Wireless Networks

Series Editor

Xuemin Sherman Shen, University of Waterloo, Waterloo, ON, Canada

The purpose of Springer's Wireless Networks book series is to establish the state of the art and set the course for future research and development in wireless communication networks. The scope of this series includes not only all aspects of wireless networks (including cellular networks, WiFi, sensor networks, and vehicular networks), but related areas such as cloud computing and big data. The series serves as a central source of references for wireless networks research and development. It aims to publish thorough and cohesive overviews on specific topics in wireless networks, as well as works that are larger in scope than survey articles and that contain more detailed background information. The series also provides coverage of advanced and timely topics worthy of monographs, contributed volumes, textbooks and handbooks.

** Indexing: Wireless Networks is indexed in EBSCO databases and DPLB **

Lin Cai • Brian L. Mark • Jianping Pan
Editors

Broadband Communications, Computing, and Control for Ubiquitous Intelligence

Springer

Editors
Lin Cai
Department of Electrical &
Computer Engineering
University of Victoria
Victoria, BC, Canada

Brian L. Mark
Department of Electrical & Computer
Engineering
George Mason University
Fairfax, VA, USA

Jianping Pan
Department of Computer Science
University of Victoria
Victoria, BC, Canada

ISSN 2366-1186 ISSN 2366-1445 (electronic)
Wireless Networks
ISBN 978-3-030-98063-4 ISBN 978-3-030-98064-1 (eBook)
https://doi.org/10.1007/978-3-030-98064-1

This Springer imprint is published by the registered company Springer Nature Switzerland AG
The registered company address is: Gewerbestrasse 11, 6330 Cham, Switzerland

Preface

This book is a collection of chapters on some of the latest developments in broadband communications for ubiquitous connectivity and caching, computing, and control for ubiquitous intelligence, which were contributed by former PhD students, postdoctoral fellows, and visiting scientists who have worked under the supervision of Professor Jon W. Mark during his long and distinguished academic career in the Department of Electrical & Computer Engineering at the University of Waterloo. For years, several former students and colleagues of Professor Mark have been contemplating some way of honoring him. This volume is a tribute to Professor Mark for his scientific contributions to the academic communities, and his enormous impact on their lives and professional careers.

This volume showcases leading-edge work on broadband communications, computing, and control that Professor Mark's former students and academic descendants are involved in, and should be a valuable resource for readers interested in pursuing research in these areas. The book has two parts: Part I contains chapters related to broadband communications, while Part II is related to caching, computing, and control. These chapters reflect the breadth and depth of Professor Mark's research interests in these areas, as well as the new directions branched out by his former students. Chapter 1 of the book contains brief stories, well wishes, and greetings by the alumni of Professor Mark's research group at the University of Waterloo.

Brief Biography of Jon W. Mark

Jon Wei Mark was born and raised in Toisan, a city-level county in Guangdong province. His father had emigrated to Canada not long after Jon's birth to find work and send money back home to the family in a time of economic hardship. Jon showed an aptitude for academics early on, breezing through primary school in record time. In his early teens, seeing little prospects available to him in China, Jon wrote a letter to his father asking him to sponsor him to emigrate to Canada. A couple of years later, Jon arrived in Toronto and reunited with his father.

Due to his father's work arrangement in a small town outside of Toronto, Jon lived independently, working a job in a restaurant while attending school in Toronto. He started in middle school, mainly to learn English. Soon after gaining sufficient proficiency in the language, he moved to a commercial school where he learned practical skills such as shorthand, typing, and accounting. Jon's father had a dream that the two of them could own and operate their own restaurant one day. Jon, however, found that he had other dreams. He decided to transfer from commercial school to Jarvis Collegiate Institute to pursue an academic path. Although such a transfer was uncommon, Jon's outstanding talents in mathematics and science were recognized, and he was admitted to Jarvis Collegiate, where he continued to excel academically. A couple of years later, he was admitted to the University of Toronto and then graduated with a BASc degree in electrical engineering in 1962.

Upon graduation, Jon took a position as an engineer with Canadian Westinghouse Company, Ltd., in Hamilton, Ontario, and quickly rose to the rank of senior engineer. While working full-time at Westinghouse, he graduated with an MEng degree in electrical engineering from McMaster University in March 1968. His MEng thesis was entitled *Optimal Detection for Echo Ranging in a Randomly Fading Environment*. From October 1968 to August 1970, he was on a leave of absence from Westinghouse to pursue PhD degree in electrical engineering at McMaster University under the auspices of a National Research Council Post Industrial Experience fellowship. Jon's doctoral dissertation, entitled *Adaptive Signal Processing for Digital Communication over Dispersive Unknown Channels* and finished in a record-breaking speed, was conducted under the supervision of Professor Simon Haykin.

After completion of his PhD Jon began his distinguished academic career in the Department of Electrical Engineering (later renamed the Department of Electrical & Computer Engineering) at the University of Waterloo in September 1970. While at Waterloo, he took several sabbatical leaves at various places. He was on sabbatical leave from 1976 to 1977, with the IBM Thomas Watson Research Center, Yorktown Heights, New York, USA, as a visiting research scientist; from 1982 to 1983, at AT&T Bell Laboratories, Murray Hill, New Jersey, USA, as a resident consultant; from 1990 to 1991, at the Laboratorie MASI, Université Pierre et Marie Curie, Paris, France, as an invited professor; and from 1994 to 1995, with the Department of Electrical Engineering, National University of Singapore, Singapore, as a visiting professor. Along the way, he supervised over 40 PhD students, postdocs, and visiting researchers at the University of Waterloo, continuing well past his official retirement as a regular faculty member in 2001. He currently holds the title of Distinguished Professor Emeritus at the University of Waterloo.

At U. Waterloo, he was a founding member of the Computer Communications Networks Group (CCNG) and later established the BroadBand Communications Research (BBCR) group. In 1996, he established the Centre for Wireless Communications as the Founding Director. Throughout his career, he was very active in the IEEE Communications Society. Dr. Mark was an editor for the *IEEE Transactions on Communications* from 1983 to 1989. He served as the Technical Program Chair of the Eighth Annual Joint Conference of the IEEE Computer and Communications Societies (IEEE INFOCOM) in 1989. He was a member of the Inter-society Steering Committee of the *IEEE/ACM Transactions on Networking* from 1992 to 2003, an editor for the *ACM/Baltzer Wireless Networks Journal* from 1997 to 2004, and an associate editor for *Telecommunication Systems* from 1996 to 2004.

Research Areas of Interest
Professor Mark's research interests have broadly been in the areas of communications, communication networks, and signal processing. He began his professional career as an engineer at Westinghouse working on problems related to satellite communications, radar, and sonar. He went on to work in the areas of adaptive equalization, spread-spectrum communications, anti-jamming secure communications over satellites, image coding, and broadband communication networks. His more recent research interests have included broadband and wireless communications and networks, including power control, resource allocation, mobility management, and end-to-end quality-of-service provisioning in hybrid wireless/wireline networks.

Acknowledgements
We would like to thank Professor Sherman Shen for reaching out to the publisher, Springer, and for all of his encouragement for this project. We thank everyone who helped make this volume possible by contributing a chapter or a personal message in Chapter 1 or sending a note of encouragement. It has been a pleasure interacting with alumni from Professor Mark's research group. We feel like an extended family of brothers and sisters sharing the common bond of having been mentored and inspired by Professor Mark. We take this opportunity to thank Professor Mark for teaching us, sharing his technical expertise with us, and, more importantly, serving as a role model for being a world-class academic researcher, mentor, teacher, and visionary leader.

Victoria, BC, Canada — Lin Cai
Fairfax, VA, USA — Brian L. Mark
Victoria, BC, Canada — Jianping Pan

Contents

Contributors

Xuehan Bai Department of Automation, Shanghai Jiao Tong University, Shanghai, China

Jun Cai Department of Electrical and Computer Engineering, Concordia University, Montreal, QC, Canada

Lin Cai Department of Electrical and Computer Engineering, University of Victoria, Victoria, BC, Canada

Lin X. Cai Department of Electrical and Computer Engineering, Illinois Institute of Technology, Chicago, IL, USA

Cailian Chen Department of Automation, Shanghai Jiao Tong University, Shanghai, China

Jie Gao School of Information Technology, Carleton University, Ottawa, Canada

Xinping Guan Department of Automation, Shanghai Jiao Tong University, Shanghai, China

Arvin Hekmati Department of Computer Science, University of Southern California, Los Angeles, CA, USA

George Karakostas Department of Computing and Software, McMaster University, Hamilton, ON, Canada

Taeyoon Kim Department of Mobile System Engineering, Dankook University, Yongin-si, South Korea

Gang Li Department of Electrical and Computer Engineering, Concordia University, Montreal, QC, Canada

Mushu Li Department of Electrical and Computer Engineering, University of Waterloo, Waterloo, ON, Canada

Xinhua Ling XLNTec Inc., Waterloo, ON, Canada

Kuang-Hao Liu Institute of Communications Engineering, National Hsing Hua University, Hsinchu, Taiwan

Yehan Ma John Hopcroft Center, Shanghai Jiao Tong University, Shanghai, China

Brian L. Mark Department of Electrical and Computer Engineering, George Mason University, Fairfax, VA, USA

Duc Minh (Aaron) Nguyen College of Engineering and Computing, Miami University, Oxford, OH, USA

Hien Nguyen Distilled Identity, Boston, MA, USA

Sangheon Pack School of Electrical Engineering, Korea University, Seoul, South Korea

Jianping Pan Department of Computer Science, University of Victoria, Victoria, BC, Canada

François Patenaude Communications Research Centre, Ottawa, ON, Canada

Cheng Ren Department of Automation, Shanghai Jiao Tong University, Shanghai, China

Humphrey Rutagemwa Communications Research Centre, Ottawa, ON, Canada

Xuemin (Sherman) Shen Department of Electrical and Computer Engineering, University of Waterloo, Waterloo, ON, Canada

Taewon Song Department of Internet of Things SCH MediaLabs, Soonchunhyang University, Asan, South Korea

Ming Tang Department of Electrical and Computer Engineering, The University of British Columbia, Vancouver, BC, Canada

Yujie Tang School of Computer Science and Technology, Algoma University, Sault Ste. Marie, ON, Canada

Peyvand Teymoori Department of Electrical and Computer Engineering, McMaster University, Hamilton, ON, Canada

Terence D. Todd Department of Electrical and Computer Engineering, McMaster University, Hamilton, ON, Canada

Miao Wang Miami University, Oxford, OH, USA

Xiaojing Wen Department of Automation, Shanghai Jiao Tong University, Shanghai, China

Vincent W. S. Wong Department of Electrical and Computer Engineering, The University of British Columbia, Vancouver, BC, Canada

Wen Wu Pengcheng Laboratory, Shenzhen, P.R. China

Lei Xu Department of Automation, Shanghai Jiao Tong University, Shanghai, China

Peng Yang School of Electronic Information and Communications, Huazhong University of Science and Technology, Wuhan, P.R. China

Jiale Ye Department of Automation, Shanghai Jiao Tong University, Shanghai, China

Qiang Ye Department of Computer Science, Memorial University of Newfoundland, St. John's, NL, Canada

Hui-Chi Yu Department of Electrical Engineering, National Cheng Kung University, Tainan City, Taiwan

Ning Zhang Department of Electrical and Computer Engineering, University of Windsor, Windsor, ON, Canada

Ran Zhang College of Engineering and Computing, Miami University, Oxford, OH, USA

Weiting Zhang School of Electronic and Information Engineering, Beijing Jiaotong University, Beijing, P.R. China

Dongmei Zhao Department of Electrical and Computer Engineering, McMaster University, Hamilton, ON, Canada

Lian Zhao Department of Electrical, Computer and Biomedical Engineering, Toronto Metropolitan University, Toronto, ON, Canada

Chapter 1
Tribute to Professor Jon W. Mark

Lin Cai

Personal Stories

Stephen Ng (Ph.D. 1977)

"One phone call changed my life.

Back in 1974, while doing my summer job in Toronto I decided to do a Ph.D. in computer communication and called Professor Mark at the University of Waterloo, who was very active in that field. He invited me to see him. We had a very good talk and I decided to do my graduate studies with him. Professor Mark was excellent in coaching me and worked very hard with his students; after one year he went to IBM Thomas J. Watson Research Center at Yorktown Heights, New York for sabbatical; however, during his sabbatical we managed to write a few good papers. We became good friends; his students and I visited his family a couple of times a year, his wife Betty was very kind and welcoming. I am very grateful that I came across such a good teacher and friend; we are still keeping in touch with each other since my graduation."

George Freeman (Ph.D. 1984)

"There is one thing which invariably causes me to think of Prof Jon Mark. It's when I hear talk that we should avoid doing something because it is too hard. I can look back on almost forty-five years of relationships with Jon: as my instructor in undergraduate and graduate courses, as my (co-)supervisor in graduate studies, as

L. Cai (✉)
Department of Electrical and Computer Engineering, University of Victoria, Victoria, BC, Canada
e-mail: cai@ece.uvic.ca

L. Cai et al. (eds.), *Broadband Communications, Computing, and Control for Ubiquitous Intelligence*, Wireless Networks,
https://doi.org/10.1007/978-3-030-98064-1_1

my chair and mentor while starting out as a new faculty member, as my colleague and sometimes research collaborator throughout most of my career at Waterloo. Always, what stands out in my recollection is authenticity. The state of the practice or the research or the department is what it is. We can observe that. The real question is what should we do about it? Is the task important? Then, if it's difficult to understand, we work hard to understand it! If it needs mathematics we don't know, we learn the mathematics! If it's hard to do, we work hard! This is the path to advances beyond small increments.

It was a conversational nudge from Prof Mark which led me into applying for scholarship support and entering the world of graduate studies. I loved it from the first day, as I felt the rush of seeing the deeper mathematics and science and learning the frontiers of knowledge through journal papers and conferences. I remember the care Jon took with onboarding new students. Before we began, I had to meet many other faculty members to ensure I was hooking up with the best supervisor for my interests. Then, I had to read widely and critically in the research literature to ensure I found a topic which sparked my curiosity. It's not that common anymore to experience such a degree of autonomy and ability to follow one's curiosity, even though, in my opinion, these are the most effective known ways to advance knowledge. I treasure that experience.

Because Prof Mark was heavily involved in administration while I was his graduate student, I feel I got some early exposure to academic politics and the mismatches between achievement and recognition. From that time, I've carried in my mind a distinction between recognition for a specific advancement to a field and recognition for accumulated incremental works (what I've since heard called "famous for being famous").

I can't help but cherish Jon's passion as an educator. There are so many examples, but crystal clear in my mind is sitting in comprehensive or thesis defense examinations where a student is not living up to their full potential. I would sense the intensity from Prof Mark as he almost levitated an inch above his chair, impatient to delve into the issues with incisive questions to open the student's mind to wider or deeper perspectives.

So how to summarize Jon's early influence on me? Perhaps the following. Excellence is a choice. The choice may lead to hard work. The result is its own best recognition. I've tried to carry that authenticity forward in my own career and believe it has served me well. I feel honored to have worked with Jon in so many ways and enthusiastically add my congratulations to this celebration of his achievements!"

Sanqi Li (Ph.D. 1985)

"Not only did Professor Mark serve as my Ph.D. thesis advisor, but he was also an incredible mentor and role model who always inspired me to do my best and helped me strive for my goals. His guidance, trusted confidence, and friendship were essential to the nurturing and cultivation of my career.

When I first wrote a letter to Professor Mark while applying for graduate studies at the University of Waterloo in 1980, I was part of the "lost generation" that grew up during China's chaotic 10-year cultural revolution. I had never received a middle or high school education, had only five courses on record through the entirety of my college life, and lived as a so-called worker-peasant-soldier student. So, without a transcript and without professor recommendations or references, I submitted my application... and beyond one's imagination, I was admitted into the graduate program at University of Waterloo. I will never forget the first time I met Professor Mark in his office. I understood less than 20% of his English and my oral English was not much better. He calmed me down and encouraged me to take some advanced courses to help. Back then, I had no confidence in myself and my only goal was to survive and not be kicked out of school and have to return to China. I still remember how panicked I had felt after taking my first midterm exam and making some stupid errors. I almost collapsed in the elevator after expressing my frustration and concerns to Professor Mark in the hallway. Professor Mark then proceeded to call me the next morning and inform me that I had gotten the highest score in the class. Throughout my first two semesters, Professor Mark often informed me of my course exam scores where I continued to rank among the highest. Although he shared some of my concerns given my previous education, he always patiently guided me and encouraged me through these lifetime changing moments, which helped me fundamentally in building my own self-confidence and has motivated me to always strive for the best.

As the first generation of graduate students from mainland China, many of us experienced some level of mental loneliness and self-isolation, largely due to extreme financial constraints and the slow adaptation of western culture during that time. During the Christmas holiday season, my memories often bring me back to my first ever Christmas party at Professor Mark's house. At night during a heavy snowfall, I started walking from the university campus to Professor Mark's house using his hand-drawn map but got lost somewhere along the way while I was completely absorbed in the bright and colorful neighborhood Christmas light displays. By the time a nice gentleman offered me a ride in the freezing weather, I realized I was already nearly a mile away from Professor Mark's house. This was my first time being invited to a "western" social gathering, which not only made me feel more accepted but also encouraged me to come out of my shell. Ever since that night, I was often invited back to Professor Mark's house for more gatherings and it is where we built our lasting and trusted friendship.

I am deeply grateful to Professor Mark for inspiring and encouraging me throughout my Ph.D. thesis work. His infectious enthusiasm for research and exploring new ideas made a significant impact on my late professional career development. I still remember the moment right after I had made some significant analytical progress in my Ph.D. thesis work, which I found out that the same work was just published at a recent conference by Bell Labs. Afterwards, I was so nervous and struggled with whether or not I should change the direction of my research. By then Professor Mark was on sabbatical leave in the USA and during our next conference call he enthusiastically encouraged me to further expand my research scope rather than change direction. As a direct result, three journal papers were published from my Ph.D. thesis.

When I was about to graduate, Professor Mark introduced me to the leading research authorities at Bell Labs, Columbia University, and IBM Thomas J. Watson Research Center in the telecommunication network field, and I subsequently landed a position at Columbia University as a Research Scientist to continue carrying out my advanced research. This paved the way for my late career development which led to my positions as a professor at the University of Texas at Austin, as a founder at multiple startups, and then as a worldwide CTO at Huawei HQ in China. Those initial five years I spent at University of Waterloo for graduate study under Professor Mark's guidance laid out a solid foundation that I was able to build upon for my later career success."

Oliver Yang (Ph.D. 1988)

"So the world is deploying 5G and researching on 6G networks nowadays. I still remember when I joined the CCNG (Computer Communication Network Group) under the supervision of Prof. Jon Mark in the 1980s; I was working on MANs (Metropolitan Area Networks) as a bridge between LANs and WANs. During my tenure at the University of Ottawa, my CCNR (Computer Communication Network Research) Lab had evolved into and investigated various "dynamic" and "static" issues in wireless networks, photonic networks and others, with focus on their design, analysis and performance evaluations.

As Jon's Ph.D. student, I could choose my research topics under the general direction while Prof Mark was advising me on research methodology and helping on various difficulties that arose. This is the freedom I enjoyed and followed when it was my turn to supervise my own Ph.D. students. In those old days at UW, the research group activities were based in CPH (Carl Pollock Hall) which was conveniently located next to my home in the MSA (Married Student Apartment). My wife and I enjoyed the social life in the CCNG which I still cherish in my memory.

Undoubtedly, Jon is an accomplished researcher as evidenced by his publications and by the talent/achievements of the students and other researchers he had supervised. Congratulations on the publication of this book that will be a good opportunity to honor his accomplishments.

Good health forever, Prof. Mark."

Jing Fei Ren (Postdoctoral Fellow and then Research Assistant Professor, 1995)

"I went to Japan from China to attend graduate school. In summer of 1990, I was working at a Panasonic lab in Japan as a visiting researcher after receiving a Ph.D. from Kyoto University. A few months before my contract ended, I sent an email to Prof. Mark to see if I could do some research in his group. At the time, email was not widely used in Japan and was not that reliable. To my surprise, I received a positive response from Prof. Mark. He asked me to arrange a recommendation letter from one of my references which I did. A few days later, I received an email offer from Prof. Mark. The opportunity was an important turning point in my life. My first trip to North America was filled with anxiety. At Kyoto University, I tried to learn English by reading and writing research papers, but I did not have a chance to practice my English in conversation. Prof. Mark told me he booked an airport van and gave me detailed instructions at Toronto Pearson International Airport to board the van. The plane landed in the evening. I was nervous and did not even know where I would stay for the first night in Canada. The driver took me to Prof. Mark's home where I met Prof. Mark, Betty, and Brian who was a university student then. They were waiting for me and had arranged their guest room for me to stay. My first real English conversation started from Prof. Mark's home. The next day, Prof. Mark brought me to school and introduced me to Guoliang and other students in BBCR group. Two months later, my wife got her visa and joined me with our 2-year-old son. We spent almost 5 years in Waterloo. Our daughter was born in K-W hospital. The experience in Waterloo is what our family will remember and cherish forever.

I interacted with students in Prof Mark's group on daily basis and learnt a lot from Prof Mark and his students as a Postdoc then as a Research Assistant Professor. Prof Mark led BBCR to successfully complete several CITR projects. On several occasions, I met a few of Prof Mark's former students who had become influential leaders in their respective technical fields. Every year when Prof Mark was in Waterloo, he would invite his group and their families to his Post Horn Pl home to have parties. He would cook barbeque for us in addition to many gourmet dishes prepared by Betty. I remember the drumstick barbeque was one of the very popular items. We would eat, chat, and laugh for several hours. When it was time to say goodbye, Prof Mark would put everyone's shoes at the entrance in order and send them off.

I joined Texas Instruments and moved to Dallas. My wife stayed in Waterloo with our two kids to finish her last semester at WLU. In April of 1996, I went back to move my family to Dallas. Prof Mark and Betty drove their new car to send our family to Pearson Airport. I have not seen Prof Mark and Betty since Infocom 96 in San Francisco but still remember the good times our family spent in Waterloo and the care, guidance, and inspiration from Prof Mark and Betty. In the time to celebrate Prof Mark's retirement, my wife, our son, and daughter join me to say congratulations to Prof Mark and Betty and wish them a very happy retirement life."

Vincent Wong (MASc. 1996)

"I had the privilege to work under the supervision of Prof. Jon Mark for my Master's study at the University of Waterloo from 1994 to 1996. My impression of Prof. Mark is that he is full of energy, dedicated to his research, and cares a lot about his graduate students. He enjoyed having coffee break with Profs. Ian Blake and Sujeet Chaudhuri. He enjoyed playing tennis after work. I remember that before our weekly meetings at his corner office in Davis Centre (DC), I would send Prof. Mark a report. He always read the report and provided very nice hand-written comments. Prof. Mark has made seminal contributions in communications and networking. I have learnt a lot from Prof. Mark on research, ranging from choosing a topic, literature survey, problem formulation, solution approaches, and paper writing. After I joined the University of British Columbia (UBC), Prof. Mark still served as one of my mentors and provided excellent career advice. When I visited Prof. Mark and Mrs. Betty Mark at their place in Toronto several years ago, Prof. Mark gave me many useful advice on healthy diet and daily exercise. I feel very proud to be one of his graduate students and am very grateful for all of his help. I take this opportunity to thank Prof. Mark for the great work he has accomplished as a researcher and mentor to many graduate students."

Lian Zhao (Ph.D. 2002)

"I was fortunate to be one of Prof. Mark's Ph.D. students. Prof. Mark has had a significant influence on me in my career path in academia. Without his guidance and tremendous support, I would not have been able to achieve what I could achieve today.

I still clearly remember that in the early stage of my Ph.D. study, I received a review invitation from IEEE Transactions on Communications and did not know what to do with it, as I had not published any journal papers yet. I was very unsure if I could accept the review invitation because I do not know whether I could provide a qualified review for this top-rank journal. I talked with Prof. Mark. He said, "Do accept the review invitation, carefully read the manuscript, and write down your review comment. I will help you proofread it before you send the review back." So I did and Prof. Mark guided me throughout the whole review process and taught me how to become a good reviewer and also how to write good papers of my own.

In 2002, I completed my thesis defense and planned to find a job in IT industry at that time or a temporary postdoc position. I just had no confidence to apply for a faculty position. It was a difficult time due to the breakdown of the telecommunication bubble in industry. I received an offer to work as a postdoc, which required me to stay half a year in a university in Europe and half a year in a university in Canada. When I talked with Prof. Mark and expressed my concern about this arrangement. Prof. Mark said to me, "You don't need to accept this offer. You can continue to work with me as a postdoc till you find a desirable position. You can also apply for a faculty position." Encouraged by Prof. Mark's words and great support, I stayed at Waterloo and started applying for a faculty position. Towards the end of 2002, I had one interview with a university in Singapore in early 2003. Unfortunately, the outbreak of SARS in early 2003 caused the postponement of this interview. Prof. Mark talked with me a few times and asked me to be patient and keep looking. In May 2003, I was invited to have an interview at Ryerson University and I was given an offer after the interview. I really appreciated Prof. Mark's encouragement and support during this long and daunting job hunting process.

After my graduation, I continued feeling how fortunate I was to be Prof. Mark's student. Bobby Ma, who was Prof. Mark's former student and started working at Ryerson University in early 1990s, greeted me during my interview and welcomed me at Ryerson as a colleague. His greeting greatly calmed my nervousness down during my interview. In the past 18 years, I have been fortunate to have an elder-academic-brother as a colleague and a friend! Bobby's family and mine visited Prof. Mark's lakeside home in Toronto from time to time. We were all impressed by the warm welcome and hospitality from Prof. Mark and his dear wife Betty. I really cherish Prof. Mark's mentorship throughout the years and sincerely wish him the best for a wonderful and happy life to be celebrated!"

Dongmei Zhao (Ph.D. 2002)

"I first met Prof. Mark in Hangzhou, China in the spring of 1998 when he and Prof. Sherman Shen attended a Wireless ATM workshop. At the time, I just received my offer to join Prof. Mark and Prof. Shen's group as a Ph.D. student. The workshop was held in a nice hotel beside West Lake. Between sessions, we walked by the lake, and Prof. Mark and Prof. Shen talked to me about living and studying in Waterloo. I arrived at Waterloo a few days before the New Year of 1999. Two days later, Prof. Shen took me to attend the New Year party at Prof. Mark's house. That was my first party in Canada. I can still remember the delicious food that Betty made on the day. My time at Waterloo was busy and fulfilling thanks to the guidance of Prof. Mark and Prof. Shen. I used to print out my work reports and put them into Prof. Mark's mail slot. His hand-written comments often filled up all the line spaces when I had the reports back. He encouraged me to pursue an academic position, helped me polish my interview presentation, and recommended me to people when we were in conferences. I am very fortunate to have worked with him. His kindness, guidance, and encouragement have helped me so much during my graduate study and after graduation. I am extremely grateful for the help Prof. Mark has given me over the years. I thank him for his wonderful supervision and wish him a long, healthy life."

Jun Cai (Ph.D. 2004)

"To my most beloved and respected supervisor: Prof. Jon W. Mark

When I sat in front of the computer to write this message, I suddenly didn't know where to start. For six and a half years of life in Waterloo, all scenes suddenly appeared in front of me. 20 years ago, I sat in the classroom listening to Prof. Mark's lectures, discussed research with him in his office, and time-to-time consulted him for my future career. Under his patient supervision, I learned how to do scientific research, how to write papers, and more importantly, how to be a man and a scholar: with integrity, meticulous, and diligent. I really want to share a little story, which I often share with my own students.

I remember once, I racked my brains and came up with an idea that I thought was perfect. So I eagerly found Prof. Mark, hoping to get his opinion. After listening to my thoughts, Prof. Mark only said to me "you haven't understood the fundamental knowledge yet." You can imagine how frustrated I was at that time. However, soon, I understood the meaning of this sentence. How can tall buildings be built without a good foundation? Without solid basic knowledge, there is no decent research.

A thousand words cannot express my gratitude to Prof. Mark. Perhaps only if I work hard and study hard can I repay the help and support he has given me.

Best wishes to Prof. Mark for an enjoyable and healthy retirement life!"

Lin Cai (MASc. 2002, Ph.D. 2005)

"In 1999, I quit my job in China to join my husband, Jianping, a postdoc fellow supervised by Prof. Mark, with no clue about my future. Jianping's research looked fascinating, but I had no background in communications or networking. The life-changing moment happened when Profs. Shen and Mark accepted me to their Broadband Communications Research (BBCR) lab, opening for me a door to a new, exciting world.

To make up for the missing background, I spent all of my savings to take four undergraduate courses in Waterloo, including the popular networking course taught by Prof. Mark. During the lectures, he often raised questions to challenge students, pushing me, originally a quiet and invisible student, to step out of my comfort zone, sharpen my mind, and voice my opinions in the large-size class. His lectures were very rich in content but without any sample questions similar to the assignments and exams. Many assignment questions were difficult; however, they were deliberately designed with a few sub-questions, guiding us to find the final answer step by step. Doing the assignments was such a rewarding brain-exercise experience, and I cannot wait to work on them. His exams were also full of surprises, while his marking was generous to honor any meaningful exploring efforts. His course was exceptional, not only delivering the knowledge but also training our critical thinking and research skills. Most importantly, he passed his endless passion in research to us.

Prof. Mark co-supervised me during my five-year Master's and then Ph.D. programs in Waterloo. He was a patient mentor, giving us all the freedom to formulate the problems by ourselves. Prof. Mark stopped me from working on incremental research (e.g., developing an algorithm to improve the performance by a few percentages in given experimental settings) and suggested focusing on significant ones (e.g., modeling and analysis of the system to find systematic, provable solutions). When I entered a dead-end or hit a wall during research, he encouraged me to take a few steps back, think hard, and RE-search a new direction.

After graduation, when I encounter any difficulty, Prof. Mark's voice will ring a bell in my mind loudly: "Work hard and think harder, Lin."

Jianping and I were also very lucky to live in Prof. Mark's big house in Waterloo for a few months before and during his sabbatical trip to Singapore. I took the chance to learn a lot from Mrs. Mark, particularly the simple, healthy, and green lifestyle. We are so grateful for their mentorship, teaching us how to be a better person and citizen. Wish them a happy retirement filled with fun and happiness."

Ahmed Hamza (Ph.D. 2015)

"I am profoundly indebted to my supervisor, Professor Jon W. Mark. For me, he was not only an academic supervisor but also a life coach. During my Ph.D. and after it, he was not only caring about my academic progress but also very supportive to me in my personal life.

I remember very well, in January 2011, four months after I had started my Ph.D. when I sent him an email to request to travel back to my home country. At that time, there was a revolution in my home country and communications were cut and I started to feel very worried about my family. I am still keeping his reply: "If you feel you can help your family being there, and that it is safe for you to go back there, then by all means." I was amazed by his response, kindness, and support. He was not concerned about how this could affect my academic progress; his only concern was my safety.

During my Ph.D., I got married and I had my first kid. He and his dear wife Mrs. Betty were very supportive to me and my wife. I remember the great advice that Mrs. Betty gave to my wife since this was our first child and we did not have family support.

In my Ph.D., he was a great supervisor that taught me how to conduct research and that the most important step in the research is not solving the research problem but formulating the research problem. I was so proud when I went to another university to attend my friend's defense and his supervisor knew that my supervisor was Prof. Jon Mark and he told me how lucky I was to have him as a supervisor.

Even after finishing my Ph.D., the kindness of Prof. Mark has not stopped. He visited me at work and my manager was so happy because he was a UW Alumni and remembered Prof. Mark from his undergraduate years. It was a great honor for me when Prof. Mark invited me and my family to his condo in Toronto, then he took us on a walking tour around the harbor front. We were astonished by the hospitality of him and his great wife Mrs. Betty.

Congratulations to Prof. Mark on his retirement! Wishing him the best of health and happiness."

Ning Zhang (Ph.D. 2015, Postdoc 2016)

"Back to the year of 2009, I was quite dazed and confused about the future, as a graduate student in Beijing, China. I decided to take an adventure and apply for Ph.D. programs overseas. With not much research experience, I doubted if I could even get an offer. It was an unforgettable moment when I received the letter from Professor Jon W. Mark, at University of Waterloo. I really appreciate the chance and trust that Prof. Mark gave to me, considering he actually had retired for many years. At that time, he indeed opened a new door to me, which changed my life path. Now I still keep the hard copy of the offer letter at home.

After joining Broadband Communications Research (BBCR) Lab at University of Waterloo in 2010, I had the chance to work with and learn from absolutely brilliant minds from all over the world. As I was a beginner researcher, Prof. Mark gave me plenty of patience and freedom to explore different topics. As a senior, Prof. Mark insisted on going to Campus for conducting research and meeting students. Through weekly meetings, he provided guidance/advices and taught me how to do research step by step. He also helped me build confidence little by little. I remember there was a formal forum that Prof. Mark gave a presentation in to give an introduction to his research works. He had tons of outstanding research works, but he picked my ongoing, preliminary research to talk in front of audiences, which was really encouraging to me and motivated me to work hard.

In the very first year, he always tried his best to help in my research during weekly meetings. As my English was not very good, I was recording during the meetings using my mobile phone so that I could replay them to digest what he said. One day, he spoke Cantonese to me for the first time and I still remember the sound "Wang zi cheng long" (Every parent wants their children to succeed). Then, he started to share a story about his past. During World War II, he was a child in a village in Guangdong Province, China. Facing the food shortage issue, he learnt that a Kung Fu school was recruiting students which could provide food to the students. To be admitted, all the children had to practice horse stance in a competition and those who gave up would have no chance. Probably over a night, Prof. Mark said he was one of those children who persevered until the end. Then, he joined that Kung Fu school and started learning Kung Fu. He showed me an old picture of all the students and masters in this Kung Fu school and asked if I could tell which child was him. Among all those students, I quickly recognized him with no reason. I feel his willpower, perseverance, and determination had never changed since then. Being a Ph.D. student, one might have the feeling of anxiety, frustration, and self-doubt. But it is also a rewarding journey. With Prof. Mark's helpful guidance, I completed all the works for my thesis and planned to have the defense in 2014. I printed my thesis and came to his office for further comments at a regular meeting time. However, there was no response behind the door, which was quite rare. Later on, I received an email from Prof. Mark saying that he didn't feel very well and had to rest at home. However, he still cared about my thesis and asked me to bring a hard copy to his house. When I came to his house, no one was there, and I left my thesis at the front door, wondering what happened and starting to worry about his health. Then, I met him in the hospital and at that time he had a very bad health condition. Luckily, he went through it and gradually recovered. In the beginning of 2015, Prof. Mark managed to attend my Ph.D. thesis defense and I successfully passed it. I really appreciate Prof. Mark's efforts and would also like to thank Prof. Sherman Shen for providing great support as a co-supervisor in that period.

I am very grateful for all the support that Prof. Mark provided during my time at University of Waterloo. I deeply appreciate his continuous encouragement and guidance in my research. Although new challenges rise all the time in research and life, the characteristics learned from Prof. Mark, such as critical thinking, focus, perseverance, and passion, can help conquer any difficulties and challenges and pave the way towards success."

Yujie Tang (Ph.D. 2017, Postdoc 2019)

"I would never forget the moment when I received the offer letter from Professor Mark in late 2010, and that opportunity opened a big gate for me to chase my dream since I was a young girl at the age of six, to be a professor. It was also an important turning point in my life. I started a whole new life at the other end of the world, far away from my parents and sister, explored a brand-new environment both physical and cultural, and met with new friends and colleagues. The very first challenge that I met was my poor English speaking at that time, which made me quite unconfident to talk. Prof. Mark was the first person to encourage me and give me useful advice on how to be more confident and comfortable to speak English.

Prof. Mark was an incredible mentor and a real role model who always inspired, encouraged, and guided me through every step of my academic journey, ranging from choosing topics, literature survey, problem formulation, solution approach, and paper writing. At the beginning of my Ph.D. thesis work, I was in fear of making mistakes and anxious about picking a wrong research direction. I am very grateful to Prof. Mark for being so patient and open-minded to allow me to explore different topics at that time. I am also deeply impressed by his infectious enthusiasm for research and how tirelessly he worked, even though he had retired at the time. He never missed any meeting with me, even when he had a serious eye infection, which deeply touched me. He always provided very nice hand-written comments on every report and manuscript that I sent to him. I still keep all the copies and cherish them as the true gem in this current digitized world. As a scholar and role model, his guidance, trust, and encouragement made a significant impact on my academic career and will shine for many years to come.

At a personal level, Prof. Mark and his wife Betty are very good friends and very kind and welcoming. They sent their sincere blessings through lovely cards and gifts on the two most important moments in my life, as a newly married and as a new mom. They shared a good deal of useful advice on healthy diet and living when we had dinner with them. The great time and experience in Waterloo is what my family will remember and cherish forever.

Joined by my husband and our daughter, we would like to send our sincere congratulations to celebrate Prof. Mark's retirement and wish him and Betty a retirement life full of happiness and health."

Yu Cheng

"I joined BBCR lab as a Ph.D. student under the supervision of Prof. Weihua Zhuang in 1999. In the first year of my study, I worked as a teaching assistant

(TA) for the networking course by Prof. Jon Mark. Prof. Mark used the textbook "Data Networks" by Bertsekas and Gallager, which is heavily mathematically oriented. That was a tough TA job. I had not learnt the textbook before, so Prof. Mark requested me to study the textbook, work out the homework problems independently, and then do the tutorials. Once, my solution to a homework problem was not fully correct, making Prof. Mark unsatisfied. I am now working as a Professor at Illinois Institute of Technology. The TA jobs here are much easier, e.g., homework grading according to solution manuals, or monitoring lab sessions. The difference between our expectation on a TA nowadays and Prof. Mark's expectation around 20 years ago can speak on his high standard on graduate students' capability and quality. It is no wonder why Prof. Mark could achieve such a splendid academic career and cultivated many great students. I am really honored that I can share a story here and show my deepest respect to Prof. Mark."

Greeting Messages from Alumni

Gordon L. Stuber (Ph.D. 1986) "Thank you Prof. Mark for being my Ph.D. co-advisor along with Ian Blake. You helped me greatly, during my 4 years of graduate studies. You always treated your graduate students well, helped them flourish, and also to have fun. I will always remember Globecom'85 in New Orleans, hurricanes in hand on Bourbon Street. You're truly a ball of fire, actively carrying on your research at Waterloo many years into your retirement. Waterloo was lucky to have you."

Hien Nguyen (Ph.D. 1993) "Congratulations on your retirement. It is still hard for me to process how you would be able to "retire" given that I know how much you love to work. I felt fortunate and would like to thank you for having chosen me as one of your Ph.D. students. My life had changed for the better from that point. I still have many fond memories of our discussion together (technical or not), and the warm Christmas parties that you, Betty, and Brian had organized for us. Back then, the world was much less connected and we didn't have YouTube nor cell phones. Thanks to your teaching devotion and kindness, many of us were able to learn enough and helped create computer networks that bring people much closer together for a better world. Please know that you have done much more than enough, compared to the rest of us, and that you can truly enjoy your retirement."

Robert Lehr (Ph.D. 1997) "Thank you, Dr. Mark, for all the support and encouragement you have provided to your graduate students over the years. From advice on writing papers, to helping your students financially, you always guided us with your experience and compassion. You have made a long-lasting, positive impact on our careers and lives, and we will always be grateful."

David Tung Chong Wong (Ph.D. 1999) "Thank you for your guidance and help during and after my Ph.D. I appreciate them very much. Take care and enjoy your retirement. "

Michael Cheung (Ph.D. 2000) "I am grateful to having Professor Mark supervising my doctoral studies. He is a true scholar, a good mentor, and a role model of a great person. His dedication to research and academic excellence has been passed on to countless number of outstanding researchers in both the academia and the industry, making valuable contributions to the scientific world. I am proud being your student."

Minghui Shi (Ph.D. 2006) "I feel very fortunate and privileged to be one of Prof. Mark's students. Prof. Mark was my co-supervisor at the UW. He guided me through my studies as a role model. After many years, I still remember his excellent art of lecturing, passion, encouragement, and, of course, the amount of editing remarks on my work. The care from Prof. Mark even went beyond the academic work and extended to my personal life and my family. Thank you very much, Prof. Mark, for everything you have done for me and other students. My family and I wish Prof. Mark good health and all the best and joy in the continuing adventures of life."

Qiang Ye (Postdoc 2019) and Wen Wu "As BBCR graduates, we have been deeply influenced by Prof. Mark's education, support, and help which are always our precious lifetime treasure. Congratulations on the retirement! We sincerely wish Prof. Mark the very best of health, happiness, and full of joy on the new journey in life! Enjoy the rest and relaxation!"

Part I
Broadband Communications for Ubiquitous Connectivity

Chapter 2
Network Slicing for 5G Networks and Beyond

Qiang Ye and Wen Wu

2.1 Introduction to 5G Communication Networks

With the advancement of telecommunication industry, the cellular communication system has been evolving to the fifth generation (5G). Different from previous generations, the 5G communication networks are expected to provide fine-grained and customized end-to-end (E2E) service deliveries with diverse and differentiated quality-of-service (QoS) provisioning for Internet of Things (IoT) [1]. The supported IoT service types range from smart homing, industrial automation, intelligent transportation systems to e-health care systems, smart cities, etc. The 3rd generation partnership project (3GPP) technical report (Release 15) specifies three main characteristics of 5G networking paradigm [2]:

1. *Enhanced mobile broadband (eMBB) communications:* The 5G networks are expected to provide seamless communication coverages for more and more end user terminals with enhanced mobility (e.g., vehicles on highways, high-speed trains), supporting increasing types of high data rate services, such as high-definition (HD) video streaming/conferencing, virtual reality/augmented reality (VR/AR). To accommodate largely increased communication demands, both 5G wireless and core network capacities are enhanced, through intensified network deployment for spatially multiplexing on currently employed network resources.
2. *Massive machine-type communications (mMTC):* Another key feature of 5G networks is to maintain a large number of heterogeneous machine-type com-

Q. Ye (✉)
Department of Computer Science, Memorial University of Newfoundland, St. John's, NL, Canada
e-mail: qiangy@mun.ca

W. Wu
Pengcheng Laboratory, Shenzhen, P.R. China

L. Cai et al. (eds.), *Broadband Communications, Computing, and Control for Ubiquitous Intelligence*, Wireless Networks,
https://doi.org/10.1007/978-3-030-98064-1_2

munication connections to realize massive IoT. To guarantee the IoT service quality, the utilization of network resources needs to be boosted to support the ever-increasing IoT traffic volume.

3. *Ultra-reliability and low-latency communications (URLLC):* IoT applications often require different dimensions and levels of QoS satisfaction. For example, industrial IoT services require strict and deterministic time-lines in task processing, while autonomous driving tasks demand low processing latency and ultra-high reliability [3, 4]. Therefore, the network resources need to be managed to support customized IoT services with QoS isolation [5]. The QoS isolation ensures that the minimum QoS requirements of any service are not violated when network dynamics happen due to end device mobility, channel quality variation, or network traffic load fluctuations.

The key features of 5G networks pose significant challenges on QoS-oriented resource provisioning. In the wireless network domain, a multi-tier of wireless base stations (BSs) are deployed, i.e., macro-cell BSs (MBSs) underlaid by multiple layers of small-cell BSs (SBSs), to exploit the spatial reuse of currently employed radio spectrum resources. However, the increasingly intensified network deployment leads to the rise of both capital and operational expenses (CapEx and OpEx) and also the increased levels of inter-cell interference. In the core network domain, data traffic from radio access networks (RANs) are aggregated at the edge (e.g., gateways) of the network core, and then traverse a sequence of network functions, such as access and mobility management function (AMF) and session management function (SMF) [2], instantiated on different network servers to realize E2E service deliveries. A traffic flow of certain service type refers to an aggregation of data packets traversing between two anchor points of a core network. Specifically, each flow is routed through a sequence of network functions, consuming computing resources (i.e., CPU time cycles) on servers and transmission resources on physical links for certain E2E service provisioning. With increasing traffic volume from different services carried in core networks, adding more network resources, including network servers providing different functionalities, physical transmission links, and forwarding devices (e.g., switches, routers), becomes cost-ineffective.

2.2 Network Slicing

To accommodate highly increased communication demands with differentiated QoS provisioning and low service delivery cost, the current resource management policies need improvement. Network function virtualization (NFV) is a cost-effective solution to softwarize network functions on servers using virtualization technologies, i.e., instantiating network functions on a virtualized environment (e.g., virtual machines (VMs)) [6, 7]. In such a way, the virtual network functions (VNFs) are decoupled from vendor-specific servers and can be flexibly programmed at different network server locations with low installation and operation cost, the process of which is also called *VNF placement*. The servers equipped with virtualization platforms (e.g., VMware, OpenStack) become commodity servers, also called NFV nodes, that can instantiate VNFs as software to provide different functionalities. By using NFV in core networks, the virtual computing/storage/networking resources on network servers are abstracted and flexibly distributed to different VMs on NFV nodes to host different sets of VNFs. The placement of VNFs is centrally controlled by an NFV controller with the consideration of in-network resource availability. Traffic flows from different service types are often required to traverse different sets of VNFs for E2E service deliveries. For example, video streaming traffic needs to pass through a transcoding function for raw video content compression and then a firewall function to ensure a secured E2E video delivery, while DNS request traffic goes through a firewall function and DNS server in sequence to search for domain name to IP address mapping.

A set of VNFs connected in a specific sequence for executing certain service functionalities is called *service function chain (SFC)*. There is a fundamental research issue of how to deploy an SFC over a physical substrate in a cost-effective manner, including instantiating VNFs on appropriate NFV nodes and configuring a routing path passing through these functions, to guarantee E2E service quality [8, 9]. To enable the programmability on routing configurations with simplified data plane operations, the NFV controller is also enabled with the software-defined networking (SDN) control functionality [10]. With SDN, the forwarding control functions at each network node are decoupled from the physical hardware and migrated to the centralized controller as a program. Based on the global network state information, the controller can be programmed with different routing algorithms for customized E2E service deliveries, and the routing decisions are enforced from the controller to each forwarding device through the OpenFlow protocol [11]. By decoupling the traffic forwarding control, the traffic routing performance in the data plane is significantly improved.

Under the SDN-enabled NFV architecture, when multiple SFCs are placed onto the physical network, it is likely that more than one VNF is instantiated on a same NFV node and traffic routing paths for different SFCs are (partially) overlapped to save CapEx and OpEx. However, when SFCs share a common set of network resources, including virtual computing resources on network servers and transmission resources on links, how to *slice* a pool of virtual resources among the SFCs to achieve QoS isolation is essential. This process is called *network slicing* [12, 13]. In 5G core networks, a key research issue is how to jointly slice the computing and transmission resources among traffic flows traversing different SFCs, with the consideration of the correlation of resource consumption between the two resource types, which is also termed as *bi-resource slicing* [12]. In 5G wireless networks, the baseband processing and radio resource control functions are virtualized and centrally managed by an SDN-enabled NFV controller. Radio resources on heterogeneous wireless BSs are aggregated as a resource pool, and are flexibly sliced and reserved for different BSs to improve the overall spectrum resource utilization for better QoS customization. In the wireless network domain, the network slicing is studied in terms of *radio resource slicing* [5], which mainly deals with how to determine the optimal ratios of sliced radio resources among BSs such that the overall resource utilization is maximized.

In this section, we present an E2E network slicing framework for both 5G wireless and core network domains. For 5G wireless networks, a dynamic radio resource slicing scheme is proposed in Sect. 2.2.1 to maximize the network-wide communication resource utilization by enabling resource sharing among BSs, while guaranteeing the QoS isolation between mobile broadband data services and machine-to-machine (M2M) communication services. For 5G core networks, we study in Sect. 2.2.2 how virtual computing resources and networking resources are jointly sliced among traffic flows with fairness guarantee and service quality satisfaction. We also discuss in Sect. 2.2.3 how artificial intelligence (AI) technologies are applied in automating the process of slice creation especially in beyond 5G networks with increased network complexity and dynamics.

2.2.1 Network Slicing in 5G Wireless Networks

5G wireless networks are built on multi-tiers of heterogeneous BS deployment, e.g., multiple layers of SBSs underlying MBS coverages, to explore the spatial reuse of radio spectrum resources for enhancing the network capacity. To further improve the overall resource utilization with reduced network deployment cost, the SDN-enabled NFV architecture softwarizes the radio processing and resource control functionalities on BSs to enable resource programmability and centralized control over heterogeneous BSs. In such a way, the radio resources can be sliced and reconfigured on BSs according to data traffic load conditions for differentiated QoS provisioning.

2.2.1.1 Dynamic Radio Resource Slicing Framework

Radio resource slicing has two levels of resource partitioning: *network level* and *service level*. In the network level, the aggregated resources are sliced among BSs and reserved for user/device groups of different services (i.e., service groups) under the BS coverages; in the service level, the sliced radio resources on each BS are partitioned among service groups for customized QoS provisioning. Most existing studies mainly focus on service-level slicing, with a focus on how to slice resources among different groups of users/devices under each BS for differentiated QoS satisfaction [14, 15]. By facilitating resource sharing among BSs, the network-level slicing needs investigation to maximize the overall resource utilization.

1. *Network model:* We consider a two-tier downlink heterogeneous wireless network, as shown in Fig. 2.1, where an MBS, denoted by B_0, is deployed at the center of a macro-cell to provide wide-area communication coverage. The

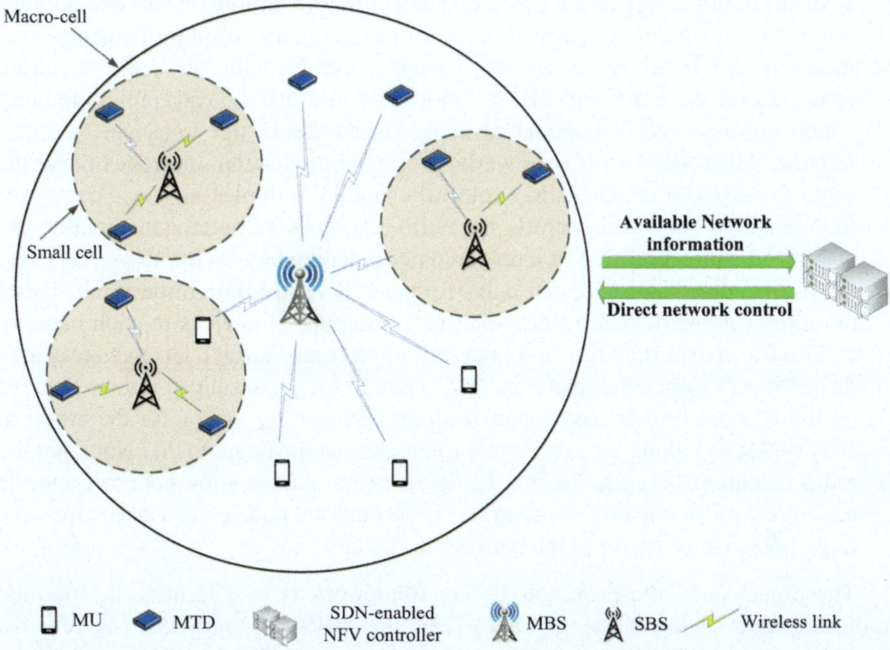

Fig. 2.1 A two-tier heterogeneous wireless network with MTDs and MUs

macro-cell is underlaid by n small-cells within the MBS coverage area, each with an SBS located at the cell center, to boost the network capacity in support of the increasing M2M communication demands. The set of SBSs under consideration is denoted by $\mathcal{B} = \{B_1, B_2, \ldots, B_n\}$. There are machine-type communication devices (MTD) and mobile users (MUs) randomly located within the BS coverages. Denote by $\mathcal{M}_{i,k}$ and $\mathcal{N}_{j,k}$ the sets of MTDs and MUs ($M_{i,k}$ and $N_{j,k}$ are the set cardinalities), respectively, from M2M communication service i ($i = 1, 2, \ldots, I$) and broadband data service j ($j = 1, 2, \ldots, J$) within the coverage of B_k ($k = 0, 1, \ldots, n$), where I and J denote the total number of supported machine-type and broadband data service types. Binary variable $x_{m,i,k}$ ($y_{l,j,k}$) is used to denote the association pattern between MTD m (MU l) from service i (j) and SBS B_k that covers the device (user), where $x_{m,i,k} = 1$ ($y_{l,j,k} = 1$) if MTD m (MU l) is associated with B_k and $x_{m,i,k} = 0$ ($y_{l,j,k} = 0$) if MTD m (MU l) is associated with MBS B_0. Since the MTDs (MUs) solely covered by the MBS have only one association pattern, we have $x_{m,i,0} = 1$ ($y_{l,j,0} = 1$). Each BS is equipped with a number of link-layer transmission queues, each for downlink packet transmissions with an associated end device/user. Let λ_i (λ_j) be the average packet arrival rate at each transmission queue of service i (j). The downlink M2M communication services and mobile broadband data services have different traffic arrival patterns. The M2M packet arrivals are often triggered by specific controlling/monitoring events and are thus event-driven, whereas the arrival process of mobile broadband traffic is periodic.

2. *Radio resource slicing optimization framework:* Initially, the radio spectrum resources on the MBS and SBSs, denoted by W_{m} and W_{s}, are pre-configured, which are assumed orthogonal to avoid inter-network-tier transmission interference. All n SBSs with non-overlapping communication coverages reuse the same set of radio resources to exploit the spatial multiplexing gain. Under the SDN-enabled NFV architecture, the radio resources on heterogeneous BSs are abstracted and aggregated as a resource pool, denoted by $W_{\mathrm{v}}(= W_{\mathrm{m}} + W_{\mathrm{s}})$, and are further sliced as $\rho_{\mathrm{m}} W_{\mathrm{v}}$ and $\rho_{\mathrm{s}} W_{\mathrm{v}}$ for the MBS and SBSs, respectively, based on current network state information (e.g., number of devices in each network cell, traffic arrival statistics, and inter-cell interference levels). Given a set of BS-device (user) association patterns, $\{x_{m,i,k}\}$ and $\{y_{l,j,k}\}$, the sliced radio resources on the BSs are further partitioned, with the ratio of $f_{i,k}$ ($g_{j,k}$), for the group of MTDs (MUs) belonging to service i (j) under the coverage of B_k. Note that the radio resource slicing ratios, the BS-device (user) association patterns, and the resource partitioning ratios among service groups are updated in a large timescale (e.g., hours) in response to the network dynamics.

The objective of the proposed slicing framework is to determine the optimal radio resource slicing ratios, ρ_{m}^* and ρ_{s}^*, on the MBS and the SBSs, respectively, with the optimal BS-device (user) association patterns, $\{x_{m,i,k}^*\}$ and $\{y_{l,j,k}^*\}$, and the optimal resource partitioning ratios, $f_{i,k}^*$ and $g_{j,k}^*$, for each service group under B_k. By enabling the network-level resource sharing, the overall radio resource utilization is maximized under different dimensions of QoS constraints (e.g., packet loss rate bound and packet transmission delay constraint for M2M communication services and minimum data rate requirement for mobile broadband services). The operation procedure of the radio resource slicing optimization includes the following three steps (refer to [5] for a detailed mathematical formulation of the optimization problem):

Step 1 Through signaling exchange between the SDN-enabled NFV controller and the BSs, the controller periodically collects the updated network state information, including the total number of MTDs and MUs, $M_{i,k}$ and $N_{j,k}$, from M2M service i and broadband data service j within the coverage area of BS B_k, traffic load statistics λ_i and λ_j, and wireless channel conditions between the BSs and end devices (users) considering the inter-cell interference.

Step 2 Based on the updated network information, the controller executes the radio resource slicing optimization to maximize the overall communication resource utilization under differentiated QoS constraints for both M2M and broadband data services, upon which the optimal slicing ratios, the optimal BS-MTD (MU) association patterns, and the optimal resource partitioning ratios for service groups under each BS are obtained.

Step 3 Repeat *Step 2* to update ρ_{m}^* and ρ_{s}^*, $\{x_{m,i,k}^*\}$ and $\{y_{l,j,k}^*\}$, and $f_{i,k}^*$ and $g_{j,k}^*$, when the network traffic load varies, to achieve consistently maximum network-level radio resource utilization.

2.2.2 Network Slicing in 5G Core Networks

In 5G core networks, to improve resource utilization with reduced network deployment and operation cost, it is likely that multiple SFCs are placed over a common (or partially overlapped) physical network path, sharing virtual computing resources on NFV nodes to instantiate multiple VNFs and transmission resources on physical links. In Fig. 2.2, we provide an illustrative example of accommodating two service traffic flows over a common network path in a core network, where packets from flow x of a DNS request service traverses in sequence a firewall function F_1 and a DNS function F_2 on NFV nodes S_1 and S_2, respectively, and traffic flow y passes through F_1 and an intrusion detection system (IDS) function F_3 on the same set of NFV nodes for a secured E2E steaming service. After traversing an NFV node, each of the traffic flows is forwarded over a number of virtual switches (vSwtiches) [16] connected by transmission links before reaching a subsequent NFV node or the destination anchor node in the core network. For a general case, a set of L traffic flows are placed on a common network path, along which there are v NFV nodes, denoted by $\{S_1, S_2, \ldots, S_v\}$, and n_u pairs of transmission links and vSwitches connecting S_u $(u < v)$ and S_{u+1}.

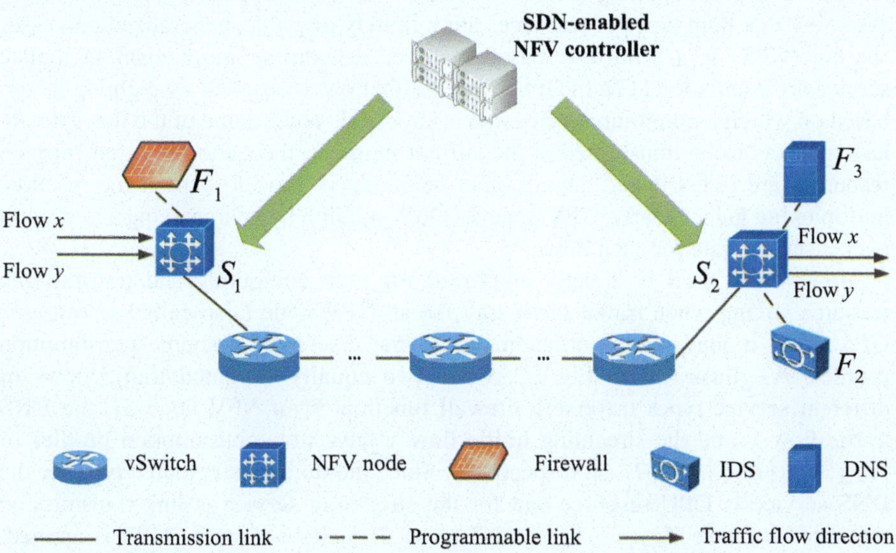

Fig. 2.2 An illustrative example of traffic flows traversing SFCs placed over a common network path in a 5G core network

2.2.2.1 Joint Computing and Transmission Resource Slicing

When each packet of a traffic flow, say flow x, passes through a VNF on an NFV node, it consumes $T_{x,1}$ amount of time for CPU processing on the NFV node and $T_{x,2}$ amount of time for packet transmission on the server output link, provided that all the computing resources on the server and transmission resources on the link are allocated to the traffic flow. This two dimensional time vector, $[T_{x,1}, T_{x,2}]$, is called *resource consumption profile* of flow x traversing a specific VNF on an NFV node. However, traffic flows of different service types often demonstrate discrepant resource consumption profiles when passing through an NFV node. For example, a DNS request packet with long packet header and short payload sizes takes more time for CPU processing, whereas a video packet consumes more resources on link transmission than processing. The type of resource that a service packet consumes more when traversing a function on an NFV node is termed a *dominant resource type*.

When multiple traffic flows from different services pass through a common NFV node, both computing resources on the server and transmission resources on the output link need to be sliced properly among the flows to ensure QoS isolation. Therefore, how to design a joint resource slicing scheme to achieve high resource utilization with fair sharing among traffic flows is a challenging research issue. By assuming that both types of resources are infinitely divisible, generalized processor sharing (GPS) is a fluid-flow based resource scheduling mechanism to realize service differentiation [17]. In GPS, every traffic flow is assigned a weighting factor, based on which a minimum service rate is allocated. When some of the flows do not have packets to be transmitted at the current moment, their allocated transmission resources are redistributed among other backlogged flows to exploit the resource multiplexing gain. Clearly, GPS achieves QoS isolation by guaranteeing a minimum packet service rate for each flow.

However, if GPS is directly employed for joint computing and transmission resource slicing when traffic flows traverse an NFV node (also called *bi-resource GPS* [12]), it may not be efficient due to the discrepant resource consumption profiles. As illustrated in Fig. 2.2, when two equally weighted traffic flows of different service types traverse a firewall function F_1 at NFV node S_1, the DNS traffic flow x and the streaming traffic flow y have time consumption profiles of $[T_{x,1}, T_{x,2}]$ and $[T_{y,1}, T_{y,2}]$, respectively. Since the dominant resource type for the DNS service is CPU resource and for the streaming service is link transmission resource, we have $T_{x,1} > T_{x,2}$ and $T_{y,1} < T_{y,2}$. If bi-resource GPS is applied, both CPU and transmission resources on S_1 are equally partitioned between the two traffic flows (applying GPS for both resource types), i.e., $h_{x,i} = h_{y,i}$ $(i = 1, 2)$, where $h_{x,i}$ and $h_{y,i}$ denote the fractions of type i resources sliced for traffic flows x and y, respectively. However, due to discrepant time consumption profiles, the equal partitioning on both resource types results in imbalanced resource provisioning, i.e., the sliced transmission resources for flow x are over-provisioned as $T_{x,1} > T_{x,2}$, leading to a certain level of resource wastage, and for flow y are insufficient, causing packet accumulation at the output link port.

To improve the overall resource utilization with low E2E delay and at the same time maintain the resource sharing fairness, we employ dominant resource GPS (DR-GPS) as our joint resource slicing scheme [18, 19]. In DR-GPS, the fractions of sliced dominant resources are equalized among all the traffic flows, and the non-dominant resources are sliced in proportion to the resource consumption profile of each flow to maximize the joint resource utilization. When some of the flows have no packets to be transmitted, their sliced resources are redistributed among other backlogged flows according to GPS. Taking the example in Fig. 2.2, when flow x and flow y pass through firewall function F_1 at S_1, we equalize the CPU resource slicing ratio for flow x with the transmission resource slicing ratio for flow y, i.e., $h_{x,1} = h_{y,2}$, to guarantee a fair resource sharing for the more demanding resource types. To reduce the packet queueing at output transmission link and improve the utilization of both types of resources, the non-dominant resources of both flows are sliced proportional to their time consumption profiles, i.e., $h_{x,2} = \frac{T_{x,1}}{T_{x,2}} h_{x,1}$ and $h_{y,1} = \frac{T_{y,1}}{T_{y,2}} h_{y,2}$. By using DR-GPS, QoS isolation among different services is achieved by guaranteeing a minimum fraction of computing and transmission resources with high resource multiplexing gain. Moreover, the scheme always maintains fair resource sharing among dominant resource types and improves the joint resource utilization with good E2E performance.

2.2.3 AI-Assisted Network Slicing in Beyond 5G Networks

As mentioned above, the advantages of network slicing have been elaborated clearly in 5G networks, and some solutions have been proposed either in the wireless network domain (Sect. 2.2.1) or in the core network domain (Sect. 2.2.2). Since 5G networks are being widely deployed starting from 2020, people are envisioning beyond 5G (B5G) networks, such as 6G networks. Seeing the great success of network slicing, it is expected that network slicing will continue evolving and play an increasingly important role in the future B5G networks [20]. In this subsection, we first introduce the potential new features in B5G networks and the corresponding new challenges. Then, we discuss how to leverage AI techniques to assist network slicing to address these challenges in the future B5G networks.

2.2.3.1 Beyond 5G Networks

Compared with the current 5G networks, the B5G networks will witness an improvement of key performance indicators (KPI) requirements, such as increased data rates, enhanced network capacity, and low latency. The detailed KPI requirements of the 5G and B5G networks are as follows: (1) *5G networks*—Aiming at supporting typical services, i.e., enhanced mobile broadband services, ultra-reliable low-latency communications services, and massive machine-type communications services, 5G is required to meet several KPI requirements, such as 20 Gbps peak data rate, 0.1 Gbps user experienced data rate, 1 ms end-to-end latency and 1 million devices/km^2. Such requirements are achieved by adopting several key enabling technologies, such as millimeter-wave (mmWave), massive multiple-input multiple-output, ultra-dense networks, and non-orthogonal multiple access [21]; (2) *B5G networks*—The B5G networks offer more than increased data rates but also expanded capacity, lower latency, and larger coverage. According to a very recent white paper, B5G networks are expected to deliver 1 Tbps peak data rate, 20–100 Gbps user experienced data rate, 0.1 ms end-to-end latency, 10 million devices/km^2 and near 100% coverage, which are an order of magnitude higher than that in 5G networks [22].

In addition to the improvement of KPI requirements, B5G networks are expected to have the following new *features*:

- *Space-air-ground integrated network (SAGIN)*—Although the current 5G networks can provide good coverage in highly populated areas, such as urban/suburban areas, B5G networks are required to deliver universal coverage, including in rural areas, remote lands, and sparsely populated areas. To achieve the goal, B5G would exploit the altitude dimension. Space networks, such as low earth orbit, medium earth orbit, and geosynchronous earth orbit satellites, aerial networks, such as unmanned aerial vehicles (UAVs) and balloons, and ground networks, such as cellular and Wi-Fi networks, can be integrated into the SAGIN, to provide global coverage, facilitate on-demand services, and support high-rate low-delay services [23, 24];
- *Abundant services with stringent QoS requirements*—A large number of emerging applications are expected to have stringent QoS requirements in multiple dimensions. Some bandwidth-consuming applications, such as mobile augmented reality, mobile virtual reality (VR), and hologram video streaming applications, usually require an extremely high data rate. Taking mobile VR as an example, the required uplink data rate is up to 5 Gbps. Some other applications, such as autonomous driving, industrial IoT networks, and UAV control systems, usually require extremely high reliability. Taking the autonomous driving as an example, the required reliability is up to 99.999%.

Due to the above features, applying existing network slicing schemes in B5G networks faces the following new challenges. Firstly, in the future SAGIN, not only the number of network components increases, but also the slice scope increases due to global network coverage. As such, it is necessary to develop *efficient* network management schemes for a large number of network slices with massive network nodes. Secondly, the stringent QoS requirements of emerging services call for *effective* network slice management schemes to guarantee stringent QoS requirements. Thirdly, since B5G networks are expected to be highly dynamic due to user mobility and time-varying service demands, the developed slice management schemes should be *adaptive* to dynamic network environments. These new challenges may undermine the feasibility of applying the traditional optimization based network slicing schemes in B5G networks. Hence, to address these challenges, it is desired to develop efficient, effective, and adaptive network slicing schemes for future B5G networks. Leveraging AI techniques to handle complicated network slicing problems is a potential solution, thereby fostering AI-assisted network slicing schemes.

2.2.3.2 AI-Assisted Network Slicing

Recently, AI techniques have achieved great success in many fields. Fuelled by powerful computing facilities and well-curated datasets, AI techniques, especially deep neural networks, have been applied in various tasks, including object detection and classification, audio recognition, and content recommendation. By directly mining from massive data, AI can handle complicated problems and adapt to the environment dynamics without requiring an accurate mathematical model beforehand. Endowed with these advantages, AI has been applied to handle many problems in wireless networks, such as beam alignment in mmWave networks [25], resource management in industrial IoT networks [26], mobility prediction in vehicular networks [27], and trajectory planning in UAV networks [28].

AI-assisted network slicing which applies AI techniques for performing network slice management can reduce the network slice management complexity [14]. In existing empirical network slicing schemes, some experienced network engineers are hired to configure a number of slice parameters according to network environments, which incurs human labor costs. Moreover, when the number of supported services is large, the number of slice parameters can be extremely high. Empirical schemes may not achieve optimal resource utilization with increased resource consumption cost. AI-assisted network slicing schemes can automatically manage network slices. This can reduce network slice management complexity and support more diverse services. By mining from extensive slice operation data, AI-assisted network slicing schemes can directly learn the optimal scheme in dynamic network environments, such that resource consumption cost can be reduced.

To better understand the detailed roles of AI-assisted network slicing, we first introduce the three stages in the network slicing lifecycle and their corresponding functions. The network slicing lifecycle consists of three phases: preparation, planning, and scheduling, as shown in Fig. 2.3. These three stages work together to facilitate network slicing for wireless networks. The functions of the three stages are detailed as follows:

- *Preparation stage*—Based on service requirements and virtual resource availability, the preparation phase is to construct and configure network slices. Specifically, service requirements are extracted to classify services based on their QoS requirements, such as service delay, service priority, throughput, and reliability. In addition, network resources (e.g., the communication, computing, and caching resources) and network functions (e.g., firewall, network name translation, domain name system) are virtualized. After virtualization, the SDN

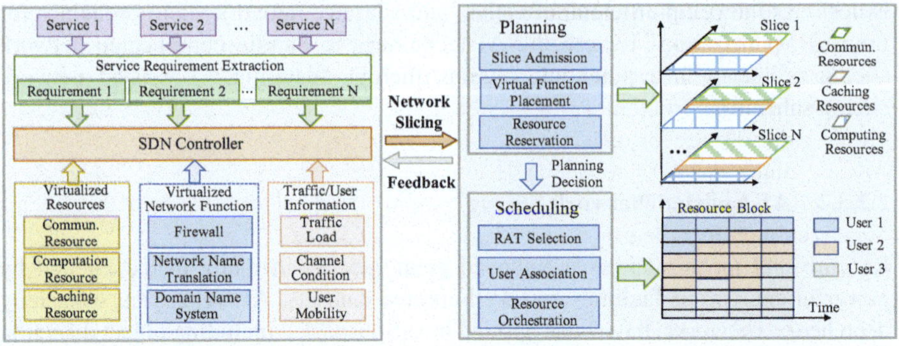

Fig. 2.3 An illustrative example for the lifecycle of network slicing. The left part is the preparation stage, the right-top part is the planning stage, and the right-bottom section is the scheduling stage

controller can efficiently manage the resources and network functions in a centralized manner.

- *Planning stage*—The main goal of the stage is to reserve network resources for different slices. To achieve this goal, extensive network information can be collected from underlying physical networks by the SDN controller, such as service demand patterns, stochastic channel conditions, and user mobility patterns. Then, the SDN controller will reserve the virtualized communication, caching, and computing resources for different slices.
- *Scheduling stage*—The main goal of the stage is to schedule the reserved resources of a slice for end users. Specifically, given the planning decision, i.e., resource reservation decision, the SDN controller dynamically allocates the reserved resource blocks to end users according to users' real-time service requests.

Note that some feedback information from the system is sent to the SDN controller, such as the resource utilization, system performance, and service-level agreement satisfaction. Based on the feedback information, the SDN controller can adjust the network slicing decisions to adapt to dynamic environments and guarantee slices' performance.

Knowing that three stages have different functions in network slicing, it is obvious that AI will play different roles in the three stages. The detailed roles of AI in the network slicing lifecycle are given as follows:

- *AI for preparation*: One exemplary task is service demand prediction. Based on the historical service traffic load, the service demand can be predicted via AI methods, such as recurrent neural networks. The predicted service demand information is utilized for decision making in the planning stage.
- *AI for planning*: There are multiple tasks in the planning stage that can be conducted via AI methods. Two examples are detailed as follows: (1) Slice admission, in which the SDN controller admits the slices to maximize system revenue, based on the network resource availability and service demand of these

slices. As the decision variable is binary, this problem can be considered as an integer optimization problem. In a large-scale network, traditional optimization methods become infeasible, and deep learning is a potential solution to address the optimization problem; (2) Resource reservation, in which the SDN controller reserves resources for different slices based on the service demand of different slices. Since the service demands are time-varying, the resource reservation should be adaptive to the service demand dynamics, which can also be addressed via a reinforcement learning method.

- *AI for scheduling*: There are also multiple tasks that can be conducted via AI methods in the scheduling stage. We take resource orchestration as an example. The reserved resource of a slice is allocated to end users after the planning decision is given. The resource orchestration decision is determined based on real-time user mobility, channel condition, etc. To efficiently utilize resources, a reinforcement learning method is desired for dynamic resource orchestration. Another example is radio access technology (RAT) selection, in which a user decides to select an optimal RAT among multiple candidate RATs. The user-perceived performance of a RAT is stochastic due to dynamic wireless environments and user mobility. Hence, the RAT selection problem can be modeled as a multi-armed bandit problem, whose objective is to identify the optimal RAT.

2.3 Case Study

In this section, a case study is presented to demonstrate the effectiveness of the proposed network slicing frameworks through computer simulations. A two-tier wireless network is created using MATLAB, where an MBS with the cell coverage radius of 600 m is deployed and is underlaid by 4 SBSs with the communication radius of 200 m. The downlink wireless transmission power levels are set as 40 dBm and 30 dBm on the MBS and each of the SBSs, respectively. The separation distance between the MBS and each SBS is set to 400 m. The pre-configured radio spectrum resources on each BS is 10 MHz, and all the SBSs reuse the same portion of spectrum resources to exploit the spatial multiplexing gain. MTDs and MUs are randomly deployed within the BS coverage areas, and we assume that all MTDs belong to one machine-type service and all MUs are subscribed to one broadband data service. The processes of downlink packet arrivals at each BS are assumed Poisson, with the rate parameter of 5 packet/s, for each MTD and periodic with the rate of 20 packet/s for each MU. The packet sizes of each machine-type packet and each data packet are set as 2000 bits and 9000 bits, respectively. For the core network, two SFCs from the DNS service and the streaming service are placed on a common network path, as shown in Fig. 2.2, with the firewall function F_1 placed on NFV node S_1 and the DNS function F_2 and the IDS function F_3 placed on NFV node S_2. There are two equally weighted traffic flows x and y traversing the respective

SFCs ($F_1 \longrightarrow F_2$ and $F_1 \longrightarrow F_3$) through the configured common traffic routing path.

The physical network path deployment and the packet-level traffic forwarding are simulated in OMNeT++ [29]. The average packet arrival rates at S_1 and the streaming service are set as 150 packet/s with the packet size of 4000 bits for the DNS service and in-between 150 packet/s and 350 packet/s with the packet size of 16,000 bits for the streaming service. The time consumption profiles for flows x and y traversing S_1 are tested over a resource virtualization platform OpenStack [30], where the VNFs of both SFCs are instantiated on different virtual machines (VMs) as customized software. By injecting each of the traffic flows to pass through the corresponding VNFs, the resource consumption profiles for flow x traversing F_1 and F_2 are [1000, 2000] packet/s and [1000, 1250] packet/s, and for flow y traversing F_1 and F_3 are [750, 500] packet/s and [800, 312.5] packet/s [19].

For the 5G wireless network, we compare the proposed network-level radio resource slicing framework with the service-level slicing scheme [15]. In the proposed framework, the radio resource slicing ratios on BSs are dynamically adjusted in response to the traffic load variations in each small-cell, as shown in Fig. 2.4, to maximize the network-wide communication resource utilization. In contrast, for the service-level resource slicing, the radio resources on each BS are partitioned among groups of service users/devices under the BS coverage. However, resource sharing among BSs is not activated.

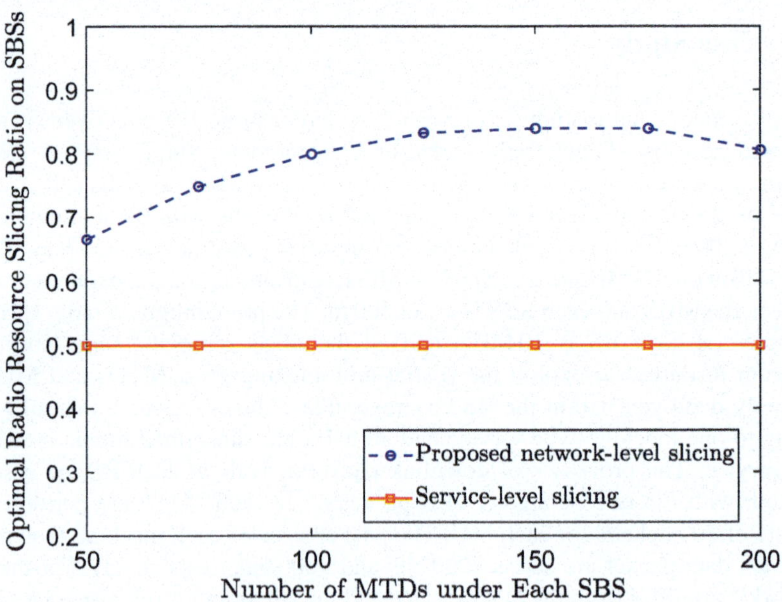

Fig. 2.4 Optimal radio resource slicing ratio on SBSs with varying network load conditions

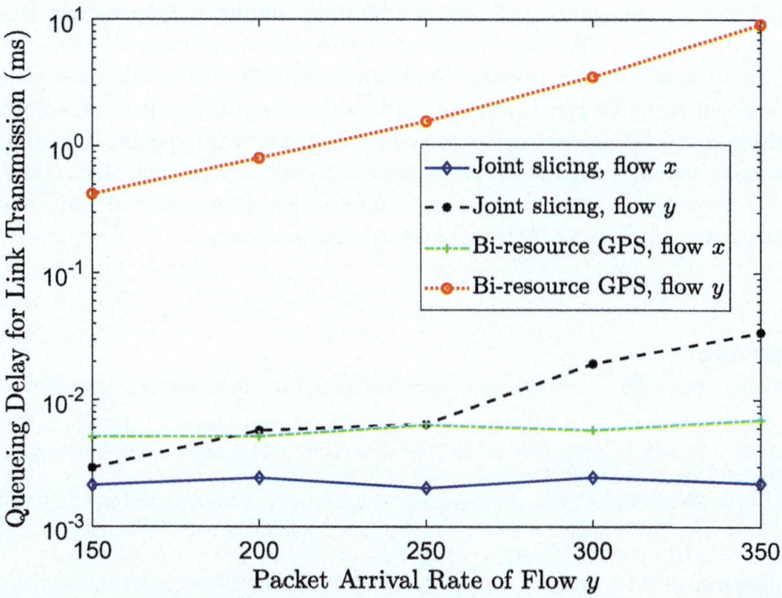

Fig. 2.5 Packet queueing delay at S_1 for link transmission

For the 5G core network, the proposed joint computing and transmission resource slicing scheme is compared with bi-resource GPS [12, 18], in terms of packet queueing delay at the output transmission link when flows x and y traverse NFV node S_1. We can see from Fig. 2.5 that the queueing delay achieved by the proposed slicing scheme is much lower than the bi-resource GPS scheme. As the proposed slicing scheme maximizes the utilization of both resource types, the queueing delay at the output link port of S_1 is minimal, whereas the transmission resource slicing is imbalanced between the traffic flows resulting in packet accumulation on the output transmission link for one of the traffic flows.

2.4 Conclusion

In this chapter, we have presented a comprehensive network slicing framework for E2E 5G networks, including both 5G wireless and core network domains. Based on the distinct features of 5G networking and service requirements, we propose new network slicing solutions to provide fine-grained resource orchestration using SDN and NFV technologies, such that the networking and computing resource utilization is improved with customized QoS satisfaction for different services. Specifically, for 5G wireless networks, a two-level dynamic radio resource slicing framework is proposed to maximize the overall communication resource utilization by enabling the network-level resource sharing among heterogeneous BSs; for 5G core networks, a joint computing and transmission resource slicing scheme is designed to achieve high resource utilization with the slicing fairness guarantee for dominant resource types and QoS isolation among different service flows. We also discuss an AI-assisted network slicing lifecycle with specific functionalities to automate the slice creation process with reduced slice management complexity for B5G networks. A case study is presented to demonstrate the effectiveness and advantages of the proposed E2E network slicing framework.

References

1. J. Ordonez-Lucena, P. Ameigeiras, D. Lopez, J.J. Ramos-Munoz, J. Lorca, J. Folgueira, Network slicing for 5G with SDN/NFV: Concepts, architectures, and challenges. IEEE Commun. Mag. **55**(5), 80–87 (2017)
2. Technical Specification Group Services and System Aspects; Summary of Rel-15 Work Items (Release 15), 3rd Generation Partnership Project, Sophia Antipolis Valbonne, France, Tech. Rep. TR 21.915 V15.0.0 (2019)
3. W. Zhuang, Q. Ye, F. Lyu, N. Cheng, J. Ren, SDN/NFV-empowered future IoV with enhanced communication, computing, and caching. Proc. IEEE **108**(2), 274–291 (2019)
4. Q. Ye, W. Shi, K. Qu, H. He, W. Zhuang, X. Shen, Joint RAN slicing and computation offloading for autonomous vehicular networks: a learning-assisted hierarchical approach. IEEE Open J. Veh. Technol. **2**, 272–288 (2021). https://doi.org/10.1109/OJVT.2021.3089083
5. Q. Ye, W. Zhuang, S. Zhang, A.L. Jin, X. Shen, X. Li, Dynamic radio resource slicing for a two-tier heterogeneous wireless network. IEEE Trans. Veh. Technol. **67**(10), 9896–9910 (2018)
6. F. Bari, S.R. Chowdhury, R. Ahmed, R. Boutaba, O.C.M.B. Duarte, Orchestrating virtualized network functions. IEEE Trans. Netw. Serv. Manage. **13**(4), 725–739 (2016)
7. R. Riggio, A. Bradai, D. Harutyunyan, T. Rasheed, T. Ahmed, Scheduling wireless virtual networks functions. IEEE Trans. Netw. Serv. Manage. **13**(2), 240–252 (2016)
8. J. Li, W. Shi, Q. Ye, N. Zhang, W. Zhuang, X. Shen, Multi-service function chain embedding with delay-guarantee: a game-theoretical approach. IEEE Internet Things J. **8**, 11219–11232 (2021). https://doi.org/10.1109/JIOT.2021.3051905
9. O. Alhussein, P.T. Do, Q. Ye, J. Li, W. Shi, W. Zhuang, X. Shen, X. Li, J. Rao, A virtual network customization framework for multicast services in NFV-enabled core networks. IEEE J. Sel. Areas Commun. **38**(6), 1025–1039 (2020)
10. Q. Duan, N. Ansari, M. Toy, Software-defined network virtualization: an architectural framework for integrating SDN and NFV for service provisioning in future networks. IEEE Netw. **30**(5), 10–16 (2016)
11. N. McKeown, T. Anderson, H. Balakrishnan, G. Parulkar, L. Peterson, J. Rexford, et al., OpenFlow: enabling innovation in campus networks. ACM SIGCOMM Comput. Commun. Rev. **38**(2), 69–74 (2008)
12. Q. Ye, J. Li, K. Qu, W. Zhuang, X.S. Shen, X. Li, End-to-end quality of service in 5G networks: examining the effectiveness of a network slicing framework. IEEE Veh. Technol. Mag. **13**(2), 65–74 (2018)
13. X. Shen, J. Gao, W. Wu, K. Lyu, M. Li, W. Zhuang, X. Li, J. Rao, AI-assisted network-slicing based next-generation wireless networks. IEEE Open J. Veh. Technol. **1**, 45–66 (2020)
14. W. Wu, N. Chen, C. Zhou, M. Li, X. Shen, W. Zhuang, X. Li, Dynamic RAN slicing for service-oriented vehicular networks via constrained learning. IEEE J. Sel. Areas Commun. **39**(7), 2076–2089 (2021)

15. C. Liang, F.R. Yu, H. Yao, Z. Han, Virtual resource allocation in information-centric wireless networks with virtualization. IEEE Trans. Veh. Technol. **65**(12), 9902–9914 (2016)
16. J. Chen, Q. Ye, W. Quan, S. Yan, P.T. Do, P. Yang, W. Zhuang, X. Shen, X. Li, J. Rao, SDATP: an SDN-based traffic-adaptive and service-oriented transmission protocol. IEEE Trans. Cogn. Commun. Netw. **6**(2), 756–770 (2019)
17. A.K. Parekh, R.G. Gallager, A generalized processor sharing approach to flow control in integrated services networks: the single-node case. IEEE/ACM Trans. Netw. **1**(3), 344–357 (1993)
18. W. Wang, B. Liang, B. Li, Multi-resource generalized processor sharing for packet processing, in *Proceedings of ACM IEEE/ACM International Symposium on Quality of Service (IWQoS)* (2013), pp. 1–10
19. Q. Ye, W. Zhuang, X. Li, J. Rao, End-to-end delay modeling for embedded VNF chains in 5G core networks. IEEE Internet Things J. **6**(1), 692–704 (2018)
20. W. Wu, C. Zhou, M. Li, H. Wu, H. Zhou, N. Zhang, W. Zhuang, X. Shen, AI-native network slicing for 6G networks (2021). arXiv:2105.08576
21. J.G. Andrews, S. Buzzi, W. Choi, S.V. Hanly, A. Lozano, A.C. Soong, J.C. Zhang, What will 5G be? IEEE J. Sel. Areas Commun. **32**(6), 1065–1082 (2014)
22. N. Rajatheva, I. Atzeni, E. Bjornson, A. Bourdoux, S. Buzzi, J.-B. Dore, S. Erkucuk, M. Fuentes, K. Guan, Y. Hu, et al., White paper on broadband connectivity in 6G (2020). arXiv:2004.14247
23. N. Zhang, S. Zhang, P. Yang, O. Alhussein, W. Zhuang, X. Shen, Software defined space-air-ground integrated vehicular networks: challenges and solutions. IEEE Commun. Mag. **55**(7), 101–109 (2017)
24. C. Zhou, W. Wu, H. He, P. Yang, F. Lyu, N. Cheng, X. Shen, Deep reinforcement learning for delay-oriented IoT task scheduling in space-air-ground integrated network. IEEE Trans. Wireless Commun. **20**(2), 911–925 (2021)
25. W. Wu, N. Cheng, N. Zhang, P. Yang, W. Zhuang, X. Shen, Fast mmwave beam alignment via correlated bandit learning. IEEE Trans. Wireless Commun. **18**(12), 5894–5908 (2019)
26. W. Wu, P. Yang, W. Zhang, C. Zhou, X. Shen, Accuracy-guaranteed collaborative DNN inference in industrial IoT via deep reinforcement learning. IEEE Trans. Ind. Informat. **17**(7), 4988–4998 (2021)
27. Y. Tang, N. Cheng, W. Wu, M. Wang, Y. Dai, X. Shen, Delay-minimization routing for heterogeneous VANETs with machine learning based mobility prediction. IEEE Trans. Veh. Technol. **68**(4), 3967–3979 (2019)
28. C. Zhou, H. He, P. Yang, F. Lyu, W. Wu, N. Cheng, X. Shen, Deep RL-based trajectory planning for AoI minimization in UAV-assisted IoT, in *International Conference on Wireless Communications and Signal Processing (WCSP)* (2019), pp. 1–6
29. OMNeT++ 5.0, http://www.omnetpp.org/omnetpp. Accessed June 2021
30. Openstack (Release Pike), https://www.openstack.org. Accessed June 2021

Chapter 3
Responsive Regulation of Dynamic UAV Communication Networks Based on Deep Reinforcement Learning

Ran Zhang, Duc Minh (Aaron) Nguyen, Miao Wang, Lin X. Cai, and Xuemin (Sherman) Shen

3.1 Introduction

Unmanned aerial vehicles (UAVs) have been attracting increasing attention as a key component in the future wireless communications [1]. Compared to the terrestrial base stations (BSs) [2, 3], UAVs equipped with wireless transceivers can serve as mobile BSs and stand out in providing highly on-demand services, better wireless connectivity to the ground users, and much lower deployment cost due to the almost infrastructure-free network construction [4]. As reported in [5], the UAV market is estimated at USD 27.4 billion in 2021 and is projected to reach USD 58.4 billion by 2026. In this booming market, UAVs have been exploited in many applications such as mobile edge computing [6, 7], crowd/traffic surveillance [8], emergency rescue [9], cached content delivery [10, 11], network coverage enhancement and extension [12], etc.

R. Zhang (✉) · D. M. (Aaron) Nguyen
College of Engineering and Computing, Miami University, Oxford, OH, USA
e-mail: zhangr43@miamioh.edu; nguyendm@miamioh.edu

M. Wang
Miami University, Oxford, OH, USA
e-mail: wangm64@miamioh.edu

L. X. Cai
Department of Electrical and Computer Engineering, Illinois Institute of Technology, Chicago, IL, USA
e-mail: lincai@iit.edu

X. (Sherman) Shen
Department of Electrical and Computer Engineering, University of Waterloo, Waterloo, ON, Canada
e-mail: sshen@uwaterloo.ca

L. Cai et al. (eds.), *Broadband Communications, Computing, and Control for Ubiquitous Intelligence*, Wireless Networks,
https://doi.org/10.1007/978-3-030-98064-1_3

Various aspects of UAV-based communications have been extensively studied, ranging from radio resource allocation and trajectory design to energy management and computing offloading [13, 14]. Conventional approaches typically formulate the studied problems into (mixed integer) non-convex optimization problems. The original NP-hard problem is generally decoupled into a set of sub-problems and solved by iterative algorithms [6, 15–22]. This conventional methodology is a better fit where the network parameters are fixed. In UAV-based communications, parameters such as the network topology, wireless channel conditions, and user distributions are usually time-varying due to the mobility of UAVs, topographic relief and the temporality of the on-demand services. As a result, the above methods need to be re-executed each time the parameters are updated. With the exponentially increasing network scale and heterogeneity in the future, it will be increasingly difficult for conventional approaches to handle the network dynamics.

Thanks to recent advances in machine learning [23], reinforcement learning (RL) [24, 25] is becoming a promising solution to UAV communication problems. By constantly interacting with the environment and learning from the interaction experiences, RL agents are strongly capable of making sequential decisions in time-varying environments free of the environment models. Existing RL-based studies on UAV communications focus mainly on control policy development given a fixed set of UAVs [26–37]. Few works have investigated how the network should be regulated considering the dynamic lineup change of the serving UAVs. Due to the ad hoc nature of the UAV networks, the serving UAV lineup can dynamically change at times. UAVs have to quit the network when their batteries are depleted; supplementary UAVs can also join the serving lineup whenever needed. Either case will inevitably create fluctuations in the network performance, thus calling for responsive regulation strategies when such changes happen. When regulated, the network should not passively react after the change, but identify the upcoming change and take actions in advance to minimize (or maximize) the performance loss (or gain) during the transition to the new optimal UAV positions. Such procedure is referred to as proactive self-regulation (PSR) in this chapter. A major challenge of using RL for PSR include is that the dimensions of the state and action space both change during the training process, which is irregular for RL. Another challenge is how to promote the learning exploration around the time of change so that the agent is able to take actions in advance. In addition, most of the existing works only consider stationary user distribution, whereas the distribution can be dynamic in practice. The works [29] and [38] considered user mobility in the RL framework, but the UAV trajectories are limited to a mesh grid.

Motivated by the above considerations, PSR of a UAV communication network is investigated in this chapter with dynamic change in UAV lineup and user distributions. We aim to achieve an optimal UAV control policy via deep RL (DRL) which relocates the UAVs in advance when (1) at least one UAV is about to quit or join the network, or (2) the user distribution changes, rather than passively relocates the UAVs after the change. To the best of our knowledge, this is the first work on optimal regulation of a UAV communication network that jointly considers the dynamic UAV lineup and user distribution. The contributions are given as follows.

- A DRL-based approach for PSR of UAV communication network is developed. The approach aims to maximize the accumulated user satisfaction (US) score of the considered time horizon where the change in UAV lineup happens. To accommodate the continuous state space and action space, the state-of-the-art actor-critic learning method, deep deterministic policy gradient (DDPG) [39], is selected among all the DRL variants so that the UAV battery status, positions, and movements can be accurately recorded.
- To promote the learning exploration around the lineup change and achieve better training performance on both the actor and critic networks, an asynchronous parallel computing (APC) structure is proposed. The proposed PSR approach under APC is referred to as PSR-APC.
- The PSR-APC approach is further extended to the case of dynamic user distribution. Time is integrated as one of the learning states to achieve a time-dependent control policy.
- Extensive simulations are conducted to demonstrate the convergence and efficacy of the proposed PSR-APC approach. Compared to a passive reaction method, the proposed approach achieves surpassing accumulated US scores during the transition period.

The remainder of the chapter is organized as follows. Section 3.3 describes the system model and formulates the problem. Section 3.4 introduces preliminary knowledge on DRL and the adopted DDPG algorithm. Section 3.5 elaborates the detailed design of the proposed APC-PSR approach. Section 3.6 extends the APC-PSR from fixed user locations to dynamic user distributions. Numerical results are presented in Sect. 3.7. Finally, Sect. 3.8 concludes the chapter.

3.2 Related Works

The conventional optimization or rule-based methods have been extensively applied to UAV communications. For instance, Nasir et al. [12] and Zeng et al. [15] studied the resource allocation (RA) and trajectory design problem of a single-UAV to maximize the minimum user rate and the mission completion time, respectively. When multiple UAVs are present, UAVs need to be coordinated in interference management, trajectory design, and user association. Mozaffari et al. [19] studied the joint UAV positioning, frequency planning, and user association problem to minimize the UAV-user latency. Wu et al. [20] additionally considered UAV trajectory design and power control. Mozaffari et al. [21] focused on energy consumption and minimized the total propulsion energy of UAVs.

When the network environment is time-varying and sequential decisions need to be made, RL-based UAV control approaches have been studied. Many existing works rely on a centralized agent to learn optimal joint policies for all the network entities. For instance, Singh et al. [31], Challita et al. [32], and Tang et al. [33] applied deep Q-learning (QL) to optimize the RA, interference management, and trajectory design of UAVs, respectively, in UAV-assisted cellular networks. Liu et al. [35] employed double QL to design optimal UAV trajectories that maximize the number of satisfied users with time-constrained requirements. To accommodate large action space and expedite convergence, actor-critic (AC) based deep RL (DRL) is applied. Cheng et al. [34] proposed an AC RL approach to optimize RA and task scheduling in UAV-assisted computing offloading. Khairy et al. [37] studied the joint altitude control and channel access problem of a solar-powered UAV network by employing actor-critic RL. Liu et al. [36] targeted the energy concerns of UAVs and exploited the up-to-date AC variant, i.e., deep deterministic policy gradient (DDPG) algorithm, to jointly maximize the energy efficiency, user fairness and network coverage.

Multi-agent RL (MARL) has been exploited in a few existing works to make the learning distributed and scalable to the network size. For instance, Klaine et al. [26] proposed a distributed QL approach to find the UAV positions that maximize the total amount of covered users. Cui et al. [27] and Hu et al. [28] applied multi-agent QL to optimize RA and trajectory design, respectively. Liu et al. [29] developed a multi-agent QL framework to optimize the UAV trajectory and power control, considering the ground user mobility. Hu et al. [30] proposed a value decomposition based MARL solution coupled with a meta-training mechanism to accelerate the learning of multi-UAV trajectories while generalizing the learning to unfamiliar environments. Pham et al. [40] and Chen et al. [41] integrated game theories into MARL to solve the complex dynamic of the joint UAV actions and simplify the complex interactions between multiple objectives and multiple UAVs, respectively.

All the above works consider a fixed set of serving UAVs. The proposed work will be among the first to fill this research gap.

3.3 System Model and Problem Formulation

In this section, the system model is first depicted, followed by the problem formulation.

3.3.1 Network Environment

As illustrated in Fig. 3.1, we consider a target area $\mathbf{A}$ with a set $\mathbf{S}_{ur}$ of N_u ground users served by a lineup $\mathbf{S}_{UAV}$ of N_{UAV} UAVs. The target area is an L-by-L square. A large percentage of the users are randomly distributed around several separate hot spots while the remaining are uniformly distributed throughout $\mathbf{A}$. The UAVs fly within $\mathbf{A}$ at a fixed altitude H to serve the ground users with guaranteed minimum throughput. The antennas of each UAV are strongly directional such that the transmit power is concentrated within an aperture angle of θ right below the UAV. As a result,

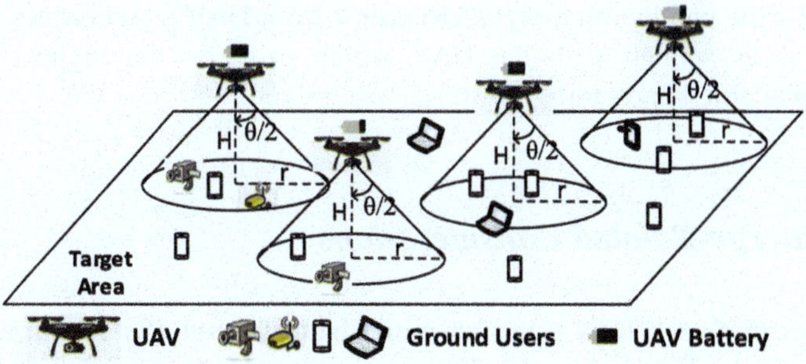

Fig. 3.1 UAV coverage range as a disk area

the coverage of a UAV on the ground is a disk area with radius $\mathbf{r} = H\tan(\frac{\theta}{2})$, as shown in Fig. 3.1. Users will not be interfered by a given UAV if they are outside its coverage disk.

3.3.2 Spectrum Access

All the UAVs are connected to external networks via back-haul links (e.g., satellite links). There is no spectrum overlapping between the UAV back-haul links and the UAV-user links so that there is no mutual interference. We denote the path loss from UAV i to ground user u as PL_{iu} which follows a commonly adopted model by Al-Hourani et al. [42]:

$$PL_{iu} = 20\log_{10}\left(\frac{4\pi f_c d_{iu}}{c}\right) + \eta \quad (\text{dB}), \tag{3.1}$$

where f_c denotes the center frequency of the spectrum assigned to user u, d_{iu} denotes the 3-D distance between UAV i and user u, c denotes the speed of the light, and η denotes extra loss taking different values for LoS and non-LoS links. Given Eq. (3.1), the signal-to-interference-and-noise ratio (SINR) from UAV i to user u is calculated as

$$SINR_{iu} = \frac{P_t G_{iu}}{n_0 + \sum_{j \in \mathbf{S}_u^{UAV} \setminus \{i\}} P_t G_{ju}}, \tag{3.2}$$

where $G_{iu} = 10^{-PL_{iu}/20}$. In Eq. (3.2), P_t is the power spectrum density (psd) of UAV transmissions, n_0 is the psd of the environment noise, $\mathbf{S}_u^{UAV}$ is the set of UAVs that cover user u.

Each user requires a minimum throughput of R_u. Thus, user u can be served by UAV i only when $i \in \mathbf{S}_u^{UAV}$ and the following condition is met:

$$W_{iu} \log_2(1 + SINR_{iu}) \geq R_u, \tag{3.3}$$

where W_{iu} is the bandwidth assigned to user u from UAV i. According to Eq. (3.3), each user is associated with the UAV which provides the highest SINR with sufficient available bandwidth.

3.3.3 Energy-Related Considerations

We consider that each UAV i has an initial battery level E_0^i. The time horizon is divided into time slots of duration T. In time slot t, UAV i spends time up to $T_1 < T$ to move a distance of $d_t^i \in [0, d_{max}]$ at a constant speed v in the direction of $\alpha_t^i \in [0, 2\pi)$ and then hovers in the new position for the remaining time to serve. The power consumption of level flight is given as follows according to [43],

$$P_{level} = \frac{W}{\sqrt{2}\rho A} \frac{1}{\sqrt{v^2 + \sqrt{v^4 + 4V_h^4}}}, \tag{3.4}$$

where $V_h = \sqrt{\frac{W}{2\rho A}}$, W is the weight of UAV in Newtons (N), ρ is the air density, and A is the total area of UAV rotor disks. From Eq. (3.4), it can be inferred that due to speed v, the power of level flight is interestingly less than that of hovering. The energy consumption of UAV i in time slot t (denoted as EC_t^i) is then represented as

$$EC_t^i = E_{FLT}(v, d_t^i, T) + E_{TX} + E_{OP}(T). \tag{3.5}$$

According to Eq. (3.5), the energy consumption of a UAV has three components: (1) energy spent on flying as a function of level speed v, flying distance d_t^i, and slot duration T, (2) energy consumed by signal transmission on both UAV-user and UAV back-haul links, and (3) energy used for operational cost which is assumed to be proportional to T.

Denote the battery residual of UAV i at the end of time slot t as E_t^i. When E_t^i is below a threshold E_{Thre}, UAV i will quit the network immediately for charging. Denote the altitude of UAV i at the end of time slot t as $H_t^i \in [H_{min}, H]$, where H_{min} is the altitude of the UAV charging point. A UAV will stop elevating when it reaches the serving altitude H. The variable H_t^i will be used in the case of UAV join-in.

3.3.4 Problem Formulation

The learning agent aims to find an optimal multi-UAV relocation policy which maximizes the accumulated US scores within a time horizon of N_T time slots, where UAVs may quit or join in the network. The optimization problem is given as follows.

$$\begin{aligned} &\max_{x_t^i, y_t^i} \sum_{t=1}^{N_T} SC_t \\ &s.t. \quad 0 \le x_t^i \le L, \forall i \in \{1, 2, \cdots, N_{UAV}\} \\ &\qquad\ \ 0 \le y_t^i \le L, \forall i \in \{1, 2, \cdots, N_{UAV}\}, \\ &\text{where} \\ &SC_t := \left(\textstyle\sum_{u \in \mathbf{S}_{ur}} X_t^u\right)^{\beta}. \end{aligned} \tag{3.6}$$

In Eq. (3.6), (x_t^i, y_t^i) represents the horizontal coordinates of UAV i. The US score in time slot t is denoted as SC_t and defined as the function of the total number of users that get served with satisfied throughput requirement. Define $X_t^u \in \{0, 1\}$ as an indicator which takes value 1 when user u is successfully served and 0 when not. The value of X_t^u is jointly determined by UAV parameters (i.e., the number of serving UAVs, the positions, battery status, altitude), user distribution dynamics, and spectrum access policy. The exponent $\beta > 0$ is a factor weighing how much the agent cares about the overall user satisfaction based on the number of users successfully served.

With the above formulation, when one UAV is about to be depleted and needs to quit soon, the agent is expected to relocate the serving UAVs ahead of the quit to reduce service holes as much as possible, rather than to react after the UAV quits. When one UAV is joining in the network, the agent is expected to determine its horizontal positions while elevating to the serving height.

3.4 Preliminaries

In the context of general RL, the agent interacts with the environment by taking an action A_t for the environment state (or observation) S_t at time step t. A reward r_{t+1} is then obtained for taking A_t at S_t. The learning target is an optimal policy π which determines the best action A for every state S that maximizes the expected future return R defined as

$$R = \sum_{t=0}^{\infty} \gamma^t r_{t+1}, \ \gamma \in [0, 1]. \tag{3.7}$$

There are generally two basic categories of RL approaches: value-based and policy-based RL. Q-learning (QL) [44] is a basic and representative value-based method. QL achieves the optimal π by estimating the value of taking action A at state S which is quantified by the function $Q(S, A)$. The optimal policy π^* is obtained as the collection of $A^* = \arg\max_A Q(S, A), \forall S$. The $Q(S, A)$ function is iterated to a guaranteed convergence according to the following formula,

$$
\begin{aligned}
&Q_{t+1}(S_t, A_t) \\
&= Q_t(S_t, A_t) + \alpha \left[r_{t+1} + \gamma \max_{A} Q_t(S_{t+1}, A) - Q_t(S_t, A_t) \right],
\end{aligned} \tag{3.8}
$$

where α is the learning rate of the RL agent. Nevertheless, QL has a major drawback that it suffers from the "curse of dimensionality." A Q-matrix needs to be maintained for each state-action pair, which is prohibitive when the state space is extremely large or infinite. This is often the case in communication and networking. To tackle this issue, deep QL (DQL) was developed which exploits a deep neural network (DNN), referred to as deep Q-network (DQN), to approximate the $Q(\cdot)$ function [45]. The number of inputs of the DQN is equal to the dimension of the state space, and the number of outputs is equal to the cardinality of the action set. Compared to the Q-matrix, a DQN reduces the input count to the dimension of the state space and consequently solves the memory anxiety by avoiding a large state space. The DQN is trained by minimizing the loss function below [46]:

$$
\mathcal{L}(\theta_Q) = \mathbb{E}[y_t - Q(S_t, A_t|\theta_Q)]^2, \tag{3.9}
$$

where θ_Q denotes the tunable weights of the DQN, y_t is the label value obtained as follows,

$$
y_t = \begin{cases} r_{t+1}, \text{ if } S_t \text{ is a terminal state;} \\ r_{t+1} + \gamma \max\limits_{A_{t+1}} Q(S_{t+1}, A_{t+1}|\theta_Q), \text{ otherwise.} \end{cases} \tag{3.10}
$$

DQL solves the dimension anxiety in state space, but the value-based methods may only apply to problems with low-dimensional discrete action space. The reason is that the value-based methods need to exhaustively search all possible actions to determine the best for a state. Such exhaustive search is difficult to achieve for a large or infinite action space. When power control or UAV mobility control (as considered in this chapter) are involved, the action space is continuous. Discretizing the action space is one possible option, but will lead to prohibitive training complexity and/or non-negligible loss in accuracy.

Policy-based methods can well solve the dimension anxiety in the action space. Instead of determining the optimal policy via the $Q(\cdot)$ values, the methods parameterize and optimize the policy $\pi(\theta\mu)$ itself. The $Q(\cdot)$ values may still be used to update the policy parameters θ_μ, but not for selecting actions directly. The actor-critic (AC) method stands out among all the policy-based methods due to the merit of reducing variance of the policy gradients. A basic AC agent is shown in Fig. 3.2. The agent consists of a critic and an actor, both being DNNs. The critic uses the collected experiences to update the $Q(\cdot)$ function via updating θ_Q. The actor combines the updated $Q(\cdot)$ values and the experiences to update π via updating θ_μ. The new action A' to be performed is determined by the actor network.

Among all the AC variants, DDPG is one of the best to handle the problem of convergence instability [39]. Specifically, DDPG exploits target networks for both the critic network ($Q(S, A|\theta_Q)$) and actor network ($\mu(S|\theta_\mu)$). The target networks,

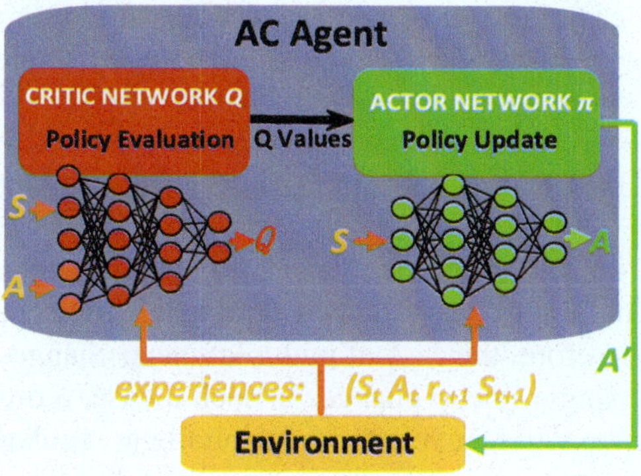

Fig. 3.2 The diagram for AC method

denoted as $Q'(S, A|\theta_{Q'})$ and $\mu'(S|\theta_{\mu'})$, have the same setup and initialization as $Q(S, A|\theta_Q)$ and $\mu(S|\theta_\mu)$, respectively, but are updated much more slowly in each time step:

$$\theta_{Q'} = \tau\theta_Q + (1-\tau)\theta_{Q'}, \qquad \theta_{\mu'} = \tau\theta_\mu + (1-\tau)\theta_{\mu'}, \tag{3.11}$$

where $\tau \ll 1$. The target networks are used to update the label value y_t in Eq. (3.9). Correspondingly, Eq. (3.10) is re-written as

$$y_t = \begin{cases} r^{t+1}, \text{ if } S_t \text{ is terminal state;} \\ r_{t+1} + \gamma Q'(S_{t+1}, \mu'(S_{t+1}|\theta_{\mu'})|\theta_{Q'}), \text{ otherwise.} \end{cases} \tag{3.12}$$

The slow update of the target networks prevents a bad y_t from being generated due to a bad deviation in θ_Q or θ_μ, thus significantly stabilizing the convergence. Using the updated Q values, the actor network is updated as follows:

$$\begin{aligned} &\nabla_{\theta_\mu} J \approx \mathbb{E}[G_a G_\mu], \\ &\text{where } G_a = \nabla_{\mu(S|\theta_\mu)} Q(S, A|\theta_Q), \\ &\qquad\quad G_\mu = \nabla_{\theta_\mu}\mu(S|\theta_\mu). \end{aligned} \tag{3.13}$$

3.5 Learning Algorithm Design for Proactive Self-Regulation Strategy

The design of the proposed PSR-APC approach is detailed in this section. The DDPG agent is implemented in a centralized server, which communicates regularly with all the UAVs via their back-haul links. During training, the agent keeps collecting the interaction experiences between UAVs and the network environment and updating both critic and actor networks. When the training is complete and the strategy is executed, the well-trained actor network sequentially collects the UAV information (states) as inputs and outputs flying instructions (actions) to each UAV in each time step. These movements collectively result in an optimal set of UAV trajectories to maximize the accumulative US score within the considered time horizon.

In addition, we consider that after one UAV quits or joins in, all the serving UAVs have sufficient time to reach the new optimal positions before another lineup change happens. Therefore, the case of multiple lineup changes can be regarded as multiple cases of a single lineup change. In such a case, a different trained agent for each lineup change will be exploited sequentially to regulate the network in the considered time horizon.

In the following, we elaborate the design from the aspects of states, actions, reward function, state transitions, tune-ups, and parallel computing. The cases of UAV quit and join-in are both considered.

3.5.1 State Space

3.5.1.1 Case of UAV Quit

The timing of UAV quit and the resultant UAV movements are closely dependent on the battery level of the UAVs. Hence, the learning states will include UAV positions and battery residual of each UAV.

The UAV positions directly determine the number of users that get successfully served in each time step. As the UAVs fly at a fixed height when serving, only the 2-D coordinates (x_t^i, y_t^i), $\forall i \in \mathbf{S}_{UAV}$ need to be considered at time step $t \leq N_T$. The movements of UAVs are limited within the target area $\mathbf{A}$, i.e., $x_t^i,\ y_t^i \in [0, L]$.

The battery residual of UAVs $\{E_t^i\}$ is a key conditional factor. It has little impact on the UAV movements when the battery level of all the UAVs is adequate. Yet when any E_t^i falls close to E_{Thre} (i.e., any UAV is running out of battery and about to quit), this factor should have significant impact on the UAV movements. The best timing of enabling the significance of $\{E_t^i\}$ will be learnt by the DDPG agent. Moreover, E_t^i is bounded within $[E_{Thre}, E_0^i]$.

Collectively, the formal state vector of the designed learning approach is defined as $S_t = [x_t^1, \cdots, x_t^{N_{UAV}}, y_t^1, \cdots, y_t^{N_{UAV}}, E_t^1, \cdots, E_t^{N_{UAV}}]$, with cardinality $3N_{UAV}$.

3.5.1.2 Case of UAV Join-In

A UAV is considered to start serving only when it reaches the serving altitude H. Similar to the battery residual of UAVs in the case of UAV quit, UAV altitude is the key factor in this case which determines the timing of proactive UAV relocation. The existing UAVs will bide their time until the joining UAV is about to reach the serving altitude. While elevating to the serving altitude, the joining UAV needs to adjust its horizontal position since where to join the UAV network is critical to maximizing the accumulative US score.

Hence, the formal state vector of the designed learning approach is defined as $S_t = [x_t^1, \cdots, x_t^{N_{UAV}}, y_t^1, \cdots, y_t^{N_{UAV}}, H_t^1, \cdots, H_t^{N_{UAV}}]$, with cardinality of $3N_{UAV}$. Based on the collected experiences, the agent will learn the best period for $\{H_t^i\}$ to take effect.

3.5.2 Action Definition

The action set of the APC-PSR approach is the same for both cases. As a centralized agent controls the movements of all the UAVs, the collective actions from all the UAVs form the agent action A_t in time step t. The action A_t^i of UAV i has two dimensions: moving direction $\alpha_t^i \in [0, 2\pi)$ and moving distance $d_t^i \in [0, d_{max}]$. In each time step, one UAV could either keep hovering still or move in any direction for a maximum distance d_{max}. Thus, the formal action vector of the proposed APC-PSR approach is defined as

$$A_t = [\alpha_t^1, \cdots, \alpha_t^{N_{UAV}}, d_t^1, \cdots, d_t^{N_{UAV}}],$$

with cardinality $2N_{UAV}$.

3.5.3 Reward Function Design

Both the cases of UAV quit and join-in share the same reward function design. Let r_t denote the reward at time step t. To align with the maximization objective in Eq. (3.6), r_t is designed as a function of the instantaneous US score in step t, i.e., SC_t:

$$r_t = \left(\frac{\sum_{u \in \mathbf{S}_{ur}} X_t^u}{N_u} \right)^\beta = \frac{SC_t}{(N_u)^\beta}. \tag{3.14}$$

In (3.14), the instantaneous US score SC_t is divided by $(N_u)^\beta$. Empirically speaking, keeping the absolute value of the instantaneous reward within 1 may result in better convergence. In addition, when $\beta > 1$, the reward difference between different $(\sum_{u \in \mathbf{S}_{ur}} X_t^u)$ values is amplified. This promotes the agent to act in advance when the UAV lineup is about to change. However, β cannot be too large as $\beta \geq 3$ has been shown to end up with lower converged values in our preliminary simulations. Moreover, under this design, maximizing the accumulated reward is equivalent to maximizing the accumulated US scores within N_T time steps.

An alternative design of reward function is to give negative rewards as a punishment when any UAV move out of the boundaries [36]. The reward function will be something like:

$$r_t = \begin{cases} \left(\sum_{u \in \mathbf{S}_{ur}} X_t^u / N_u\right)^\beta, & \text{if inside boundaries} \\ P, & \text{otherwise} \end{cases} \tag{3.15}$$

where P can be a negative constant or variable proportional to the number of UAVs crossing the boundaries. During training, when one UAV moves out of boundaries, its current movement will be cancelled. A negative reward will be issued for taking the current action A_t at the current state S_t. With such a design, all the episodes will have a fixed number N_T of time steps. Reasonable as this design is, it may make convergence more difficult. This is because in a good reward design with both positive and negative rewards, the negative rewards need to "combat" the positive ones closely during the training for better convergence performance and speed. However, the relative ratio between the positive and negative rewards keeps changing during the training, making it more challenging and computationally complex to achieve satisfying convergence.

3.5.4 *State Transition Definition*

In either case, a state is a terminal state if at least one of the two conditions is met: i) when any UAV moves outside the boundaries of the target area, i.e., $x_t^i < 0$, $y_t^i < 0$, $x_t^i > L$, or $y_t^i > L$; or ii) when N_T time steps are completed. The current episode will end when the terminal state is reached and a new one will start.

Due to the dynamic UAV lineup change, the number of UAVs in the network may change accordingly. This causes the dimension of the *actual* state-action space to explore during the training to vary after one UAV quits or joins the network.

3.5.4.1 Case of UAV Quit

Consider that UAV i quits the network at time step t_q. Then x_t^i, y_t^i, and E_t^i will stay unaltered for any $t > t_q$. Whatever actions α_t^i and d_t^i $(t > t_q)$ are selected, the positions and battery residual of UAV i will never be updated. In other words, the actual dimension of the explorable state space is reduced from $3N_{UAV}$ to $3(N_{UAV} - 1)$. At the same time, UAV i will not be considered in the reward calculation after t_q.

3.5.4.2 Case of UAV Join-In

Suppose UAV i completes charging and is ready to take off to join the network at time step t_c. When $t < t_c$, x_t^i, y_t^i, and H_t^i will stay unaltered. Suppose UAV i elevates to the serving altitude at time step t_s. When $t_c \leq t < t_s$, UAV i is excluded from the instantaneous reward calculation, but its horizontal positions (x_t^i, y_t^i) will change towards the optimal position to maximize the instantaneous reward when formally joining the network. A constant elevation distance h per time step will be used.

3.5.5 *Training Tune-Ups*

3.5.5.1 Tune-Ups for Neural Network Training

Both the critic and actor networks are DNNs. We design both networks to be just complex enough to accurately learn the nonlinear mappings between inputs and outputs while preventing overfitting. Both DNNs contain 2 fully connected hidden layers with 400 and 300 hidden nodes, respectively. To bound the actions as designed in Sect. 3.5.2, we employ *tanh* and *scaling* layers in the actor network. In both networks, ReLU function is used as the activation function, and L_2 regularization is adopted to suppress overfitting. The learning rates for updating both θ_Q and θ_μ is 10^{-4}. Although a larger learning rate may expedite convergence, it more likely leads to convergence instability or sub-optimum. We set the mini-batch size for DNN training to 512, which is a compromise between computational complexity and variance reduction of the gradients of the loss functions. Input normalization is also enabled for faster convergence.

3.5.5.2 Tune-Ups for RL Training

During RL training, both target networks $Q'(S, A|\theta_{Q'})$ and $\mu'(S|\theta_{\mu'})$ are updated slowly at a rate $\tau = 0.001$. The discount factor γ is set to 0.9. A higher γ will force the agent to account more of the future rewards, thus making convergence more difficult. In addition, DDPG adopts an exploration algorithm where the output of the actor network is added with a random noise of zero mean and decaying variance over time steps. In our implementation, the initial variance is 0.6 and decays at a rate of 0.9995. Experience replay is used with sufficient buffer to contain all the experiences. Insufficient buffer may make the agent lose valuable experiences at an early stage if one does not know well which experiences to drop, which will cause notable convergence instability or even divergence.

3.5.6 Parallel Computing

One major challenge during the training is how to fully explore the state-action space to promote action-taking ahead of the lineup change. Failing to do so will lead to convergence to a sub-optimal (sometimes even bad) result, which is often the case in our early simulations. The reason of insufficient exploration is mainly two-fold. First, although experience replay randomly sample experiences from the entire buffer to train the DNNs, the sampled experiences in one mini-batch will inevitably have some correlation due to the Markov nature of RL. Correlation among training examples of DNN will harm the learning accuracy. Second, the dimension of the explorable state space changes accordingly when the UAV lineup changes. Such a change during training often leads to no UAV relocation after the change or no proactive movement ahead of the change.

Increasing the random noise added to the output of the actor network does not help in our case. Inspired by the asynchronous advantage actor-critic (A3C) [47] algorithm, we propose to use the structure of asynchronous parallel computing (APC) to promote exploration, as shown in Fig. 3.3. The structure contains a host client and multiple parallel workers. The host client maintains a unified pair of critic and actor networks and periodically updates both networks from the collected experiences. The parallel workers are mutually independent, each interacting with an independent copy of the same environment. In A3C, each worker has its own set of network parameters and sends the *gradients* of policy loss to the host client. But in APC, the parallel workers share the same set of policy parameters from the host client and upload their own *experiences* to the host client to update the unified

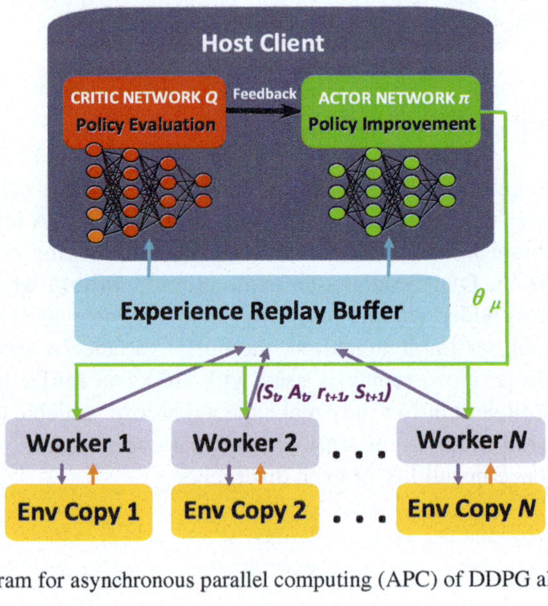

Fig. 3.3 The diagram for asynchronous parallel computing (APC) of DDPG algorithm

neural networks. In our implementation, each worker uploads the experiences upon completion of the current episode and immediately receives updated policy parameters from the host, thus being asynchronous from each other.

Algorithm 1 PSR-APC approach: host client side (UAV quit/join-in)

/***Host Client***/
Randomly initialize critic network $Q(S, A|\theta_Q)$ and actor network $\mu(S|\theta_\mu)$;
Initialize the target networks $Q'(S, A|\theta_{Q'})$ and $\mu'(S|\theta_{\mu'})$ with the same weights: $\theta_{Q'} := \theta_Q, \theta_{\mu'} := \theta_\mu$;
while Not all workers complete all episodes **do**
    if Receive experience set **Ep** from worker k **then**
        Store **Ep** into experience replay buffer B;
        Send θ_μ to worker k;
    end if
    Sample a mini-batch of experiences from B;
    Update θ_Q according to Eqs. (3.9) and (3.12);
    Update θ_μ according to Eq. (3.13);
    Update $\theta_{Q'}$ and $\theta_{\mu'}$ according to (3.11);
end while

Since independent workers interact with different copies of the same environment, the experiences from each worker will be independent. In this way, the correlation among the sampled experiences in a mini-batch are significantly diluted, thus leading to potentially better network training performance. Note that APC itself does not increase the convergence speed in our case as the computational complexity of neural network training is much higher than that of simulating the environment. As a summary, Algorithms 1 and 2 display the pseudo-codes of the proposed APC-PSR approach for both cases of UAV quit and join-in.

3.6 Proactive Self-Regulation with Dynamic User Distribution

The above section considers a *fixed* user distribution in the entire time horizon. But it may be more practical if a *dynamic* user distribution is considered. With dynamic user distribution in the considered time horizon, the optimal UAV positions may have to change from time to time in order to maximize the accumulated US score. This is more challenging compared to the case of the fixed distribution. In such a situation, the optimal self-regulation of UAVs is more towards the optimal UAV trajectory design, with additional proactive response to the UAV lineup change.

In this problem, how to determine the optimal trajectories for all the UAVs according to the dynamically changing user distributions is the key. Some existing works [10, 29] have considered time-varying user distributions under the RL framework. In these works, time-varying user mobility patterns are first predicted using

Algorithm 2 PSR-APC approach: parallel worker side (UAV quit/join-in)

/***Parallel Worker***/
for episode := $1, \cdots, N$ **do**
  Obtain the initial state S_1, IsTerminal := False;
  for epoch $t := 1, \cdots, N_T$ **do**
    $A_t = \mu(S_t|\theta_\mu) + \mathcal{N}$, where $\mathcal{N}$ is stochastic noise with zero mean and decaying variance over t;
    Execute A_t and observe next state S_{t+1};
    for UAV $i := 1, \cdots, N_{UAV}$ **do**
      /*Case of *UAV Quit**/
      if $E_t^i <= E_{Thre}$ **then**
        $S_{t+1}^i := S_t^i$, where S_t^i=$\{x_t^i, y_t^i, E_t^i\}$;
        Exclude UAV i when calculating r_{t+1};
      else
        Obtain S_{t+1}^i according to S_t^i and A_t;
      end if
      /*Case of *UAV Join-In**/
      if $H_t^i < H$ **then**
        Exclude UAV i when calculating r_{t+1};
        $H_{t+1}^i = \min\{H_t^i + h, H\}$;
      end if
      Obtain $\{x_{t+1}^i, y^{i_{t+1}}\}$ according to S_t^i and A_t;
      /*Shared codes begin*/
      if UAV i goes out of boundaries **then**
        Cancel the movement of UAV i;
        IsTerminal := True;
      end if
    end for
    Calculate r_{t+1};
    if IsTerminal==True **then**
      Break;
    end if
  end for
  Send experiences in this episode to host client;
  Obtain updated θ_μ from host client;
end for

either echo state networks (ESNs) or Long-Short Term Memory (LSTM), based on which (multi-agent) QL is employed to achieve the optimal UAV trajectories. Nevertheless, the trajectories are obtained in such a way that optimal positions in each time slot are first derived by RL algorithms given a slot-specific user distribution, and then stringed together into the trajectories. These methods have to train the RL agent(s) once every time slot, instead of training once for the entire time horizon with dynamic user distribution. This may incur high training complexity if a large number of time slots are involved. Different from the above works, we incorporate the time slot t as one dimension of the state space, so that the obtained policy is determined not by one specific distribution in a particular time step, but by a dynamic distribution along the entire time horizon. In this manner, the agent only needs to be trained once over the entire time horizon and ends up with a time-aware

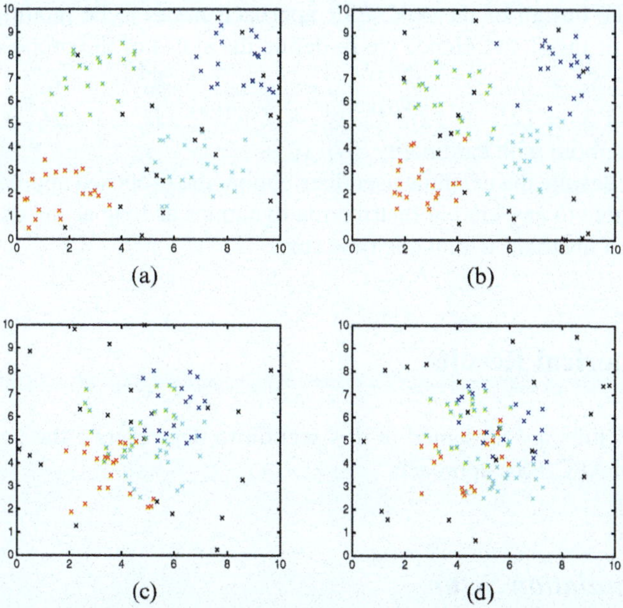

Fig. 3.4 Illustration of time-varying user distribution: Snapshots of different time slots. Users are represented by dots of different colors. Black-dot users are uniformly distributed throughout **A**, and users of each other color belong to one hot spot. From snapshots 1 to 4, hot spots move from being scattered to being overlapped. (**a**) Snapshot 1. (**b**) Snapshot 2. (**c**) Snapshot 3. (**d**) Snapshot 4

optimal policy that may take different actions at different time slots even if the rest of the states are the same.

A simplified model on dynamic user distribution is exploited to investigate the learning performance of the APC-PSR approach on the time-variability of the user distribution. Instead of specifying mobility models for individual users, a different trace is considered for each hot spot center. That is, the center of each hot spot follows a different race to move in the target area **A** with time t. The percentage of users in proximity to each hot spot can be either fixed or moderately varying. Figure 3.4 illustrates an example of moving hot spots and the corresponding user distribution. It can be seen that there are 4 hot spots initially located at the 4 corners of **A**. During N_T time steps, the 4 hot spots first move towards the center of **A**, i.e., $(L/2, L/2)$, stop when reaching a certain distance from the center, stay for a period, and finally move back to where they were initially. Such a disperse-gather-disperse procedure can be used to simulate some realistic scenarios such as when users commute between residences and a central business district during workdays.[1]

As the user distribution is changing, the agent may take different actions even if the UAVs are in the same positions and energy/altitude status at different time steps.

[1] In such a scenario, the duration of one time step needs to be scaled up to the order of minutes.

Therefore, the design of the APC-PSR approach needs to be modified to include time as one of the states. Hence we re-define the states of the proposed APC-PSR approach as: $S_t = [x_t^1, \cdots, x_t^{N_{UAV}}, y_t^1, \cdots, y_t^{N_{UAV}}, E_t^1, \cdots, E_t^{N_{UAV}}, t]$ (case of UAV quit), or $S_t = [x_t^1, \cdots, x_t^{N_{UAV}}, y_t^1, \cdots, y_t^{N_{UAV}}, H_t^1, \cdots, H_t^{N_{UAV}}, t]$ (case of UAV join-in), both with cardinality $3N_{UAV} + 1$.

Note that despite the example user distribution, the proposed approach is deemed to be applicable to any kind of distribution dynamics as long as the distribution can be considered invariant within one time step.

3.7 Numerical Results

Numerical results are presented in this section to demonstrate the performance of the proposed APC-PSR approach.

3.7.1 *Simulation Setup*

The target area is a 10×10 unit square with each unit being 100 meters. The simulations are conducted using Reinforcement Learning Toolbox of Matlab 2020a on a Windows 10 server with Intel Core i7-7700 CPU @ 3.60 GHz and 16 GB RAM. The training has a maximum 10000 episodes, each having up to 100 time steps. The trained agents are tested for a period of $N_T = 100$ time steps. In addition, we consider the transmission-related power of UAVs negligible compared to the propulsion power [48]. Table 3.1 summarizes the main parameters below. Note that $unit \cdot s$ in the table indicates that the value is a product of power (1 power unit = 9.428 W according to Eq. (3.4)) and time (unit is second).

The learning converges to a narrow range instead of a fixed value in most of the simulated cases. For the sake of better presentation, we smooth the convergence curve of the episode reward by averaging over the latest 100 episodes. Intermediate agents with good episode rewards are saved during training and compared during tests to determine the best trained agent.

3.7.2 *Simulation Results*

3.7.2.1 Case Without UAV or User Dynamics

We first simulate the cases without any UAV lineup or user distribution change to get a reference of optimal UAV positions under different N_{UAV}. Figure 3.5 shows the convergence performance of the accumulated US scores. The initial positions of all the UAVs are the evenly separated dots on a circle centered at (5,5). It

Table 3.1 Summary of main parameters

Parameters	Values
Default number of users N_u	100
Default number of UAVs N_{UAV}	5
UAV level speed v	40 km/h
UAV max. elevation speed	14.4 km/h
UAV weight W, air density ρ	4 kg×9.8 m/s^2, 1.225 kg/m^3
Total area of rotor disks A	0.18 m^2
UAV height H, aperture angle θ	3 units, 60°
Max. distance per epoch d_{max}	1 unit
Spectrum center frequency f_c	2 GHz
Spectrum access technology	LTE with resource blocks (RBs)
Spectrum and RB bandwidth	4.5 MHz and 180 kHz
psd of transmission and noise	−49.5 dBm, −174 dBm
Required user throughput R_u	250 kbps
LOS path loss parameter η	1 dB
Time duration per epoch T	10 s
Maximum UAV moving time per epoch T_1	9 s
Maximum UAV communication time per epoch $T - T_1$	1 s
Factor of US score β	2
Energy threshold to quit E_{Thre}	150 unit·s

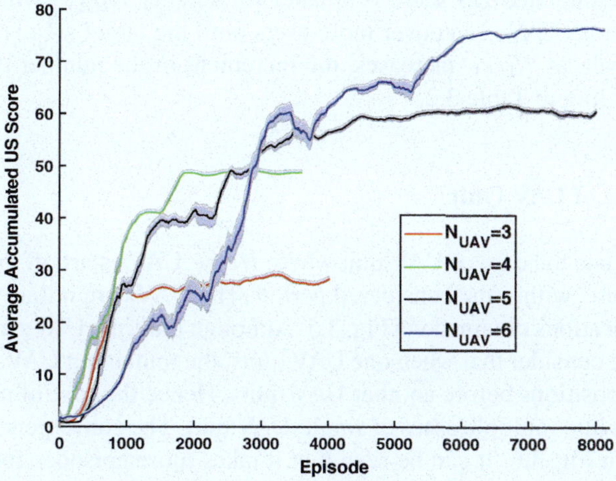

Fig. 3.5 Convergence with 95% credit interval for different N_{UAV} without UAV or user dynamics as a benchmark

can be seen that for all the simulated N_{UAV} values, the accumulated US scores eventually converge, with larger N_{UAV} taking more training episodes. The reason is straightforward: the more UAVs there are, the larger dimension of the state-action space has, thus requiring more time to fully explore and exploit. In addition, the

Table 3.2 Maximum number of served users with N_{UAV}

N_{UAV}	3	4	5	6
Number of served users	56	71	80	88
Increment	–	15	9	8

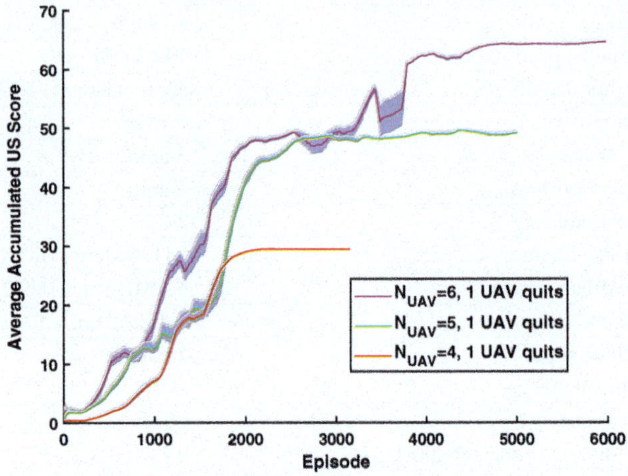

Fig. 3.6 Convergence performance with 95% credit interval for the case of UAV quit

converged accumulated US score is smaller for smaller N_{UAV}, which aligns with the fact that more UAVs can cover more users until the target area is saturated with UAVs. However, as N_{UAV} increases, the increment in the number of served users reduces according to Table 3.2.

3.7.2.2 Case of UAV Quit

We then simulate the case of UAV quit where N_{UAV} UAVs start off in the beginning and 1 UAV quits within the considered period. The UAVs are initially positioned at the optimal locations obtained via Fig. 3.5. Although multiple UAVs may quit during the period, we consider that when one UAV quits, the remaining UAVs will reach the new optimal positions before another UAV quits. Hence the case of multi-UAV quit can be broken into multiple cases of single-UAV quit. The convergence performance is presented in Fig. 3.6. It can be seen that it takes more episodes for larger N_{UAV} to converge. Then, the optimal epoch-wise US scores are shown in Fig. 3.7. As a comparison to the proposed PSR-APC approach, a passive reaction approach is also simulated, which only relocates the remaining UAVs passively after one UAV quits the network.

The epoch-wise US scores in Fig. 3.7 are obtained by combining the agents achieved in Figs. 3.5 and 3.6. The UAVs start from the circular positions, then to the optimal positions maximizing the epoch US score; a UAV then quits the network and the remaining UAVs are finally relocated to the new optimal positions. It can be observed that for all the simulated N_{UAV}, the epoch US score first increases to a

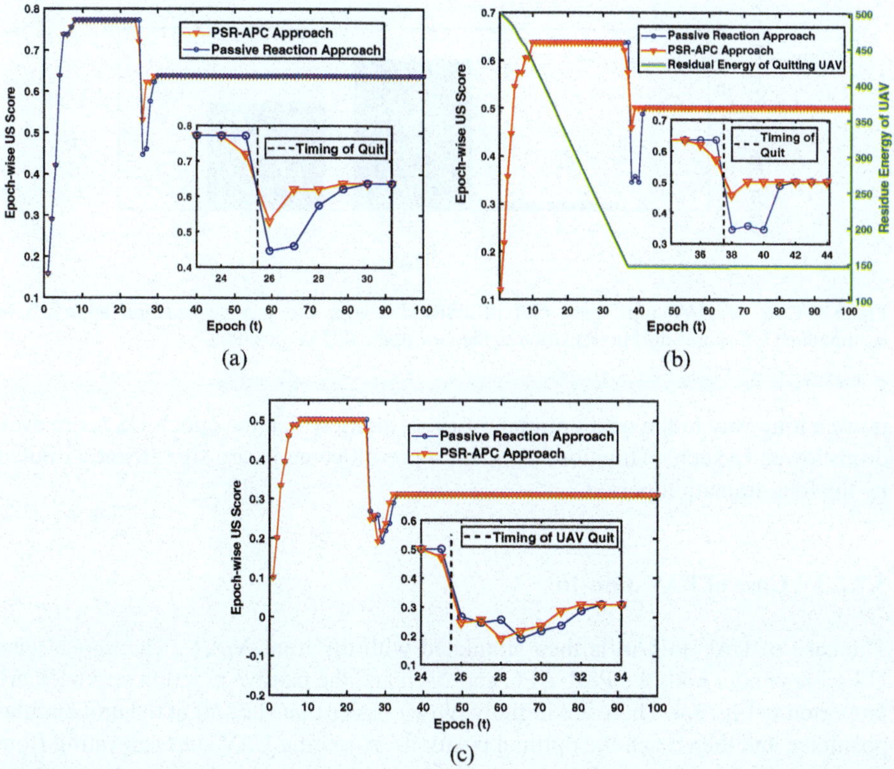

Fig. 3.7 Case of UAV quit: Epoch-wise reward comparison between the PSR-APC approach and the passive reaction approach. (**a**) $N_{UAV} = 6$. (**b**) $N_{UAV} = 5$. (**c**) $N_{UAV} = 4$

maximum as the UAVs are heading to the optimal positions. When a UAV quits the network, the epoch US scores drop dramatically due to the service holes caused by the quit. After a short period of self-regulation, the scores rise up to a new maximum smaller than the previous one when the remaining UAVs reach the new optimal positions. The proposed PSR-APC and the passive reaction approach differ around the timing of UAV quit. The passive reaction approach has no reaction before UAV quit and thus experiences dramatic drop in US scores. On the contrary, the PSR-APC approach monitors the UAV battery status and starts moving the UAVs one or two epochs before the UAV quit. Although the epoch US scores may drop early due to pre-movements, they will not drop that low when the UAV quits as those under the passive reaction approach. Besides, the transition process will be completed earlier. As a result, the accumulated US scores during the transition are higher than those under the passive reaction approach, as shown in Fig. 3.8. However, proactive movement is not always considerably beneficial since the gain depends on specific user distribution and user-to-UAV ratio. When $N_{UAV} = 4$, the gain is marginal. The reason is that before one UAV quits, the 4 UAVs are separately positioned over 4 hot spots far away from each other. When one UAV quits, at least one UAV needs to

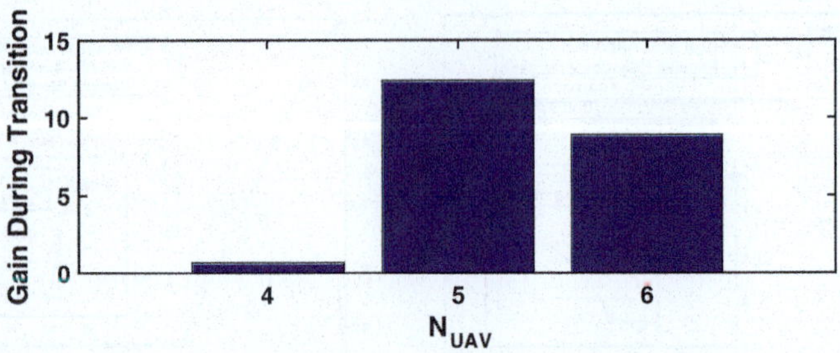

Fig. 3.8 Case of UAV quit: gain (%) of PSR-APC over the passive reaction approach in accumulated US scores during transition to the new optimal UAV positions

move a long way to the next optimal position, along which the epoch US score even drops lower. In such a situation, the gain of pre-movements are significantly diluted by the long transition period.

3.7.2.3 Case of UAV Join-In

The case of UAV join-in is then simulated with different N_{UAV}. The epoch-wise US scores under both the PSR-APC approach and the passive reaction approach are presented in Fig. 3.9. There are initially N_{UAV} UAVs that start off at the unit circular positions, and then reach the optimal positions. A joining UAV starts elevating from the ground in the center (5,5) at epoch 11, and reaches the serving altitude (i.e., formally join the network) at epoch 19. The passive reaction approach in this case relocates UAVs only after the joining UAV joins the network right above (5,5), while the PSR-APC approach starts tuning the horizontal positions of the joining UAV right after it starts off. This ensures that when the joining UAV reaches the serving altitude, all the UAVs are already near the new optimal positions. This is confirmed by the curves in Fig. 3.9. It can be observed that under the PSR-APC approach, all the UAVs are at the new optimal positions in the very first epoch after the new UAV joins in, while it takes the passive approach couple of epochs to dispatch the UAVs to the new optimal positions. In addition, there is no pre-movement of the existing UAVs when the new UAV is about to join. This is because the new optimal positions of the existing UAVs are within 1 epoch reach to the previous optimal positions in our user distribution settings. The gain in accumulated US score during the transition period introduced by the PSR-APC approach is shown in Fig. 3.10, achieving at least 10% for all simulated N_{UAV}.

3.7.2.4 Case of UAV and User Dynamics

At last, the case with dynamic user distribution is simulated. The disperse-gather-disperse procedure shown in Fig. 3.4 is employed. We divide the 100 epochs evenly into 10 segments. The centers of the hot spots are updated (i.e., user distribution is updated) at the beginning of each segment. The order of update is snapshot

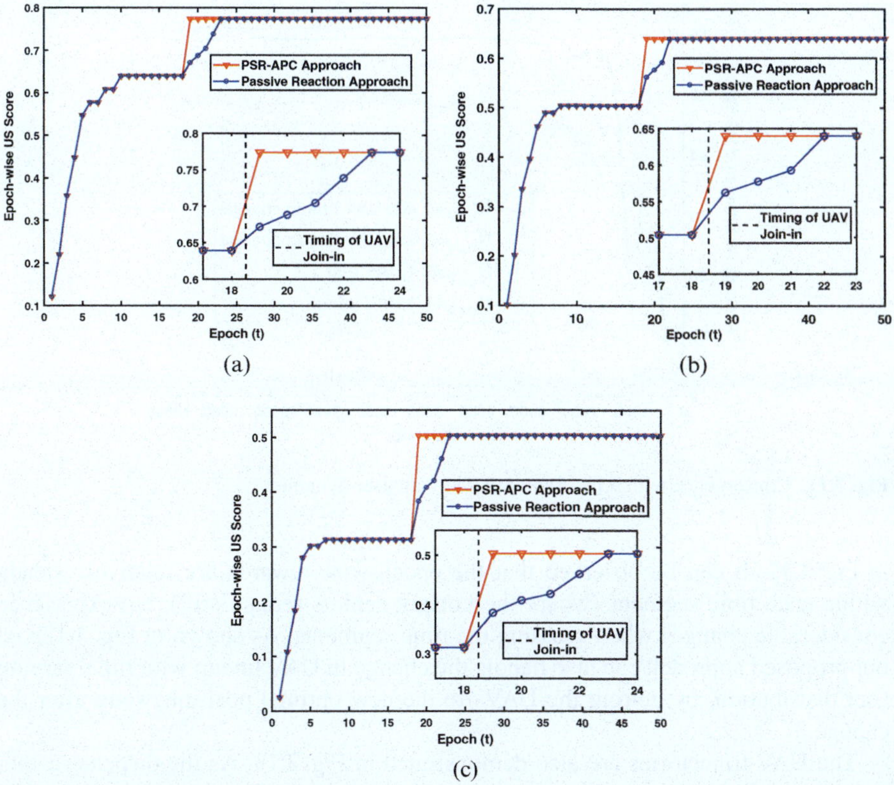

Fig. 3.9 Case of UAV join-in: epoch-wise reward comparison between the PSR-APC approach and the passive reaction approach. (**a**) $N_{UAV} = 5$. (**b**) $N_{UAV} = 4$. (**c**) $N_{UAV} = 3$

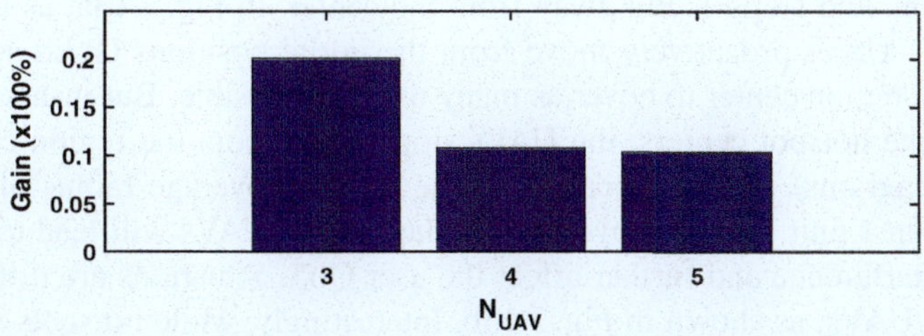

Fig. 3.10 Case of UAV join-in: gain (%) of PSR-APC over the passive reaction approach in accumulated US scores during transition to the new optimal UAV positions

1→2→3→4→4→4→4→3→2→1. The N_{UAV} UAVs start off initially from the optimal positions of snapshot 1, and move accordingly while the user distribution changes. The convergence of 4 situations is shown in Fig. 3.11: 4 UAVs and 5 UAVs with no UAV quit or join-in, 5 UAVs with 1 UAV quit, and 4 UAVs with 1 UAV join-in. In addition, the epoch-wise reward of each situation is presented

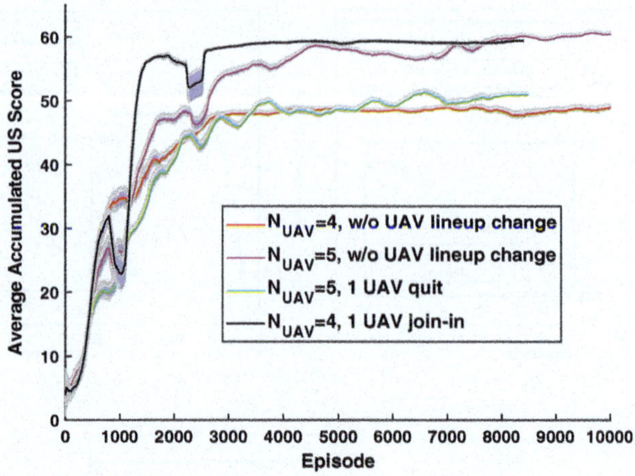

Fig. 3.11 Convergence with 95% credit interval with user dynamics

in Fig. 3.12. It can be observed that the epoch-wise rewards are relatively steady within each time segment (where the hotspot centers remain still), but experience considerable changes when crossing the time segments. As shown in Fig. 3.12b, c, our proposed approach can also handle the change in UAV lineup with time-varying user distribution, by getting the UAVs to the new optimal positions soon after the change.

The UAV trajectories are also demonstrated in Fig. 3.13. As the disperse-gather procedure is just the opposite mirror of the gather-disperse procedure, we only present the first 50 epochs. In all 4 subfigures, the black dashed lines represent the traces of the hotspot centers, moving from corners towards the center of the target region, and stops 1 unit away from the center. In Fig. 3.13a, as the hotspots move, the 4 UAVs proactively move from the initial positions (solid aqua circles) towards the region center to cover as many users as possible. But instead of exactly following the hotspot centers, the UAVs stop farther from the region center (solid brown pentagrams). This is because (1) the ground coverage radius of each UAV is larger than 1 unit, and (2) coverage overlapping of UAVs will lead to significant intercell interference and further affect the user QoS. Situations are different when there are 5 UAVs, as shown in Fig. 3.13b. Interestingly, while hotspots move, UAV 1 and UAV 3 almost stay still because of the existence of UAV 2. As UAV 2 moves towards the region center, it is able to cover most of the user flows from hotspots in the top and bottom right corners, so that UAV 1 and UAV 3 can stay put to cover more uniformly distributed users. Accordingly, the trace of UAV 4 leans a little towards the center to help cover the center users, and the trace of UAV 5 backs off a little to reduce overlapping.

Figure 3.13c shows the UAV traces when UAV 1 quits the network around epoch 16. It can be observed that after UAV 1 quits, the traces of UAV 2 and 5 turn more towards the hotspot in the top right corner to cover more users. UAV 4 goes

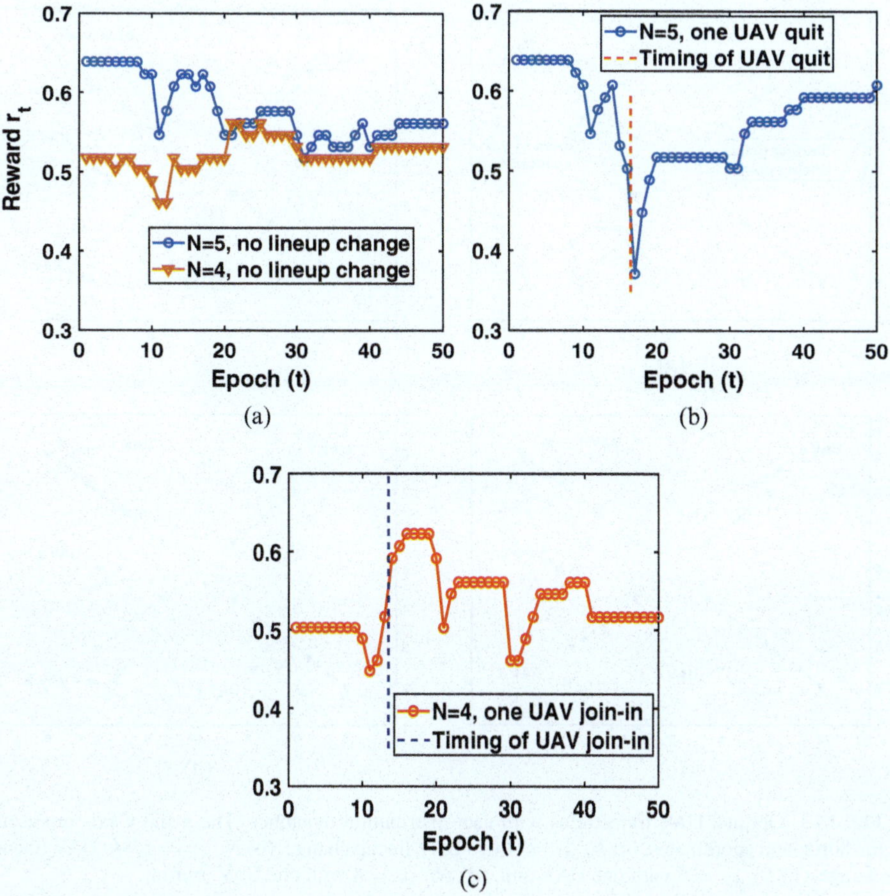

Fig. 3.12 Epoch-wise reward of different situations with dynamic user distributions. (**a**) No UAV lineup change. (**b**) $N_{UAV} = 5$ with one UAV quit. (**c**) $N_{UAV} = 4$ with one UAV join-in

deeper towards the region center to cover the users missed by UAV 2 and 5 while UAV 3 stays put. Figure 3.13d shows the UAV traces when UAV 2 starts off at the region center and formally joins the network at epoch 14. The dashed part of UAV 2 represents horizontal trace before it reaches the serving altitude. It can be observed that before UAV 2 joins the network, the 4 existing UAVs move very similarly to Fig. 3.13a. After UAV 2 joins, UAV 1 and 3 gradually back to the proximity of their original starting points, while UAV 4 and 5 start to move like Fig. 3.13b where there are 5 existing UAVs.

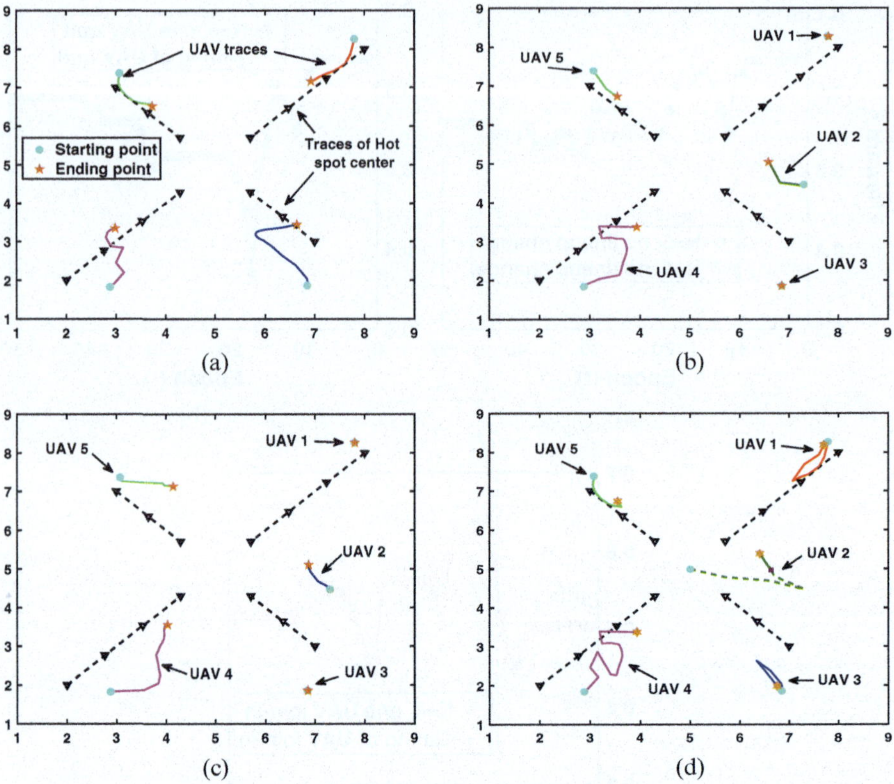

Fig. 3.13 Optimal UAV trajectories with user distribution dynamics. The x and y axis represent the horizontal coordinates. (**a**) $N_{UAV} = 4$ w/o UAV lineup change. (**b**) $N_{UAV} = 5$ w/o UAV lineup change. (**c**) $N_{UAV} = 5$ with one UAV quit. (**d**) $N_{UAV} = 4$ with one UAV join-in

3.8 Conclusions

In this chapter, an RL-based regulation strategy of a UAV communication network has been investigated with dynamic UAV lineup and user distribution. The learning approach, i.e., PSR-APC, has been designed to responsively control the UAV trajectories when the group of serving UAVs or the user distribution changes within a considered time horizon. Simulation results have demonstrated that compared to the passive reaction approach, the proposed approach can achieve up to 20% higher accumulated US scores during the transition process. In addition, when the user distribution is dynamically changing, the proposed approach has been shown to be able to capture the dynamics and move UAVs accordingly.

References

1. Q. Zhang, M. Jiang, Z. Feng, W. Li, W. Zhang, M. Pan, IoT enabled UAV: network architecture and routing algorithm. IEEE Internet Things J. **6**(2), 3727–3742 (2019)
2. X. Shen, J. Gao, W. Wu, K. Lyu, M. Li, W. Zhuang, X. Li, J. Rao, Ai-assisted network-slicing based next-generation wireless networks. IEEE Open J. Veh. Technol. **1**, 45–66 (2020)
3. W. Zhuang, Q. Ye, F. Lyu, N. Cheng, J. Ren, SDN/NFV-empowered future IOV with enhanced communication, computing, and caching. Proc. IEEE **108**(2), 274–291 (2019)
4. Y. Zeng, R. Zhang, T.J. Lim, Wireless communications with unmanned aerial vehicles: opportunities and challenges. IEEE Commun. Mag. **54**(5), 36–42 (2016)
5. Unmanned aerial vehicle (UAV) market (2019). https://www.marketsandmarkets.com/Market-Reports/unmanned-aerial-vehicles-uav-market-662.html
6. M. Li, N. Cheng, J. Gao, Y. Wang, L. Zhao, X. Shen, Energy-efficient UAV-assisted mobile edge computing: resource allocation and trajectory optimization. IEEE Trans. Veh. Technol. **69**(3), 3424–3438 (2020)
7. D. Shi, H. Gao, L. Wang, M. Pan, Z. Han, H.V. Poor, Mean field game guided deep reinforcement learning for task placement in cooperative multi-access edge computing. IEEE Internet Things J. **7**, 9330–9340 (2020)
8. N.H. Motlagh, M. Bagaa, T. Taleb, UAV-based IoT platform: a crowd surveillance use case. IEEE Commun. Mag. **55**(2), 128–134 (2017)
9. N. Zhao, W. Lu, M. Sheng, Y. Chen, J. Tang, F.R. Yu, K.-K. Wong, UAV-assisted emergency networks in disasters. IEEE Wireless Commun. **26**(1), 45–51 (2019)
10. M. Chen, M. Mozaffari, W. Saad, C. Yin, M. Debbah, C.S. Hong, Caching in the sky: proactive deployment of cache-enabled unmanned aerial vehicles for optimized quality-of-experience. IEEE J. Sel. Areas Commun. **35**(5), 1046–1061 (2017)
11. H. Wu, F. Lyu, C. Zhou, J. Chen, L. Wang, X. Shen, Optimal UAV caching and trajectory in aerial-assisted vehicular networks: a learning-based approach. IEEE J. Sel. Areas Commun. **38**(12), 2783–2797 (2020)
12. A.A. Nasir, H.D. Tuan, T.Q. Duong, H.V. Poor, UAV-enabled communication using NOMA. IEEE Trans. Commun. **67**(7), 5126–5138 (2019)
13. M. Mozaffari, W. Saad, M. Bennis, Y.-H. Nam, M. Debbah, A tutorial on UAVs for wireless networks: applications, challenges, and open problems. IEEE Commun. Surv. Tuts. **21**(3), 2334–2360 (2019)
14. Q. Wu, L. Liu, R. Zhang, Fundamental trade-offs in communication and trajectory design for UAV-enabled wireless network. IEEE Wireless Commun. **26**(1), 36–44 (2019)
15. Y. Zeng, X. Xu, R. Zhang, Trajectory design for completion time minimization in UAV-enabled multicasting. IEEE Trans. Wireless Commun. **17**(4), 2233–2246 (2018)
16. Q. Wu, R. Zhang, Common throughput maximization in UAV-enabled OFDMA systems with delay consideration. IEEE Trans. Commun. **66**(12), 6614–6627 (2018)
17. Y. Zeng, J. Xu, R. Zhang, Energy minimization for wireless communication with rotary-wing UAV. IEEE Trans. Wireless Commun. **18**(4), 2329–2345 (2019)
18. H. Guo, J. Liu, UAV-enhanced intelligent offloading for internet of things at the edge. IEEE Trans. Ind. Inf. **16**(4), 2737–2746 (2019)
19. M. Mozaffari, A.T.Z. Kasgari, W. Saad, M. Bennis, M. Debbah, Beyond 5G with UAVs: foundations of a 3D wireless cellular network. IEEE Trans. Wireless Commun. **18**(1), 357–372 (2018)
20. Q. Wu, Y. Zeng, R. Zhang, Joint trajectory and communication design for multi-UAV enabled wireless networks. IEEE Trans. Wireless Commun. **17**(3), 2109–2121 (2018)
21. M. Mozaffari, W. Saad, M. Bennis, M. Debbah, Mobile unmanned aerial vehicles (UAVs) for energy-efficient internet of things communications. IEEE Trans. Wireless Commun. **16**(11), 7574–7589 (2017)
22. Z. Yang, C. Pan, K. Wang, M. Shikh-Bahaei, Energy efficient resource allocation in UAV-enabled mobile edge computing networks. IEEE Trans. Wireless Commun. **18**(9), 4576–4589 (2019)

23. E. Alpaydin, *Introduction to Machine Learning* (MIT Press, Cambridge, 2020)
24. R.S. Sutton, A.G. Barto, *Reinforcement Learning: An Introduction*, 2nd edn. (Bradford Books, Cambridge, 2018)
25. N.C. Luong, D.T. Hoang, S. Gong, D. Niyato, P. Wang, Y.-C. Liang, D.I. Kim, Applications of deep reinforcement learning in communications and networking: a survey. IEEE Commun. Surv. Tuts. **21**(4), 3133–3174 (2019)
26. P.V. Klaine, J.P. Nadas, R.D. Souza, M.A. Imran, Distributed drone base station positioning for emergency cellular networks using reinforcement learning. Cognit. Comput. **10**(5), 790–804 (2018)
27. J. Cui, Y. Liu, A. Nallanathan, Multi-agent reinforcement learning-based resource allocation for UAV networks. IEEE Trans. Wireless Commun. **19**(2), 729–743 (2019)
28. J. Hu, H. Zhang, L. Song, Z. Han, H.V. Poor, Reinforcement learning for a cellular internet of UAVs: protocol design, trajectory control, and resource management. IEEE Wireless Commun. **27**(1), 116–123 (2020)
29. X. Liu, Y. Liu, Y. Chen, L. Hanzo, Trajectory design and power control for multi-UAV assisted wireless networks: a machine learning approach. IEEE Trans. Veh. Technol. **68**(8), 7957–7969 (2019)
30. Y. Hu, M. Chen, W. Saad, H. V. Poor, S. Cui, Distributed multi-agent meta learning for trajectory design in wireless drone networks (2020). arXiv:2012.03158
31. S. Singh, A. Kumbhar, I. Güvenç, M.L. Sichitiu, Distributed approaches for inter-cell interference coordination in UAV-based LTE-Advanced HetNets, in *2018 IEEE 88th Vehicular Technology Conference (VTC-Fall)* (IEEE, Piscataway, 2018), pp. 1–6
32. U. Challita, W. Saad, C. Bettstetter, Interference management for cellular-connected UAVs: a deep reinforcement learning approach. IEEE Trans. Wireless Commun. **18**(4), 2125–2140 (2019)
33. F. Tang, Y. Zhou, N. Kato, Deep reinforcement learning for dynamic uplink/downlink resource allocation in high mobility 5G HetNet. IEEE J. Sel. Areas Commun. **38**(12), 2773–2782 (2020)
34. N. Cheng, F. Lyu, W. Quan, C. Zhou, H. He, W. Shi, X. Shen, Space/aerial-assisted computing offloading for IoT applications: a learning-based approach. IEEE J. Sel. Areas Commun. **37**(5), 1117–1129 (2019)
35. X. Liu, M. Chen, C. Yin, Optimized trajectory design in UAV based cellular networks for 3D users: a double Q-learning approach (2019). arXiv:1902.06610
36. C.H. Liu, Z. Chen, J. Tang, J. Xu, C. Piao, Energy-efficient UAV control for effective and fair communication coverage: a deep reinforcement learning approach. IEEE J. Sel. Areas Commun. **36**(9), 2059–2070 (2018)
37. S. Khairy, P. Balaprakash, L. X. Cai, Y. Cheng, Constrained deep reinforcement learning for energy sustainable multi-UAV based random access IoT networks with NOMA (2020). arXiv:2002.00073
38. Y. Huang, X. Mo, J. Xu, L. Qiu, Y. Zeng, Online maneuver design for UAV-enabled NOMA systems via reinforcement learning, in *2020 IEEE Wireless Communications and Networking Conference (WCNC)*. (IEEE, Piscataway, 2020), pp. 1–6
39. T.P. Lillicrap, J.J. Hunt, A. Pritzel, N. Heess, T. Erez, Y. Tassa, D. Silver, D. Wierstra, Continuous control with deep reinforcement learning (2015). arXiv:1509.02971
40. H.X. Pham, H.M. La, D. Feil-Seifer, A. Nefian, Cooperative and distributed reinforcement learning of drones for field coverage (2018). arXiv:1803.07250
41. D. Chen, Q. Qi, Z. Zhuang, J. Wang, J. Liao, Z. Han, Mean field deep reinforcement learning for fair and efficient UAV control. IEEE Internet Things J. **8**(2), 813–828 (2020)
42. A. Al-Hourani, S. Kandeepan, A. Jamalipour, Modeling air-to-ground path loss for low altitude platforms in urban environments, in *2014 IEEE Global Communications Conference* (IEEE, Piscataway, 2014), pp. 2898–2904
43. J.M. Seddon, S. Newman, *Basic Helicopter Aerodynamics*, vol. 40 (Wiley, Hoboken, 2011)
44. M. Han, S. Khairy, L. X. Cai, Y. Cheng, R. Zhang, Reinforcement learning for efficient and fair coexistence between LTE-LAA and Wi-Fi. IEEE Trans. Veh. Technol. **69**(8), 8764–8776 (2020)

45. D. Shi, J. Ding, S.M. Errapotu, H. Yue, W. Xu, X. Zhou, M. Pan, Deep Q-network-based route scheduling for TNC vehicles with passengers' location differential privacy. IEEE Internet Things J. **6**(5), 7681–7692 (2019)
46. T. Hester, M. Vecerik, O. Pietquin, M. Lanctot, T. Schaul, B. Piot, D. Horgan, J. Quan, A. Sendonaris, I. Osband, et al., Deep Q-learning from demonstrations, in *Thirty-Second AAAI Conference on Artificial Intelligence* (2018)
47. V. Mnih, A.P. Badia, M. Mirza, A. Graves, T. Lillicrap, T. Harley, D. Silver, K. Kavukcuoglu, Asynchronous methods for deep reinforcement learning, in *International Conference on Machine Learning* (2016), pp. 1928–1937
48. Y. Zeng, R. Zhang, Energy-efficient UAV communication with trajectory optimization. IEEE Trans. Wireless Commun. **16**(6), 3747–3760 (2017)

Chapter 4
Utility-Based Dynamic Resource Allocation in IEEE 802.11ax Networks: A Genetic Algorithm Approach

Taewon Song, Taeyoon Kim, and Sangheon Pack

4.1 Introduction

Wireless local area networks (WLANs) have grown extensively over decades as advanced services such as ultra HD and 4K video, multimedia streaming, and rapid file transfer have become widespread among the general public. As a result, the number of personal devices, including smartphones, laptops, and high-definition multimedia devices, dramatically increases. As the number of devices increases, it leads to severe congestion, and the devices can hardly be connected to the Internet. Because of the congestion, the latest WLAN standard, IEEE 802.11ax [1], is primarily aimed at improving efficiency in high-density WLANs.

One of the most promising techniques in IEEE 802.11ax to deal with the dense deployment scenario is orthogonal frequency division multiple access (OFDMA), which has been adopted in various existing standards such as IEEE 802.16e WiMAX [2], long-term evolution (LTE), and 5G new radio (NR). In the OFDMA technique adopted in IEEE 802.11ax, the entire bandwidth is divided into several resource units (RUs). By allowing multi-user channel access and multi-user data

T. Song
Department of Internet of Things, SCH Media Labs, Soonchunhyang University, Asan, South Korea
e-mail: twsong@sch.ac.kr

T. Kim
Department of Mobile System Engineering, Dankook University, Yongin, South Korea
e-mail: 2000kty@dankook.ac.kr

S. Pack (✉)
School of Electrical Engineering, Korea University, Seoul, South Korea
e-mail: shpack@korea.ac.kr

L. Cai et al. (eds.), *Broadband Communications, Computing, and Control for Ubiquitous Intelligence*, Wireless Networks,
https://doi.org/10.1007/978-3-030-98064-1_4

transmission through orthogonal RU allocation, OFDMA can significantly reduce contention and preamble overhead.

On the other hand, as the number of access points (APs) is rapidly increased, each basic service set (BSS) becomes seriously overlapped. Therefore, a network-wide optimization for OFDMA resource allocation should be considered. Only a few studies considered adjacent BSSs; however, they only investigated interference mitigation issues through directional transmissions, and OFDMA resource allocation was not taken into consideration in detail.

In this chapter, we propose a utility-based dynamic resource allocation (UDRA) scheme in which a network-wide utility maximization problem is formulated to consider AP throughput and fairness among associated stations jointly. Since the formulated problem is an NP-hard problem, we map the optimization problem onto the genetic algorithm for a realistic WLAN environment. Extensive simulation results demonstrate that the proposed genetic algorithm has much lower complexity than the exhaustive search algorithm, while its performance in terms of throughput and fairness is nearly identical to the exhaustive search algorithm.

This chapter's key contribution is twofold. The first is that the frequency resource can be dynamically optimized using an interaction without any wired connectivity among APs. The simulation shows that depending on the given parameters, the network throughput of UDRA is 38% higher than conventional algorithms, or Jain's fairness index [3] of UDRA is higher than that of other algorithms. The second is that UDRA exhibits nearly the same performance thanks to the genetic algorithm compared to an exhaustive searching algorithm while the running time of UDRA is significantly reduced.

The remainder of this chapter is organized as follows. Sections 4.2 and 4.3 summarize the related works on OFDMA resource management. Sections 4.4 and 4.5 describe OFDMA operation in 802.11ax, which is a foundation for UDRA and demonstrates the formulated problem and the genetic algorithm to solve it. Simulation results and concluding remarks are given in Sects. 4.6 and 4.7, respectively.

4.2 Related Works

A fundamental problem for OFDMA resource management is how to allocate limited resources to the stations efficiently. That is, to increase spectral efficiency, stations should be allocated to appropriate time and frequency resource. Unfortunately, the problem of finding the optimal allocation was shown to be NP-hard in [4]. Therefore, many studies have focused on reducing computational complexity for resource management. The approaches to overcome this computational complexity can be categorized mainly into (1) solving sub-optimal solutions relying on relaxation [5–8] and (2) introducing alternative frameworks [9, 10].

A mixed integer nonlinear problem (MINLP) was formulated in [5], and the authors proposed a sub-optimal solution to the MINLP problem relying on convex relaxation. In [6], an optimal problem of assigning users to RUs while maximizing their sum rate was formulated, and a relaxed scheduling and resource allocation problem utilizing the divide-and-conquer approach was introduced. One approach assumes a more realistic and practical assumption. Since OFDMA is a technique for enabling multi-user transmission, the authors of [7] developed a resource allocation algorithm in which a scheduled duration is optimally determined to minimize the padding overhead occurring in the stations that are not transmitting at that time. In [8], the authors defined resource allocation as a selecting block for services, in order to meet some requirements such as the latency or traffic demands. The authors then proposed a sub-optimal and low-complexity algorithm to perform the assignment of blocks to services.

In [9], the authors defined a welfare function that reflected the total benefit covering all players and formulated the resource allocation problem as a game-theoretical framework. In [10], an auction-theoretic approach is proposed for the resource allocation problem to reduce the computation time. In this literature, several resource allocation schemes have been proposed, but most of them attempt to optimize OFDMA resource allocation within BSS without considering resource allocation information from adjacent BSSs.

Recent works on resource management consider a more challenging environment with multiple and densely deployed APs. In [11, 12], transmit beamforming was considered to mitigate the effect of inter-cell interference. The authors also investigated an achievable rate when transmit beamforming is applied. Indeed, this approach, such as transmit beamforming, is challenging to adopt in a WLAN because it requires a directional transmission. Although these studies considered adjacent BSSs, they only investigated interference mitigation issues through directional transmissions, and OFDMA resource allocation was not taken into consideration in detail.

4.3 Background on OFDMA and RU Allocation in IEEE 802.11ax

The basic OFDM principle is to utilize orthogonal subcarriers in frequency for data transmissions. Thus, broadband wireless radio channels with frequency-selective fading are replaced by a set of narrow-band channels (subcarriers) with flat fading. Each data symbol is then transmitted in one subcarrier, which is robust for multipath propagation. Additional advantages of OFDM are its highly efficient use of frequencies, its cost-effective and flexible digital signal processing, and its low complexity of MIMO principles.

802.11ax, which is the most widely used standard for WLAN, supports bands of 20 MHz, 40 MHz, 80 MHz, 80+80 MHz (combining two 80 MHz channels), and 160 MHz (single 160 MHz channel) [1]. In OFDMA transmission, the spectral band is divided into several resource units (RUs). In the time domain, the RU spans the

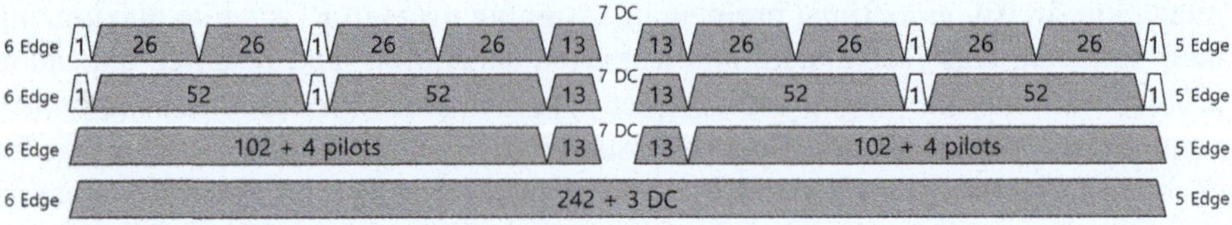

Fig. 4.1 Subdividing 20 MHz channel using OFDMA in IEEE 802.11ax

Table 4.1 Total number of RUs by channel bandwidth

RU type	20 MHz	40 MHz	80 MHz	160 and 80+80 MHz
26-subcarrier RU	9	18	37	74
52-subcarrier RU	4	8	16	32
106-subcarrier RU	2	4	8	16
242-subcarrier RU	1	2	4	8
484-subcarrier RU	N/A	1	2	4
996-subcarrier RU	N/A	N/A	1	2
2x996-subcarrier RU	N/A	N/A	N/A	1

entire data portion of the High Efficiency (HE) PLCP Protocol Data Unit (PPDU). In the frequency domain, it consists of a subset of successive subcarriers. In the frequency domain, RUs can be 26, 52, 106, 242, 484, or 996. RUs in HE multi-user (MU) PPDUs that use OFDMA transmissions can be one of these sizes. The position of the RU in the HE PPDU is fixed. Each RU, larger than 26, can be divided into two smaller RUs. The entire bandwidth can be used as a single 484-tone RU, or divided into two 242-tone RUs, each of which can be split into smaller RUs until a 26-tone RU is reached. When an RU is created, the AP assigns one RU to each user or a group of users for transmission. When bandwidth is split into RUs and is allocated to each user, the transmission is pure OFDMA, which can also be used for MU-MIMO if the RU is a 106 or higher subcarrier, then referred to as a joint transmission between MU-MIMO and OFDMA. Figure 4.1 illustrates how an 802.11ax system multiplexes a 20 MHz channel using different resource unit (RU) sizes. The smallest division of the channel can support up to 9 users simultaneously for every 20 MHz of bandwidth. The number of users that can be supportable for the RU type and various available channels are listed in Table 4.1. In this chapter, we follow the existing parameters regarding OFDMA.

4.4 System Model

In this section, UDRA, an optimization problem that jointly considers the network throughput and the fairness index in OFDMA resource allocation, is addressed. To this end, we first describe a system model on which UDRA is based and demonstrate

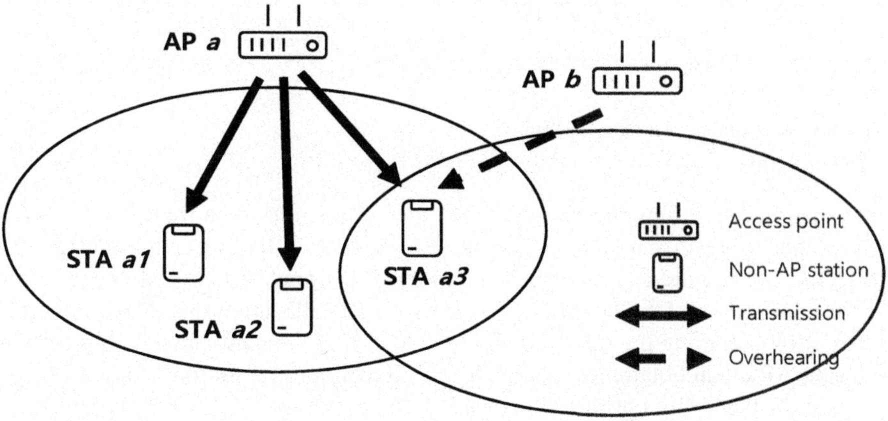

Fig. 4.2 Schematic representation of UDRA

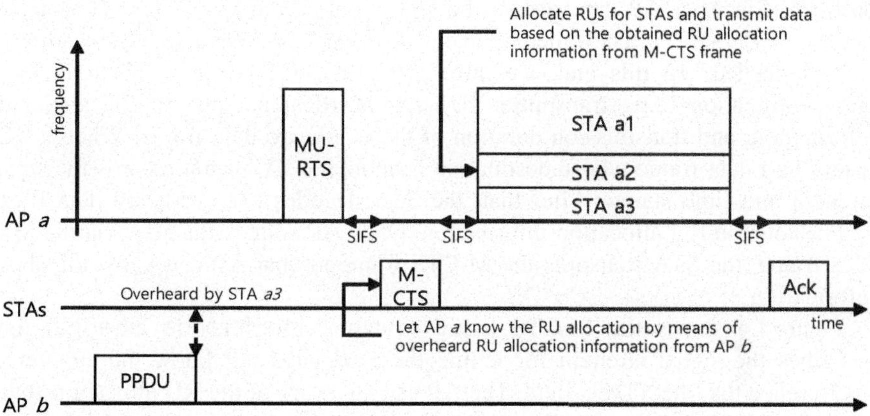

Fig. 4.3 The timing diagram of UDRA in the system model

an illustrative example of UDRA. Next, a modified CTS (M-CTS) frame structure, which is essential to operate UDRA, is followed.

In our model, each BSS consists of one AP and one or more STAs. Only downlink traffic from APs to STAs is considered since it occupies a dominant portion of traffic for WLAN applications. Furthermore, we assume that the AP's buffer to be transmitted to associated stations is in a saturated state, which means the APs always have frames to be sent.

Figures 4.2 and 4.3 show a schematic example of how UDRA works. In this example, as shown in Fig. 4.2, there are two APs, APs a and b, where STAs $a1$, $a2$, and $a3$ are connected to AP a. STA $a3$ is on the area where two transmission ranges are overlapped. The solid arrows and the dotted arrows stand for determined transmissions and overheard transmissions, respectively. Once STA $a3$ that resides in the overlapped area overhears a data frame from AP b, it will know RU

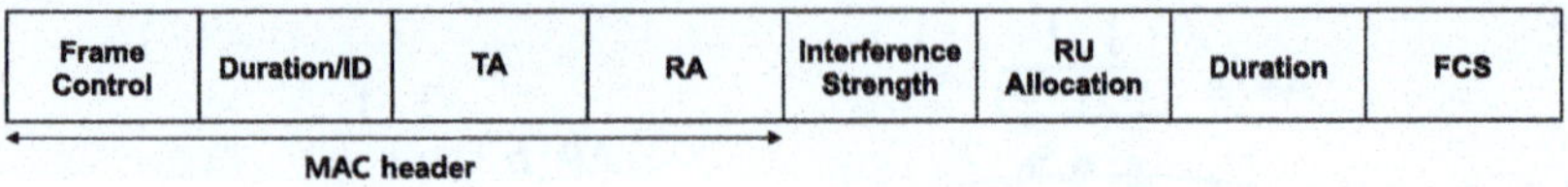

Fig. 4.4 A modified CTS (M-CTS) frame

information of AP b by means of the PHY preamble in the data frame. When STAs $a1$, $a2$, and $a3$ receive a multi-user request-to-send (MU-RTS) frame from AP a, all the STAs are required to respond with an M-CTS frame, which will be addressed in the next section. Besides, STA $a3$ reports the RU allocation information of AP b to AP a by including the information in the M-CTS frame, as shown in Fig. 4.3. As a result, AP a can obtain network-wide RU allocation information, which will then be used for the utility optimization problem in Sect. 4.5.

Some information, such as transmission duration and transmission signal power from adjacent BSSs, is required to formulate a network-wise utility maximization problem. Since an AP can receive the signal only from associated STAs in the existing IEEE 802.11, a new method to deliver the information collected by adjacent APs is needed. To this end, we introduce the M-CTS frame, which includes the identification (e.g., transmitter ID), overheard signal power, RU allocation information, and transmission duration of the overheard data frame. When a STA listens to a data frame whose destination is another STA, it first records the signal strength and time stamp. After that, the STA decodes the overheard data frame and detects the RU allocation information. Once AP solicits the STA via the MU-RTS frame, the STA transmits the M-CTS frame so that AP can know the above information.

Figure 4.4 shows the M-CTS frame structure. Interference Strength field describes the signal strength for letting the associated AP know the amount of interference that the STA suffers. The AP can be aware of the RU allocation status of adjacent BSS utilizing RU Allocation field. Duration field lets the AP know how long the transmission of adjacent BSS lasts. Although Interference Strength, RU Allocation, and Duration fields have not been defined in the existing CTS frame, we can modify some existing fields to describe them or define a newly designed frame format. Formatting these fields is beyond the scope of this chapter, and the formats of these fields are not restricted.

4.5 Utility-Based Dynamic Resource Allocation Scheme

In this section, an optimization problem that jointly considers the network throughput and the fairness [3] in OFDMA resource allocation is defined. Also, as a practical solution to the problem, we mapped UDRA to genetic algorithm in the following subsection.

4.5.1 Optimal Resource Allocation Problem Formulation

Let N and M be the number of STAs and the number of sub-channels, respectively. $Y = \{0, 1\}^{N\times M}$ represents the assignment matrix where an element of Y, $y_{n,m}$, is 1 if STA n occupies sub-channel m; otherwise, $y_{n,m} = 0$. Meanwhile, $I = \mathbb{R}^{N\times M}$ is the interference matrix where an element of I, $i_{n,m}$, is the value included in the Interference Strength field of the M-CTS frame. $S = \mathbb{R}^{N\times M}$ is the signal power matrix where its element, $s_{n,m}$, refers to the signal power of STA n in sub-channel m.

To jointly consider both throughput and fairness index, we need to calculate the normalized throughput and fairness. For the normalized throughput, signal-to-interference-plus-noise ratio (SINR) needs to be first defined. When STA n occupies sub-channel m and the assignment matrix is given by Y, SINR can be expressed as

$$SINR_{n,m}(Y) = \frac{t}{T}\cdot \frac{s_{n,m}[\text{mW}]\cdot y_{n,m}}{i_{n,m}[\text{mW}]\cdot N_0[\text{mW}]}, \tag{4.1}$$

where t is the length interfered by the adjacent AP, which can be obtained by overhearing data frames from adjacent AP, while T is the frame length in bytes. N_0 is the thermal noise, which is expressed as $-174 + 10\log_{10}\frac{B}{M}$ [13, 14] and $f[\text{mW}]$ represents that f is in a milli-watts scale. Then, the attainable throughput of STA n from sub-channel m is expressed as

$$c_{n,m}(Y) = \frac{B}{M}\cdot \log_2(1 + SINR_{n,m}(Y)). \tag{4.2}$$

Since there are M resource units, the total attainable throughput of STA n is given by:

$$c_n(Y) = \sum_{m=1}^{M} c_{n,m}(Y) = \sum_{m=1}^{M} \frac{B}{M}\cdot \log_2(1 + SINR_{n,m}(Y)). \tag{4.3}$$

Using Eq. (4.3), the attainable network throughput per transmission and the fairness index can be derived. First of all, the attainable network throughput per transmission can be expressed as

$$c(Y) = \sum_{n=1}^{N} c_n(Y). \tag{4.4}$$

Meanwhile, the Jain's fairness index can be computed as

$$f(Y) = \frac{\left(\sum_{n=1}^{N} c_n(Y)\right)^2}{N \cdot \sum_{n=1}^{N} c_n(Y)^2}. \tag{4.5}$$

Using Eqs. (4.4) and (4.5), the utility to balance the throughput and fairness can be defined as

$$u(Y) = \alpha \cdot c(Y) + (1 - \alpha) \cdot f(Y), \tag{4.6}$$

where α is a weighting factor to prioritize either the attainable network throughput or the fairness between STAs. For example, once α approaches to one, the attainable network throughput will be prioritized. On the contrary, when α approaches to zero, the fairness among STAs is preferred and the fairness will be emphasized.

Finally, the utility optimization problem can be expressed as

$$\begin{aligned} &\max_{Y \in \{0,1\}^{N \times M}} \quad u(Y), \\ &s.t. \sum_{n=1}^{N} y_{n,m} \leq 1, \forall m \in \{1, 2, \ldots, M\}, \end{aligned} \tag{4.7}$$

where the constraint represents that two or more STAs cannot occupy the same sub-channel.

Assuming the signal powers of sub-channels that suffer channel fading are independently and identically distributed (i.i.d.), the complexity of the utility optimization problem in (4.7) can be $O(2^{M \cdot N})$ with big-O notation, which is known as NP-hard. Thus, we will explain a practical genetic algorithm for this problem in the next subsection.

4.5.2 Genetic Algorithm

Genetic Algorithm (GA) is a meta-heuristic popular in computer science [15]. GA applies the principle of survival of the fittest to produce a better and better approximation to the solution of the problem that GA is trying to solve. For each generation, a new set of approximations is created through the process of selecting individuals according to their level of fitness in the problem domain, and propagating the individuals together using operators borrowed from genetic processes carried out in nature (e.g., crossover and mutation) is created. This process, as occurs in natural adaptation, leads to the evolution of groups of individuals who adapt better to the environment than the individuals from which they were created.

Genetic algorithm is a well-known heuristic algorithm to deal with the NP-hard problem, which emulates the evolution process in nature and the process consists of natural selection, reproduction, and mutation.

In the genetic algorithm, the following concepts are employed to find the optimal solution. First of all, a *chromosome* is a set of parameters, which defines a proposed solution to the problem that the genetic algorithm is trying to solve. For standard optimization algorithms, this can be the domain of the objective function. This set of chromosomes is called *population*. On the other hand, the *fitness function* represents a function to be optimized, i.e., objective function, and the *fitness value* is the output of the fitness function when one of chromosomes is given by an input.

For each *generation*, a predetermined number of chromosomes are arbitrarily selected and their fitness values are compared among them. After the comparison, the natural selection process starts with the selected chromosomes that have greater fitness value than others. This selected chromosomes are then *reproduced* for the next generation. Above-mentioned procedure continues for a certain number of generations.

A problem mapping the optimization algorithm on genetic algorithm is depicted in Fig. 4.5. In our RU allocation problem, we denote an assignment matrix, Y, as a *chromosome*. A *fitness value* is calculated for each chromosome. In this problem, the fitness value can be a utility, $u(Y)$, as seen in Eq. (4.7). Chromosomes are randomly generated within the universal set, which is a binary matrix with M rows and N columns that satisfy constraints.

A detailed procedure for solving the RU allocation problem can be represented as follows. This procedure is depicted in Algorithm 3. First of all, the generation number, i, is initiated (see line 1). Once an AP transmits MU-RTS to associated STAs, some respond to MU-RTS by transmitting M-CTS if they overheard any data frames from adjacent APs. In so doing, the AP can obtain the interference matrix I and the signal power matrix S (see line 2). At the first generation, the AP randomly generates j_{max}, a predefined population size, assignment matrices. After that, the AP calculates the utility for each matrix (see lines 6 to 9). Next, some "winner" matrices are survived, and the next generation will be triggered. This procedure continues until the difference between the utility of the ith generation and the $(i-1)$th is smaller than a predefined threshold, $u_{threshold}$, or the maximum running time, $T_{maxstall}$, elapsed.

Even though the exhaustive search requires exponential processing time, the genetic algorithm has a polynomial executing time because it runs up to $i_{max} \cdot j_{max}$ cycles at most and each cycle requires polynomial processing time. In the literature, several studies have been conducted to determine the optimal number of populations, j_{max}. Since the derivation of the optimal number is beyond the scope of this chapter, j_{max} is set to $10 * M \cdot N$ according to [16].

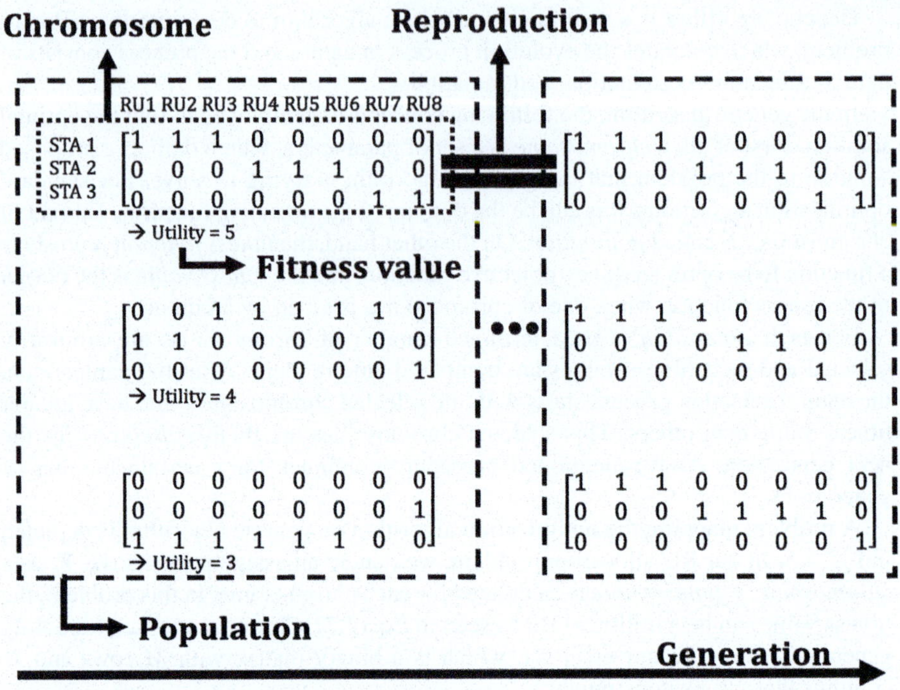

Fig. 4.5 Mapping from UDRA to genetic algorithm

4.6 Simulation Results

Extensive simulations have been conducted by a MATLAB simulator to evaluate the performance of UDRA. Simulation parameters are based on the IEEE 802.11ax standard. Defined parameters are summarized in Table 4.2. First of all, we show how UDRA performs and how fast it runs compared to an exhaustive search. After that, we analyze the throughput and fairness aspects of UDRA compared to conventional algorithms for α and examine the effect of α in detail.

4.6.1 UDRA vs. Exhaustive Search

Complexity and the resulting performance degradation between exhaustive search and UDRA are presented in Fig. 4.6. Specifically, this figure shows how much can UDRA reduce its running time compared to exhaustive search and how much does UDRA underperform exhaustive search. Since the complexity of exhaustive search grows exponentially, the simulations are conducted in a small-scale environment with 2 to 4 stations and 8 RSUs.

Algorithm 3 Genetic algorithm for UDRA

$i = 1$
Obtain the interference matrix I and the signal power matrix S
Randomly generate a set of j_{max} assignment matrices $\mathbb{Y}_i$ which meet the constraint of the optimization problem in Eq. (4.7)
while $|u_i - u_{i-1}| > u_{threshold}$ for $T_{maxstall}$ times **do**
  $j = 1$
  for each assignment matrix $Y_{i,j}$ in $\mathbb{Y}_i$ **do**
    Calculate a utility for the assignment matrix by means of Eqs. (4.1)–(4.7)
    $j = j + 1$
  end for
  Calculate the best utility, $u_i = \max_{\forall j} u(Y_{i,j})$
  Select a portion of the assignment matrices and leave them for the next population
  Randomly generate assignment matrices and make a set of assignment matrices with the survived assignment matrices for the next generation $\mathbb{Y}_{i+1}$
  $i = i + 1$
end while
Determine the assignment matrix that makes the utility maximum value
return assignment matrix Y

Table 4.2 Simulation parameters

Parameter	Value
Multiple access scheme	OFDMA
Channel bandwidth	80 MHz
RU type	106-subcarrier RU
Number of RUs	8
Noise model	Thermal noise
Optimization methodology	Genetic algorithm
Number of populations	$10 * M \cdot N$
Max stall generations	150

As shown in Fig. 4.6, UDRA exhibits the same performance when the number of STAs is 2. Meanwhile, UDRA shows degraded throughput compared with the exhaustive search by 3.56 and 3.77% when the numbers of STAs are 3 and 4, respectively. Even though the genetic algorithm has slightly reduced throughput, it significantly reduces running time compared with the exhaustive search. For example, UDRA can achieve 31.25, 2.93, and 0.24% of the exhaustive search for the running time for the cases with 2, 3, and 4 STAs, respectively.

4.6.2 *Network-Wise Throughputs and Fairness Indexes*

Figures 4.7 and 4.8 show the total network throughput and the Jain's fairness index for UDRA, round-robin algorithm, and randomly allocation algorithm as the number of stations varies. From Fig. 4.7, it can be seen that the throughputs of round-robin

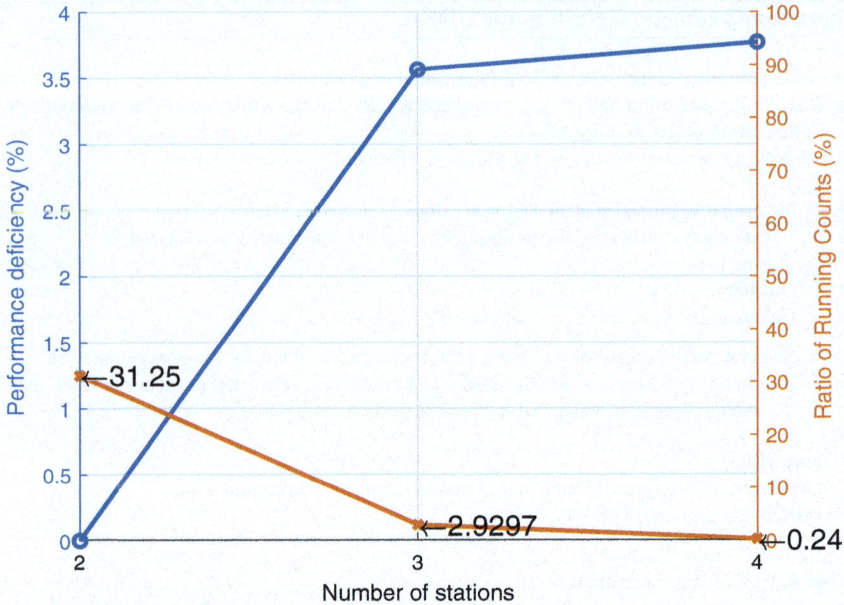

Fig. 4.6 UDRA vs. exhaustive search

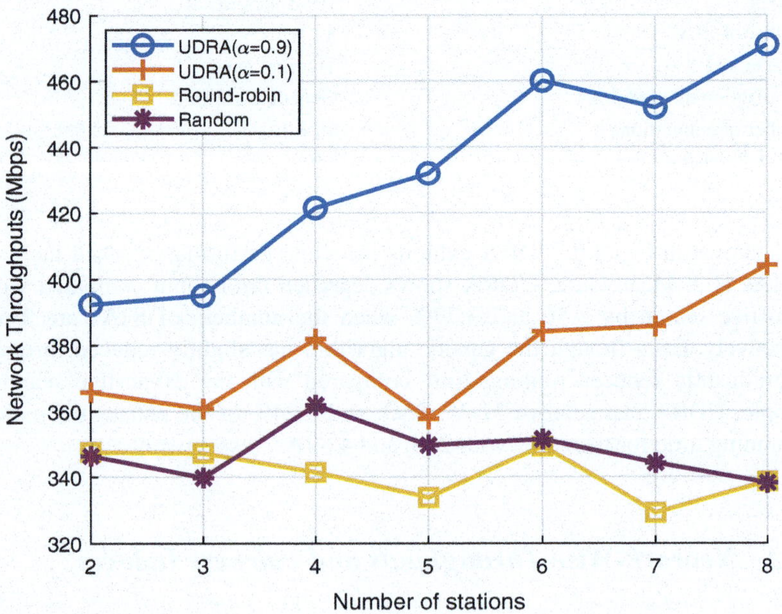

Fig. 4.7 Throughputs

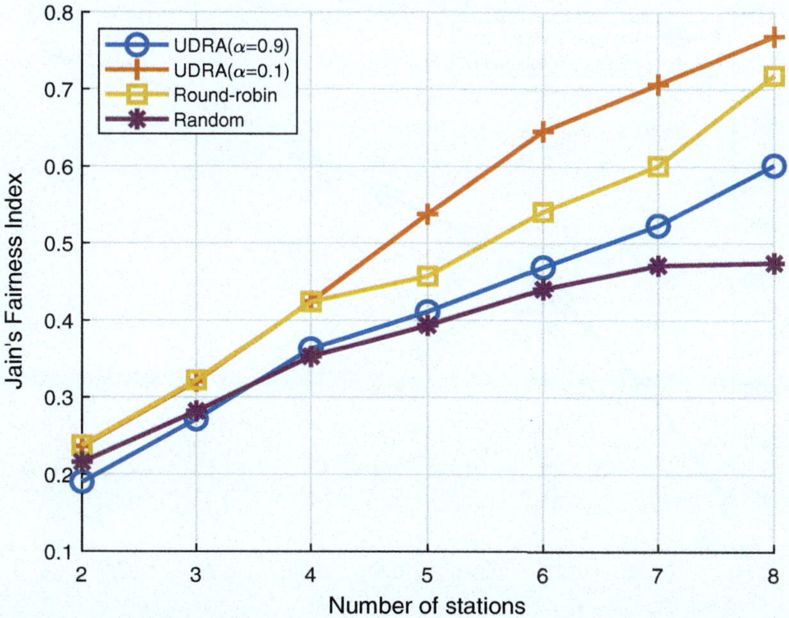

Fig. 4.8 Jain's fairness indexes

and random allocation hardly increase as the number of STAs increases. This is because the total resource, i.e., the channel bandwidth, to be allocated is identical regardless of the number of stations, and they allocate the resources evenly to the stations in the long term. On the other hand, the AP in UDRA can finely allocate RUs to STAs by solving the formulated optimization. Therefore, it can be found that the network throughput of UDRA increases with the increase in the number of STAs, although the total resource does not vary. Specifically, UDRA with $\alpha = 0.1$ and with $\alpha = 0.9$ exhibits from 5.6 to 19.5% and from 13.2 to 39.2% higher throughputs compared to conventional algorithms as the number of stations varies, respectively. Meanwhile, throughput of UDRA with $\alpha = 0.1$ is 16.9–27.0% lower than that with $\alpha = 0.9$. This is because UDRA with $\alpha = 0.1$ prioritizes the normalized fairness rather than the normalized throughput.

Figure 4.8 shows the Jain's fairness index depending on the number of STAs. It can be seen that overall fairness trends consistently increase as the number of stations increases. UDRA with $\alpha = 0.1$ outperforms other algorithms in terms of fairness. Also, it can be seen that the fairness indexes of the random algorithm are inferior to the other algorithms. This is because the random algorithm does not consider whether an associated station is affected by adjacent AP's transmission or not, and this causes severe collisions, especially when the number of APs becomes large.

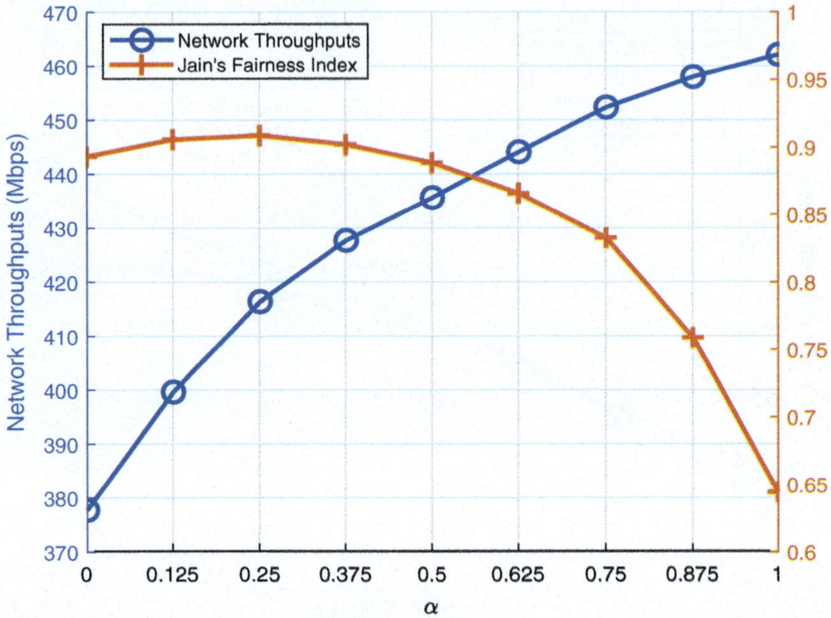

Fig. 4.9 Effect of α

Figure 4.8 also shows the Jain's fairness index depending on the number of STAs. It can be seen that overall fairness trends consistently increase as the number of stations increases for the same reason mentioned above.

As shown in Figs. 4.7 and 4.8, the performance of UDRA is highly affected by α, and thus we analyze the effect of α. From Fig. 4.9, the throughput becomes higher while the fairness index gets smaller as α increases. This can be explained as follows. When α is high, the normalized throughput is emphasized, and thus each AP tends to allocate the best sub-channel to the best STA, leading to a severe imbalance of throughput. For example, an STA whose SINR is much stronger than other STAs will occupy most of the RUs, and thus the STA can get much higher throughput for a high value of α.

On the other hand, the normalized fairness is prioritized when α is low, and therefore each AP tends to allocate sub-channels fairly to the STAs. In this case, STAs residing in the overlapped area, e.g., station $a3$ in Fig. 4.2, will have an opportunity to access the medium since it has a higher opportunity to suffer higher interference than station $b1$. Apparently, such fair resource allocation leads to degraded network throughput, and thus the optimal value of α should be carefully chosen under the service requirements.

Besides, the slope of network throughputs increases nearly linearly, whereas the slope of Jain's fairness index tends to decrease sharply as seen in Figs. 4.7 and 4.8. When $\alpha = 1$, UDRA would behave as if it allocates RUs to the stations in a greedy manner, and thus the metric of fairness becomes degraded. Also, at $\alpha = 1$, the network throughput increases linearly, while the fairness index decreases sharply, so a small amount of concern for fairness can result in a quite fair resource allocation while performing large throughputs.

4.7 Conclusion

In this chapter, a utility-based dynamic resource allocation algorithm for OFDMA-based wireless networks is proposed. By using M-CTS, stations residing in overlapped area can overhear RU allocation and interference power. Then the station can deliver information to its associated AP and thus the AP can utilize it for utilizing RUs efficiently. After that, AP operates utility maximization problem with a factor, α. By adjusting α, the throughput as well as the fairness can be achieved. We next formulate genetic algorithm that operates in polynomial running time. The simulation results demonstrated that the genetic algorithm has few or no performance drop while its running time remarkably decreases.

References

1. Institute of Electrical and Electronics Engineering (IEEE), 802.11ax-2021 - IEEE standard for information technology–telecommunications and information exchange between systems local and metropolitan area networks–specific requirements part 11: wireless LAN medium access control (MAC) and physical layer (PHY) specifications amendment 1: enhancements for high-efficiency WLAN
2. Institute of Electrical and Electronics Engineering (IEEE), IEEE standard for air interface for broadband wireless access systems
3. R. Jain, D.-M. Chiu, W.R. Hawe, *A Quantitative Measure of Fairness and Discrimination for Resource Allocation in Shared Computer System*, vol. 38 (Eastern Research Laboratory, Digital Equipment Corporation Hudson, Hudson, 1984)
4. K. Jain, J. Padhye, V.N. Padmanabhan, L. Qiu, Impact of interference on multi-hop wireless network performance. Wireless Netw. **11**(4), 471–487 (2005)
5. A. Abdelnasser, E. Hossain, D.I. Kim, Tier-aware resource allocation in OFDMA macrocell-small cell networks. IEEE Transl Commun. **63**(3), 695–710 (2015)
6. K. Wang, K. Psounis, Scheduling and resource allocation in 802.11ax, in *IEEE INFOCOM 2018 - IEEE Conference on Computer Communications*, vol. 2018 (2018), pp. 279–287
7. M. Karaca, S. Bastani, B.E. Priyanto, M. Safavi, B. Landfeldt, Resource management for OFDMA based next generation 802.11 WLANs, in *2016 9th IFIP Wireless and Mobile Networking Conference (WMNC)* (2016), pp. 57–64
8. L. You, Q. Liao, N. Pappas, D. Yuan, Resource optimization with flexible numerology and frame structure for heterogeneous services. IEEE Commun. Lett. **22**(12), 2579–2582 (2018)
9. H. Dai, Y. Huang, R. Zhao, J. Wang, L. Yang, Resource optimization for device-to-device and small cell uplink communications underlaying cellular networks. IEEE Trans. Veh. Technol. **67**(2), 1187–1201 (2018)
10. N. Tadayon, S. Aissa, Radio resource allocation and pricing: auction-based design and applications. IEEE Trans. Signal Process. **66**(20), 5240–5254 (2018)

11. K. Ishihara, T. Murakami, Y. Asai, Y. Takatori, M. Mizoguchi, Cooperative inter-cell interference mitigation scheme with downlink MU-MIMO beamforming for dense wireless LAN environment. Wireless Pers. Commun. **93**, 661–674 (2014)
12. K. Ishihara, T. Murakami, H. Abeysekera, M. Akimoto, Y. Takatori, Distributed smart antenna system for high-density WLAN system. Electron. Lett. **54**(6), 336–338 (2018)
13. J.B. Johnson, Thermal agitation of electricity in conductors. Phys. Rev. **32**, 97–109 (1928). http://link.aps.org/doi/10.1103/PhysRev.32.97
14. H. Nyquist, Thermal agitation of electric charge in conductors. Phys. Rev. **32**, 110–113 (1928). http://link.aps.org/doi/10.1103/PhysRev.32.110
15. Z. Bankovic, D. Stepanovic, S. Bojanic, O. Nieto-Taladriz, Improving network security using genetic algorithm approach. Comput. Electr. Eng. **33**(5), 438–451 (2007)
16. R. Storn, On the usage of differential evolution for function optimization, in *Proceedings of North American Fuzzy Information Processing* (2002), pp. 519–523

Chapter 5
Intelligentized Radio Access Network for Joint Optimization of User Association and Power Allocation

Hui-Chi Yu and Kuang-Hao Liu

5.1 Introduction

With the rapid growth of mobile data traffic, the traditional cellular network faces numerous challenges to fulfill the demand of higher data rates and persistent service quality. On one hand, high cell density is beneficial to reduced propagation loss due to distance. On the other hand, the inter cell interference (ICI) increases with the cell density that dramatically degrades the service quality of cellular users. To alleviate the aforementioned problem, multi-cell coordination has been considered by 3rd Generation Partnership Project (3GPP) as an effective means to mitigate strong ICI through cell coordination [1]. Different implementation choices of cell coordination have been proposed, including coordinated schedule/beamforming (CS/CB), joint transmission (JT), and dynamic point selection (DPS). The early standardization efforts aim to define the high-level requirements on the changes of the specifications to support multi-cell coordination. More recent work has been focusing on the new control signaling for enabling multi-cell coordination [2]. It is expected that the next-generation cellular network will undergo a major evolution on the network infrastructure where the notion of the cell boundary will be more vague because of cell coordination.

To enable multi-cell coordination, both the radio access network (RAN) and the core network (CN) that serve as two major entities of the cellular network will undergo some major changes. In the traditional RAN, each radio tower or the base station (BS) needs to handle both signal processing and resource management

H.-C. Yu
Department of Electrical Engineering, National Cheng Kung University, Tainan City, Taiwan

K.-H. Liu (✉)
Institute of Communications Engineering, National Hsing Hua University, Hsinchu, Taiwan
e-mail: khliu@ee.nthu.edu.tw

L. Cai et al. (eds.), *Broadband Communications, Computing, and Control for Ubiquitous Intelligence*, Wireless Networks,
https://doi.org/10.1007/978-3-030-98064-1_5

for the users within its coverage. These two functions will be split in the future RAN where the signal transmission and reception will be performed by the entity called the transmission/reception point (TRP). The resource management task will be covered by another entity called the distributed unit (DU). Thus the functionality of the traditional BS is replaced by one DU connected with several TRPs. This new infrastructure with functional split facilitates cell coordination in a cost-effective manner because now the hardware/software resources of a DU can be shared by multiple TRPs. On the other hand, the future CN will be fully configurable by software so that network operators can flexibly split their networks into several slices for satisfying diverse service requirements and upgrading their CN conveniently [3].

With the flexible infrastructure to support multi-cell coordination, the RAN still needs to be optimized to achieve the maximum performance. In order to make efficient use of radio resources, the RAN can be configured to activate a set of TRPs to serve a user equipment (UE), referred to as the user association (UA) problem. Here two critical factors that affect the network performance include: which set of TRPs should be active and what transmission power should be allocated to serve the UE? By properly activating TRPs, severe ICI can be avoided. For example, neighboring TRPs can jointly design their beamforming vectors to cancel ICI or transmit the same downlink data that turns the ICI into useful signal. The former is known as CB and the latter as JT in the categories of multi-cell coordination. The 3GPP specification mainly focuses on the control plane to support multi-cell coordination while the formation of a coordinating TRP set, or commonly referred to as the cluster, is subject to practical implementation. Also, transmission power allocation (PA) is defined only for legacy single-TRP transmission but that across multiple coordinated TRPs is open.

5.2 Related Work

Traditional RAN The joint UA and PA optimization is important to the future RAN featured by multi-cell coordination. Most of the existing works consider UA and PA separately. PA for sum rate maximization under the maximal power constraint is a well-known non-convex optimization problem and it is NP-hard. Hence the best practice is to transfer the PA problem into a convex one that facilitates efficient problem solving algorithms [4, 5] despite the obtained solution is sub-optimal in general. On the other hand, UA for load balancing has also shown to be an NP-hard problem [6], which makes it difficult to find the exact optimal solution directly. The joint UA and PA design has been addressed in [7, 8] . Due to the NP hardness of the problem, iterative algorithms have been proposed in the previous work to deliver the near-optimal solutions. However, these algorithms require global channel state information (CSI) and their performance is sensitive to the CSI accuracy.

Clustering When a set of TRPs are activated to serve the same UE, they form a cluster and exchange information to enable multi-TRP transmission. The clustering scheme is critical to the network-wide performance, but it is often considered as an implementation issue not specified in the standard. From the implementation perspective, there are two kinds of clustering schemes, including the *network-centric* and the *user-centric* clustering [9]. In the network-centric clustering, different TRPs are configured according to the network architecture. For example, the TPRs that belong to the same site may form a so-called intra-site cluster, which appears to be the easiest way to perform multi-TRP transmission because information exchange within the same site can be readily realized through an internal pipeline. It may also be possible to group the TRPs of different sites into a cluster, namely, the inter-site cluster provided with fast backhaul links for information exchange across different sites. System-level simulations reveal that in the dense urban scenario, the intra-site clustering often performs better than the inter-site clustering [10]. In practice, the backhaul link has limited capacity and non-negligible latency. Hence the implementation of inter-site clustering is more challenging than the intra-site clustering. Thanks to recent advances in optical fiber technologies, instant and reliable backhaul transmission becomes more promising. The encouraging development of backhaul transmission stimulates the standardization progress to support multi-TRP transmission by defining detailed control-plane specifications [2].

In the user-centric clustering, multiple TRPs are grouped as a cluster based on the user reference. For example, the first few TRPs that provide the strongest reference signal received power (RSRP) to the user can form a cluster. Alternatively, a cluster may be formed by the neighboring TRPs whose RSRP values are sufficiently close [11]. Clearly, the TRPs in the user-centric cluster may not belong to the same site and not even close in their locations due to potential shadowing and blockage effects. Compared with the network-centric clustering, the user-centric clustering usually achieves a higher gain than the network-centric clustering [11] despite its stringent requirement on the backhaul link.

ML for UA and PA To embrace significant system-wide capacity improvement of multi-cell coordination, fast algorithms to solve the joint UA and PA optimization problem is of paramount importance. This motivates some recent efforts on applying machine learning (ML) approaches to solve classical optimization problems in wireless networks. For example, the weighted minimum mean squared error (WMMSE) algorithm is popular for solving the PA problem [4], but it requires extensive matrix inversion and bisection search to obtain a good solution. The authors in [12] propose to approximate WMMSE by a fully connected deep neural network (DNN). Their results indicate that the sum rate achieved by the DNN is very close to that obtained by WMMSE with less computational time. Similar to the original WMMSE, the DNN used in [12] requires network-wide CSI. To alleviate the signaling overhead for collecting full CSI, the authors in [13] proposed to train a convolutional neural network (CNN) offline and then each transmitter (i.e., the agent) decides the transmission power by the trained CNN using local CSI. The PA problem for the massive multi-input multi-output (MIMO) network is studied in [14]. Considering the huge dimension of the training data due to the large number of antennas, a DNN for solving the PA problem is trained by user locations instead of the exact channel coefficients as in [12, 13]. It is demonstrated that geographical location information of UEs is a good proxy sufficient for a DNN to learn the optimal transmission power. In the aforementioned work, the employed ML model falls in the category of supervised learning, which requires a large set of training data to ensure learning accuracy and thus it encounters the scalability problem, e.g., when multi-cell coordination is required to serve a large number of users.

Deep Reinforcement Learning-Based Approaches Another line of research employs reinforcement learning (RL) to find the near-optimal solution of important problems in wireless networks. Particularly, Q-learning has been widely used for its simplicity and acceptable performance. In Q-learning, an agent interacts with the environment by taking an action based on its observation about the status of the environment and the reward it receives by taking a specific action. In [15], the joint UA and PA problem is solved by a multi-agent Q-learning approach, where each user acts as an agent. Each agent needs to maintain a Q-table, which stores the state–action pairs about the past experience of interacting with the environment. Such a lookup table-based approach is not practical to handle the problem with large action and state spaces. A promising solution is to replace the Q-table by a neural network, leading to the so-called deep Q-learning (DQL). Some attempts have successfully applied DQL for solving various problems in wireless networks. For example, DQL is used in [16] to enable distributed dynamic spectrum access for network utilization maximization. In [17], a DQL-based approach is developed to solve the PA problem in a multi-cell network. DQL is also used in [18] for transmission power allocation in dense small-cell networks. The work in [19, 20] focuses on how DQL can learn the optimal transmission power allocation in the presence of random variations and delays in the CSI. Particularly, it is shown in [19] that the past state and reward information can help the DQN learn the optimal solution when the wireless channels are time varying. Furthermore, [20] demonstrates that the DQN zero discount factor can be used as an estimator of the optimal transmission power.

The above existing work focuses on the PA problem assuming a serving cell for each user has been determined, i.e., a certain UA scheme is in place. The joint UA and PA problem is investigated in [21, 22] using the DQN as the solver to approximate the optimal solution with different objectives. Particularly, the maximum sum rate is considered in [22], while the maximum energy efficiency is addressed in [22]. Their results suggest that compared to the DQN for solving the PA problem along, the DQN trained to learn the optimal UA and PA solution jointly possesses a larger performance gap from the real optimal solution.

5.3 Main Contribution

With the support of multi-cell coordination in the future RAN, each cell can cooperate with its neighboring cells to decide its serving users and transmission power. As the UA and PA are coupled decisions, they should be decided jointly to achieve the optimum performance. Clearly, the joint UA and PA optimization is non-convex and combinatorial and thus searching for the global optimal solution is highly challenging. The classical optimization would use a certain iterative algorithm where each iteration requires to solve either the UA or PA sub-problem when the solution of the other is fixed. Many iterations are required before the algorithm delivers a converged result. Iterative algorithms often encounter two major challenges in solving the joint UA and PA problem: scalability and convergence speed. In the context of multi-cell coordination, the widely used performance indicator, such as the sum rate or energy efficiency, involves not only the channel gains within the cell but also those in neighboring cells. Besides, the UA and PA decisions taken by neighboring cells will influence the performance of each other. As a result, the classical iterative-based algorithms need to handle a large solution space and consume more iterations to converge. On the other hand, the ML-based approaches have demonstrated great success in solving complex combinatorial problems in numerous fields. Therefore, it would be interesting to explore the ML-based approaches for solving the performance optimization problem for the future RAN with multi-cell coordination. Most of the existing works reviewed in Sect. 5.2 focus on a single-cell scenario, while only very few consider the multi-cell network without cross-cell coordination. To the best of our knowledge, how to apply ML to solve the joint UA and PA problem under the coordination across different cells is still an open issue and thus this contribution is devoted to fill the gap.

5.4 System Model

Network Architecture Consider a cellular network with N sites where each site covers a geographic area consisting of three cells. Each site has K UEs uniformly located and each cell has one TRP with N_t antennas. Each UE is assumed to have one antenna for simplicity but the considered framework can be readily modified to handle the case with multi-antenna UEs. For convenience, denote the set of sites TRPs and UEs as $\mathcal{N}$, $\mathcal{M}$, and $\mathcal{U}$, respectively. Under the considered cell structure, there are $3N$ TRPs and KN UEs in the network. At time slot t, let $\zeta_{m,u} \in \{0, 1\}$ indicate whether TRP m is activated to serve a UE u, i.e.,

$$\zeta_{m,u} = \begin{cases} 1, & \text{if UE } u \text{ is served by TRP } m, \\ 0, & \text{otherwise.} \end{cases} \tag{5.1}$$

Channel Model The radio propagation over the wireless medium suffers both large-scale and small-scale fading. For UE u, the channel coefficient between this UE and TRP m can be modeled by:

$$g_{m,u} = \sqrt{\beta_{m,u}} \boldsymbol{h}_{m,u}, \tag{5.2}$$

where $\beta_{m,u}$ captures the distance-dependent path-loss and log-normal shadowing and $\boldsymbol{h}_{m,u} \in \mathbb{C}^{1\times N_t} \sim \mathcal{CN}(0, 1)$ is the Rayleigh fading coefficient due to time-varying multi-path propagation. The time variation is modeled by a first-order complex Gauss–Markov process as given by Nasir and Guo [19]:

$$\boldsymbol{h}_{m,u}^{(t)} = \rho \boldsymbol{h}_{m,u}^{(t-1)} + \boldsymbol{e}_u, \tag{5.3}$$

where ρ is the correlation coefficient and $\boldsymbol{e}_u \sim (0, 1 - \rho)$. According to the well-known Jakes' model, $\rho = J_0(2\pi d_f T)$, where $J_0(\cdot)$ is the first-kind zero order Bessel function, f_d is the maximum Doppler frequency, and T is the slot length.

Signal Model Based on the considered network structure, the received signal of user u at slot t is given by:

$$\begin{aligned} y_u^{(t)} = &\sum_{m\in\mathcal{M}} \zeta_{m,u}^{(t)} \sqrt{p_m^{(t)}} g_{m,u}^{(t)} \boldsymbol{w}_{m,u}^{(t)} x_m^{(t)} \\ &+ \sum_{m\in\mathcal{M}} (1 - \zeta_{m,u}^{(t)}) \sqrt{p_m^{(t)}} g_{m,u}^{(t)} \boldsymbol{w}_{m,u}^{(t)} s_m^{(t)} + z_u^{(t)}, \end{aligned} \tag{5.4}$$

where $p_m(t) \in \mathbb{R}$, $\boldsymbol{w}_{m,u}^{(t)} \in \mathbb{C}^{N_t\times 1}$, and $x_m^{(t)} \in \mathbb{C}$ denote the transmission power, precoding vector, and transmission signal by TRP m at slot t for UE u. The symbol $\dagger$ represents the conjugate and transpose of a complex vector. Here, the first term is the desired signal of UE u, the second term corresponds to the overall interference signal, and $z_u^{(t)}$ is the additive white Gaussian noise (AWGN) with zero mean and variance σ^2. To maximize the received signal power, TRP m employs the maximum transmission ratio (MTR) precoder to serve UE u as given by $\boldsymbol{w}_{m,u}^{(t)} = (\boldsymbol{h}_{m,u}^{(t)})^\dagger / \|\boldsymbol{h}_{m,u}^{(t)}\|$. According to (5.4), the received signal-to-interference-plus-noise ratio (SINR) of UE u at slot t is given by:

$$\mathrm{SINR}_u^{(t)} = \frac{\sum_{m\in\mathcal{M}} \zeta_{m,u}^{(t)} (G_{m,u}^{(t)})^2 p_m(t)}{\sum_{m\in\mathcal{M}} (1 - \zeta_{m,u}^{(t)}) (G_{m,u}^{(t)})^2 p_m(t) + \sigma^2}, \tag{5.5}$$

where $G_{m,u}^{(t)} = |(g_{m,u}^{(t)})^\dagger \boldsymbol{w}_{m,u}^{(t)}|$ represents the precoded channel gain. As a result, the spectral efficiency of the downlink transmission for UE u at slot t can be expressed as

$$C_u^{(t)}(\boldsymbol{p}^{(t)}, \boldsymbol{\zeta}^{(t)}) = \log_2(1 + \mathrm{SINR}_u^{(t)}), \tag{5.6}$$

Table 5.1 Table of notations

Notation	Meaning	Notation	Meaning
A	Number of quantized power levels	$s^{(t)}/a^{(t)}/r^{(t)}$	State/action/reward at slot t
$C_u^{(t)}$	Spectral efficiency of UE u at slot	T	Slot length
$\bar{C}_u^{(t)}$	Normalized spectral efficiency of UE u	$\mathcal{U}_m$	Candidate UE set of TRP m
$\mathcal{C}_m$	The set of TRPs in the same cluster as TRP m	$\boldsymbol{w}_{m,u}^{(t)}$	Precoding vector used by TRP m for UE u at slot t
f_d	Maximum Doppler frequency	$\beta_{m,u}$	Large-scale fading coefficient
$\boldsymbol{g}_{m,u}/G_{m,u}$	Channel coefficient/precoded channel gain between TRP m and UE u	γ	Discount factor
$\boldsymbol{h}_{m,u}$	Small-scale fading coefficient	ϵ	Probability of exploration
$\mathcal{I}_u$	The set of selected interfering TRPs of UE u	$\zeta_{m,u}$	Decision variable for UA
k_c	Number of candidate UEs	η	Learning rate
$P_{\max}$	Maximum transmission power	$\boldsymbol{\theta}_{\text{target}}^{(t)}/\boldsymbol{\theta}_{\text{train}}^{(t)}$	Model parameters of target/train DQN
$p_m^{(t)}$	Transmission power of TRP m at slot t	ρ	Correlation coefficient
$\mathcal{Q}_m$	The set of UEs interfered by TRP m		
$R^{(t)}$	Accumulated reward up to slot t	σ^2	Noise power

where $[p_1^{(t)}, \cdots, p_M^{(t)}]^T \triangleq \boldsymbol{p}^{(t)}$ and $[\zeta_{1,1}^{(t)}, \cdots, \zeta_{M,KN}]^T \triangleq \boldsymbol{\zeta}^{(t)}$ represent the transmitting power vector and the user association vector at slot t that need to be jointly optimized to maximize the total spectral efficiency.

5.5 Problem Formulation

In this section, we formulate the joint PA and UA problem that aims to maximize the sum rate in the downlink transmission with multi-cell coordination. Based on the described system model, the design problem can be formulated as follows:

$$
\begin{aligned}
(\mathrm{P}) \qquad & \max_{\boldsymbol{p}^{(t)},\boldsymbol{\zeta}^{(t)}} \sum_{n=1}^{N} \sum_{u\in\mathcal{U}} C_u^{(t)}(\boldsymbol{p}^{(t)}, \boldsymbol{\zeta}^{(t)}) \\
& \text{s.t. } 0 \le p_m^{(t)} \le P_{\max}, \quad \forall\, m \in \mathcal{M} \qquad (5.7)
\end{aligned}
$$

$$\sum_{m \in \mathcal{M}} \zeta_{m,u}^{(t)} \leq N_c, \quad \forall n \in \mathcal{N}, u \in \mathcal{U} \tag{5.8}$$

$$\zeta_{m,u}^{(t)} \in \{0, 1\}, \quad \forall m \in \mathcal{M}, n \in \mathcal{N}, u \in \mathcal{U}, \tag{5.9}$$

where (5.7) follows because of the maximum power of each TRP, (5.8) specifies the maximum cluster size to N_c, and (5.9) corresponds to the binary UA decision. It is known that the optimal power control that maximizes the sum rate is an NP-hard problem [23]. With the additional UA variables, it is more challenging to solve (P) even in a single-cell setup. Despite the optimal solution can be found by searching all possible combinations of $(\boldsymbol{p}^{(t)}, \boldsymbol{\zeta}^{(t)})$, the computational complexity is extremely high for a large-scale network with many TRPs and UEs. This motivates a multi-agent DQL approach, which can efficiently provide the close-to-optimal solutions with affordable complexity.

5.6 DQL Framework

Among numerous machine learning models, Q-learning falls in the category of model-free RL. Let S denote the state space of all possible states in the considered network and A denote the action space consisting of user association and power allocation decisions. By observing state $s^{(t)} \in S$ at time slot t, the agent takes action $a^{t)} \in A$. After interacting with the environment, the agent receives a reward $r^{(t)}$ and moves to the new state $s^{(t+1)}$ at slot $t+1$. Traditional RL deals with discrete state spaces. In our problem, S includes the channel status, which may change with time in different rates depending on the UE moving speed and the propagation environment. Since a discrete state space cannot fully characterize the time-varying channel status, a neural network-based formulation of RL is adopted as will be detailed in the following.

Figure 5.1 illustrates the proposed DQL framework for jointly optimizing transmission power and user association. Specifically, each TRP acts as an agent to take an action based on the observed state from the environment. The definition of states and actions critically affects the learning accuracy of DQL and they will be elaborated in Sec. 5.6.2. A policy π refers to the rule for the agent to choose an action from the action space for a given state. To evaluate the expected return for selecting action a in state s under policy π, a Q-function, also known as the DQN, is used as given by:

$$Q_\pi(s, a; \boldsymbol{\theta}) = \mathbb{E}_\pi \left[R(t) \middle| s^{(t)} = s, a^{(t)} = a \right], \tag{5.10}$$

where $\boldsymbol{\theta}$ represents the set of DQN parameters and $R^{(t)}$ is the accumulated discounted reward given as

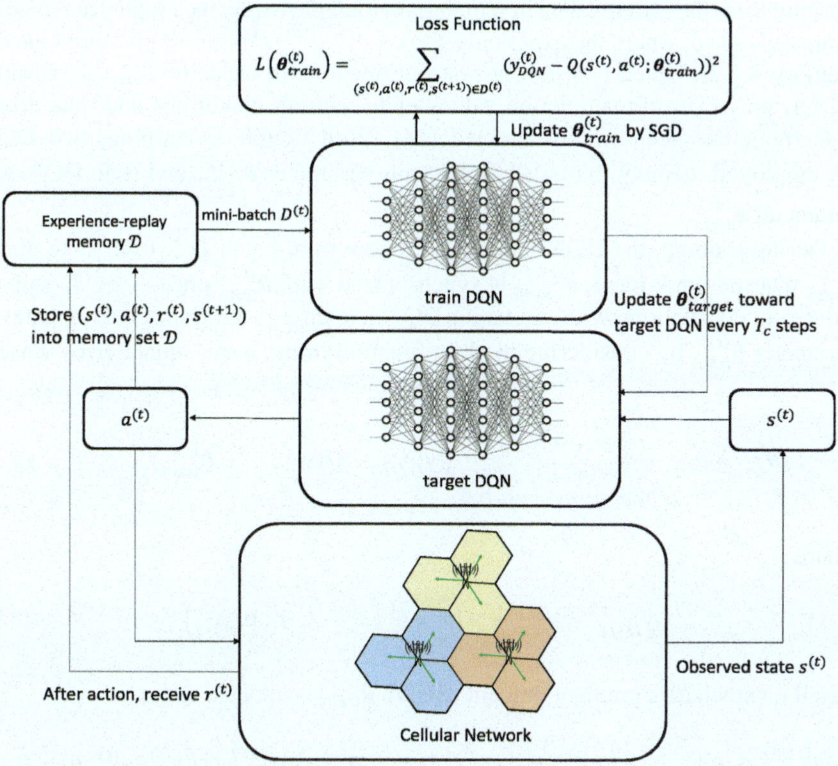

Fig. 5.1 The proposed DQL framework

$$R^{(t)} = \sum_{\tau=0}^{\infty} \gamma^{\tau} r^{(\tau+t)}, \tag{5.11}$$

where $0 \leq \gamma \leq 1$ is a discount factor that strikes the balance between the future reward and the current one by taking an action. Then the optimal policy can be expressed as

$$Q^*(s, a) = \mathbb{E}_{s'}\left[r + \gamma \max_{a'} Q\left(s', a'\right) | s' = s, a' = a\right]. \tag{5.12}$$

5.6.1 DQN

Since the Q-function is non-linear, a neural network is used to approximate the optimal Q-function, $Q(s, a; \boldsymbol{\theta}) \approx Q^*(s, a)$. Whether the approximation is sufficiently accurate relies on the independence of the training data. Instead of learning from the training data in sequential order, the experience replay technique is commonly used, where the experience tuple $\left(s^{(t)}, a^{(t)}, r^{(t)}, s^{(t+1)}\right)$ is stored into the memory set $\mathcal{D}$, which is a first-in-first-out queue with capacity C_{memory}. Besides, a DQL may overestimate action values under certain conditions and thus double Q-learning has been widely adopted [24]. With double Q-learning, two DQNs are employed, namely the target DQN with parameter $\boldsymbol{\theta}^{(t)}_{\text{target}}$ and train DQN with parameter $\boldsymbol{\theta}^{(t)}_{\text{train}}$.

During training, the DQN samples a random mini-batch $D^{(t)}$ from $\mathcal{D}$ to update $\boldsymbol{\theta}^{(t)}_{\text{train}}$. On the other hand, $\boldsymbol{\theta}^{(t)}_{\text{target}}$ is synchronized with $\boldsymbol{\theta}^{(t)}_{\text{train}}$ once every T_c steps to reduce its correlations with the target DQN. At time t, the train DQN updates its parameter $\boldsymbol{\theta}^{(t)}_{\text{train}}$ by considering the loss function in the mean square error sense as given by:

$$L(\boldsymbol{\theta}^{(t)}_{\text{train}}) = \sum_{(s^{(t)},a^{(t)},r^{(t)},s^{(t+1)})\in D^{(t)}} \left(y^{(t)}_{DQN} - Q(s^{(t)}, a^{(t)}; \boldsymbol{\theta}^{(t)}_{\text{train}})\right)^2, \tag{5.13}$$

where

$$y^{(t)}_{DQN} = r^{(t)} + \gamma \max_{a'} Q\left(s^{(t+1)}, a'; \boldsymbol{\theta}^{(t)}_{\text{target}}\right). \tag{5.14}$$

Based on stochastic gradient descent (SGD), $\boldsymbol{\theta}_{\text{train}}$ is updated as

$$\boldsymbol{\theta}^{(t+1)}_{\text{train}} = \boldsymbol{\theta}^{(t)}_{\text{train}} + \eta\left(y^{(t)}_{DQN} - Q\left(s^{(t)}, a^{(t)}; \boldsymbol{\theta}^{(t)}_{\text{train}}\right)\right) \cdot \nabla Q\left(s^{(t)}, a^{(t)}; \boldsymbol{\theta}^{(t)}_{\text{train}}\right), \tag{5.15}$$

where $0 \leq \eta \leq 1$ is the learning rate. To approximate the near-optimal solution, centralized training is considered in the training stage, while each agent trains a single DQN by using all agents' experience memory. The shared experience across all agents not only helps to stabilize the training of the DQN but also improves the learning efficiency.

5.6.2 *Design the DQN*

As mentioned, each TRP acts as an agent and chooses an action using a dedicated DQN consisting of an input layer, followed by L fully connected hidden layers each with N_l neurons and a fully connected output layer. For TRP m, the input layer of the DQN is fed by the state vector $s^{(t)}_m$. Based on the input, the DQN estimates the Q-value of each action for a given state. Then, the agent takes the action according to the adaptive ϵ-greedy algorithm. Specifically, an agent takes the action that has the maximum Q-value with probability $1 - \epsilon$ and it randomly selects an action with probability ϵ to balance exploration and exploitation. After interacting with the environment, TRP m receives a reward $r_m^{(t)}$ and moves to the next state $s_m^{(t+1)}$.

When applying the DQN to solve the joint power control and user association problem, three key elements including states, actions, and rewards need to be carefully designed. They are defined as follows:

State In multi-TRP transmission, a UE suffers two kinds of interference, including the one from the TRPs within the same cluster, referred to as the intra-cluster interference, as well as the one from the TRPs of different clusters, referred to as the inter-cluster interference. Both types of interference can be well suppressed by carefully choosing the TRP's transmission power and the serving/cooperating TRPs for each UE. This implies the state information about the TRPs within the same cluster is as important as that of other clusters. However, each TRP only has some local information about the environment and thus information exchange among different TRPs is necessary to assist a better decision making.

To avoid catastrophic bandwidth consumption for information exchange, each TRP considers a set of candidate UEs, which is denoted as $\mathcal{U}_m^{(t)}$ for TRP m at time t. Only the state information related to the candidate UEs is collected and exchanged. A simple choice of $\mathcal{U}_m^{(t)}$ is to pick those UEs that have stronger channel gains with TRP m because their superior channel conditions are beneficial to boost the sum rate. For each candidate UE in $\mathcal{U}_m^{(t)}$, the state is composed of three feature groups. For simplicity, suppose the number of candidate UEs of each TRP, denoted by k_c, is identical to all TRPs.

1. **Intra-cluster features** The intra-cluster feature group consists of five sets of information: current and the past CSI, the indication about how likely a TRP may serve a UE, and the current and the past transmission power. Each type of information has its own purpose to the DQL. For illustration, let C_m denote the set of TRPs within the same cluster as TRP m.

   - Current CSI: To support multi-TRP transmission, each TRP needs to collect not only the CSI about itself but also the CSI about other TRPs in the same cluster. Thus there are three sets of CSI collected by each TRP.

     - The first-type CSI refers to the serving channel between each TRP in C_m and its candidate UEs in $\mathcal{U}_m^{(t)}$. The serving channel gain $\boldsymbol{G}_{m,u}^{(t)}$ for $u \in \mathcal{U}_m^{(t)}$ allows TRP m to evaluate the downlink spectral efficiency .
     - Each TRP collects the channel gains between its candidate UEs and other TRPs in the same cluster, namely, $\boldsymbol{G}_{m',u}^{(t)}$ for $m' \in C_m^{(t)} \backslash m$ and $u \in \mathcal{U}_m^{(t)}$. The second-type channel has two roles depending on the transmission mode. In the multi-TRP transmission mode, TRP m' in $C_m^{(t)}$ serves UE $u \in \mathcal{U}_m^{(t)}$ together with TRP m and thus $\boldsymbol{G}_{m',u}^{(t)}$ is also the serving channel gain. In the single-TRP transmission mode, the transmission from TRP m' causes interference to UE $u \in \mathcal{U}_m^{(t)}$. In this case, $\boldsymbol{G}_{m',u}^{(t)}$ is used by TRP m to estimate the interference level for UE $u \in \mathcal{U}_m^{(t)}$.

  - The channel gain of the third type is denoted by $\boldsymbol{G}^{(t)}_{m',u'}$ for $m' \in \mathcal{U}^{(t)}_m \backslash m$ and $u' \in \mathcal{U}^{(t)}_{m'}$, which is also the first-type CSI associated with TRP m' in the same cluster as TRP m. By exchanging the third-type CSI, TRPs in $C^{(t)}_m$ can diagnose the gain and loss of performing multi-TRP transmission.

- Past CSI: Besides the CSI at the current slot t, TRP m also collects that at time $t-1$, i.e., $\boldsymbol{G}^{(t-1)}_{m,u}$, $\boldsymbol{G}^{(t-1)}_{m',u}$, $\boldsymbol{G}^{(t-1)}_{m',u'}$ to help track the variation of CSI for $m \in \mathcal{U}^{(t)}_m$, $m' \in C^{(t)}_m \backslash m$, $u \in \mathcal{U}^{(t)}_m$, and $u' \in \mathcal{U}^{(t)}_{m'}$. Although having more past measurements helps the agent to make a better decision, the complexity of the DQN also grows.
- Preference indication: With multi-TRP transmission, each TRP's transmission power and association decision will affect the interference power experienced by not only its serving UE but also other non-serving UEs. Due to this dependence, we implement a candidate indicator denoted by $\boldsymbol{V}_{u,m'} \in \{0,1\}^{k_c \times 1}$ for $m' \in C_m$ and $u \in \mathcal{U}^{(t)}_m$, which indicates whether each candidate UE of TRP m is also the candidate UE of TRP m' in the same cluster with TRP m. For example, if the candidate UE $u \in \mathcal{U}^{(t)}_m$ coincidentally is the kth candidate UE of TRP m''s in C_m, the kth element of $\boldsymbol{V}_{u,m'}$ will be one, and it is zero otherwise. Similar to CSI, the preference indicator at time $t-1$ is also included in $s^{(t)}_m$.
- Past transmission power: The last set of state information is the transmission power used by all the TRPs in the same cluster. For TRP m, $p^{(t-1)}_{m'}$, $\forall\, m' \in C_m$ will be included in $s^{(t)}_m$.

2. **Inter-cluster features** With dense deployments cells, a UE may suffer severe interference from the TRPs of other clusters. Information such as CSI and transmission power of TRPs from other clusters would be useful for each TRP to learn its best action under the interference from other clusters. Since collecting the information from all the other clusters is costly, we limit the information exchange across clusters only to the first few inter-cluster TRPs that have the largest channel gain to the candidate UE of each TRP because they serve as the strongest interferers. To this end, let $\mathcal{I}_u$ represent the set of selected inter-cluster interfering TRPs to UE $u \in \mathcal{U}^{(t)}_m$. Then the channel gains $\boldsymbol{G}^{(t)}_{i,u}$ and transmission powers $p^{(t)}_i$ at time t as well as $\boldsymbol{G}^{(t-1)}_{i,u}$ and $p^{(t-1)}_i$ at time $t-1$ for $i \in \mathcal{I}_u$ and $u \in \mathcal{U}^{(t)}_m$ are included in $s^{(t)}_m$.

5.6.2.1 Actions

Based on the state information, the actions to take by each agent (i.e., TRP) include which candidate user to serve (i.e., user association) and the transmission power. Recall that the transmission power $p^{(t)}_m$ is a continuous decision variable in (5.7). While DQN can handle continuous action spaces through methods such as actor-critic or policy gradient, we choose to use discrete actions because the transmission power is also the state information that needs to be exchanged between TRPs. Using quantized transmission power as the DQN state is commonly considered in the related literature [18, 19]. To facilitate state information exchange between TRPs, the interval $[0, P_{\max}]$ is split into A levels, corresponding to $\lceil \log_2(A) \rceil$ number of bits with $\lceil \cdot \rceil$ denoting the ceiling function. As a result, the set of possible transmission power can be represented as $\{0, \frac{P_{\max}}{A-1}, \frac{2P_{\max}}{A-1} \cdots, P_{\max}\}$. On the other hand, the set of UA variables for TRP m is $\{\zeta_{m,1}^{(t)}, \cdots, \zeta_{m,k_c}^{(t)}\}$. The action space for TRP m is thus fully characterized by:

$$A_m = \{(\zeta_{m,1}^{(t)}, 0), (\zeta_{m,1}^{(t)}, \frac{P_{\max}}{A-1}), \cdots, (\zeta_{m,k_c}^{(t)}, 0), \cdots, (\zeta_{m,k_c}^{(t)}, P_{\max})\}. \tag{5.16}$$

5.6.2.2 Reward

The proposed DQL framework is expected to solve the optimization problem (P). Thus the action adopted by the agent should maximize the sum rate as given by the objective function of (P). However, simply defining the sum rate as the reward for each agent is problematic in the considered multi-agent DQL. Specifically, the sum rate achieved by the multi-TRP transmission is always higher than that by the single-TRP transmission. Thus each agent tends to maximize its reward by persistently participating in the multi-TRP transmission with other agents. However, the gain of the multi-TRP transmission comes at the cost of excessive use of resources. As a result, less number of UEs can be served if the multi-TRP transmission is overused that does not maximize the sum rate. To avoid overestimating the contribution of a TRP to the sum rate, we introduce the notion of normalized spectral efficiency (NSE), which normalizes the spectral efficiency achieved by serving a UE to the number of TRPs used for serving that UE. For an arbitrary UE u associated with TRP m, its NSE can be expressed as

$$\bar{C}_u^{(t)} = \frac{C_u^{(t)}}{\sum_{m \in \mathcal{M}} \zeta_{m,u}^{(t)}}, \tag{5.17}$$

where the numerator is the spectral efficiency given in (5.6) and denominator indicates the number of active TRPs for serving UE u, which acts as a penalty to the active TRP due to its introduced interference to other UEs. Then we define the reward for TRP m as the difference between the gain and the cost of serving UE u. In specific, the NSE in (5.17) presents the gain that TRP m contributes to the network by serving UE u. On the other hand, TRP m causes interference to other UEs in the proximity that results in reduced sum rate. Hence the cost of serving UE u is measured as the reduced NSE due to the association between TRP m and UE u. Let $Q_m^{(t)}$ denote the set of UEs interfered by TRP m. For any UE in $Q_m^{(t)}$, the achieved NSE when TRP is not active in serving UE u is given by:

$$\bar{C}_{u' \in Q_m(t), \sim m} = \log_2 \left(1 + \frac{\sum_{m \in \mathcal{M}} \zeta_{m,u'}^{(t)} (\mathbf{G}_{m,u'}^{(t)})^2 p_m^{(t)}}{\sum_{m' \in \mathcal{M} \setminus m} (1 - \zeta_{m',u'}^{(t)}) (\mathbf{G}_{m',u'}^{(t)})^2 p_{m'}^{(t)} + \sigma^2} \right) \Big/ \sum_{m' \in \mathcal{M}} \zeta_{m',u'}^{(t)}. \tag{5.18}$$

Finally the reward of TRP m for serving UE u is

$$r_m^{(t)} = \bar{C}_u^{(t)} - \sum_{u' \in Q_m(t)} \left(\bar{C}_{u' \in Q_m(t), \sim m} - \bar{C}_{u'}^{(t)} \right), \tag{5.19}$$

where the summation term accounts for the reduced NSE of each UE in $Q_m^{(t)}$ due to TRP m's transmission.

5.7 Results and Discussions

Simulations are performed to evaluate the performance of the proposed DQN for solving (P). In simulations, the number of sites N varies from 4 to 25 with the minimum inter-site distance of 500 m. Each site has three TRP, each with two antennas. UEs are uniformly distributed in the area whose size depends on N. The small-scale fading gain follows the Rayleigh distribution with Doppler frequency $f_d = 10$ Hz. To model the path loss, the COST Hata model is considered with

$$\begin{aligned} \beta =& 46.3 + 33.9 \log_{10}(f_c) - 13.82 \log_{10}(h_{\mathrm{BS}}) \\ & - a(h_{\mathrm{UE}}) + \left[44.9 - 6.55 \log_{10}(h_{BS}) \right] \log_{10}(d) \text{ [dB]}, \end{aligned} \tag{5.20}$$

where f_c is the carrier frequency, h_{TRP} and h_{UE} denote the antenna height of the TRP and UE antennas, respectively, d (km) is the distance between a TRP and a UE. For the suburban scenario,

$$a(h_{UE}) = \left(1.1 \log_{10}(f) - 0.7\right) h_{UE} - \left(1.56 \log_{10}(f_c) - 0.8\right). \tag{5.21}$$

The rest of physical parameters are listed in Table 5.2.

The DQN architecture, as shown in Fig. 5.1, contains $L = 5$ hidden layers of fully connected neural networks with 512, 256, 256, 128, and 64 neurons in each layer, respectively. For each hidden layer, rectified linear unit (Relu) is adopted as the activation function, while the linear activation function is used for the output layer. The input layer is fed with the intra-cluster and inter-cluster feature groups. The state of the intra-cluster feature group has 128 elements and that of inter-cluster feature group has 100 elements. Thus, each state has a total dimension of 228. In addition, the number of power levels A is set to 10 as in [19]. The number of candidate UEs per TRP is set to $k_c = 5$. For each UE, the set of interfering TRPs

Table 5.2 Physical parameters considered in simulations

Parameter	Value
System frequency f_c	2000 MHz
TRP antenna height h_{BS}	32 m
UE antenna height h_{UE}	1 m
Antenna pattern $A(\theta)$	$A(\theta) = -\min[12(\frac{\theta}{\theta_{3dB}}, A_m)]$ $\theta_{3textdB} = 70°, A_m = 20\,\text{dB}$ [1]
Log-normal shadowing standard deviation	8 dB
Maximum transmit power P_{max}	38 dBm
AWGN power σ^2	−114 dBm

Table 5.3 DQN parameters

Parameter	Value
Maximum training episodes E_{max}	5000
Training interval T_{max}	10
Memory size C_{mem}	50,000
Learning rate η	10^{-3}
Initial epsilon	0.2
Final epsilon	10^{-4}

$\mathcal{I}_m$ has five TRPs and the cluster size $|C_m|$ for each TRP m is equal to three. As to each TRP, the set of interfered UEs Q_m has the size of five.

Train the DQN We implement the proposed DQUPA with TensorFlow. In the training stage, the model parameters θ_{train} and θ_{target} are randomly initialized and then trained with the maximum number of episodes denoted by E_{max}. Each episode contains T_{max} time slots during which the large-scale fading remains constant and the small-scale fading changes independently from one slot to another. At the end of an episode, the UE locations are re-drawn randomly. Each agent, namely TRP, takes actions following the adaptive ϵ-greedy policy, which determines the user association and power allocation based on the estimated Q-value to compromise between exploitation and exploration. In this way, exploration enables an agent to improve more informed decisions and avoid trapping in the sub-optimal decision. The DQN is trained by taking a random mini-batch $D^{(t)}$ from the memory $\mathcal{D}$ every T_c slots, and the Adam algorithm is adopted as the optimizer in our work. During the testing stage, the ϵ-greedy algorithm is inactivated, i.e., the agent only takes the action with the maximum Q-value from the output layer. Important DQN parameters are listed in Table 5.3.

Benchmarks In the following figures, the results obtained by jointly solving UA and PA using DQN are denoted by

DQN (joint).

In comparison, a few benchmark schemes are considered as explained below:

- Greedy UA: each UE is associated with the TRP, which offers the strongest channel gain.
- Genetic algorithm (GA): GA is a well-known meta-heuristic algorithm. We use GA to solve the UA sub-problem assuming each TRP has global CSI, i.e., every channel gain in the network is available to each TRP. With this ideal assumption, the solution obtained by GA is close to optimal.
- Max-P: Each TRP transmits with the maximum power without power allocation.
- WMMSE [23]: This algorithm delivers the optimal power allocation that maximizes the sum rate under the condition that global CSI is available at each TRP.
- DQN (Separate): Two separate DQNs are used to solve (P) sequentially. The UA sub-problem is solved by one DQN assuming each TRP transmits with the maximum power. Then the PA sub-problem is solved by the other DQN based on the UA result delivered by the first DQN.

While the high complexity of (P) makes it difficult to find the exact optimal solution, we anticipate that the solution obtained by subsequently solving the UA sub-problem and the PA sub-problem using GA and WMMSE, respectively, should be close to the optimal one. In what follows, the results obtained using GA and WMMSE is denoted as "GA+WMMSE."

5.7.1 Training and Testing Results

We first study the training performance under different DQN parameters for $M = 12$ TRPs and $K = 10$ UEs in each cell. Figure 5.2a, b shows the SE during the training stage under different discount rate γ and batch size D, respectively. One can see that there exists a good choice of the discount rate that leads to the best training performance, and same as the batch size. It is shown from Fig. 5.2 that when an agent trusts all future awards (i.e., $\gamma = 1$) or when it cares about an immediate award only (i.e., $\gamma = 0$), it does not learn the best action appropriately. Weighting the recent awards more than the future awards by choosing a smaller discount factor, say $\gamma = 0.3$, achieves a faster convergence of the training process than a larger value (e.g., $\gamma = 0.9$). Similarly, a mild choice of the batch size, e.g., $D = 128$ is promising to stabilize the training process.

The training performance of the proposed DQN in comparison with some benchmarks is depicted in Fig. 5.3. As can be seen, the proposed DQN for jointly solving UA and PA shows a linear improvement of the average SE over the first 4000 episodes. At the 4000th episode, we deactivate the adaptive ϵ-greedy algorithm such that the agents only take the action with the maximum output value of DQN output layer. Consequently, the average SE slightly fluctuates between 5.25 and 5.5 bps/Hz, indicating that the proposed DQN reaches a convergence.

Figure 5.4 shows the testing result after training. The proposed DQN (joint) achieves nearly the same SE as GA+WMMSE. This may be a sign that the proposed

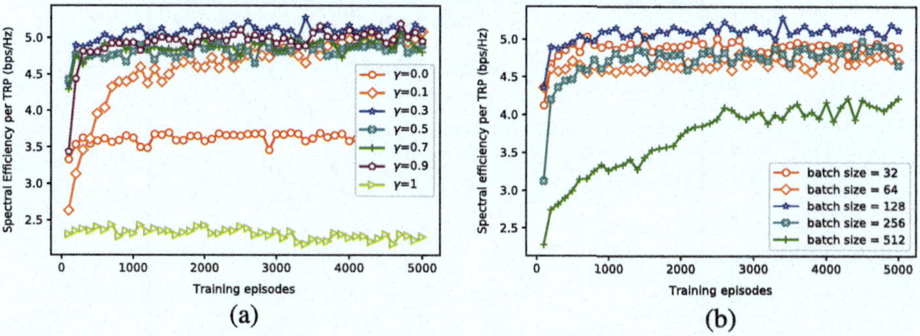

Fig. 5.2 Training results during 5000 episodes. (**a**) Impact of discount rate γ. (**b**) Impact of batch size D

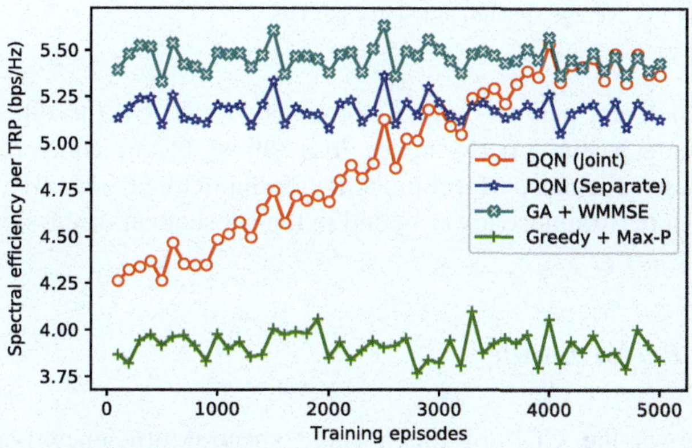

Fig. 5.3 Training results during 5000 episodes

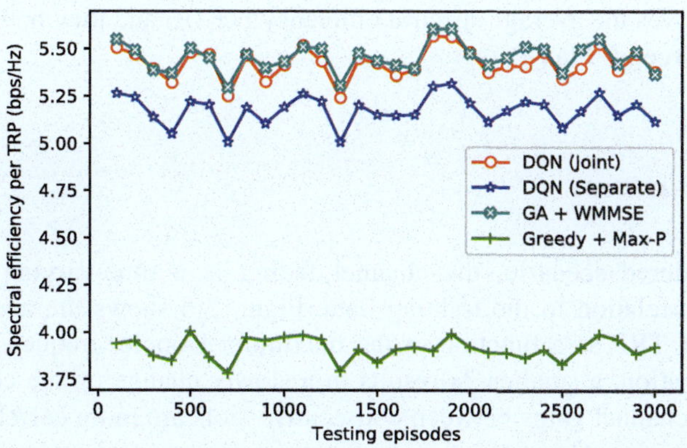

Fig. 5.4 Testing results over 3000 episodes

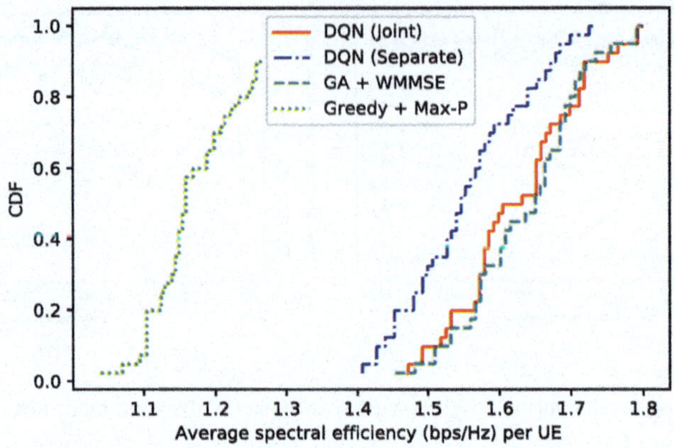

Fig. 5.5 CDF of the average spectral efficiency per UE

DQN (joint) delivers near-to-optimal solutions of (P). On the other hand, DQN (separate) incurs a slight performance loss (about 6.2%) compared with DQN (joint). All these schemes perform much better than Greedy+Max-P, which will be considered as a performance lower bound in the subsequent discussions.

5.7.2 UE Performance

Figure 5.5 shows the CDF of the average spectral efficiency per UE of different schemes. Again, the performance of DQN (joint) is very close to that of GA+WMMSE. Compared with the two DQN-based schemes, DQN (joint) slightly improves the average spectral efficiency per UE and they both significantly outperform Greedy+Max-P.

5.7.3 Robustness

In the considered scenario, the channel fading is a time-varying process that introduces correlation to the training data. Figure 5.6 shows the average spectral efficiency per TRP as a function of the maximum Doppler frequency f_d. With a fixed slot duration, a smaller f_d results in a slower change on the channel fading and thus the channel gains between consecutive slots are more correlated and vice versa. As $f_d \to \infty$, the correlation coefficient ρ is close to zero in which case the channel gains are uncorrelated variables. This case is labeled as "uncorrelation" in the figure. As can be seen, the proposed DQN is robust to the time-varying channels. We note that the SE achieved by GA+WMMSE and Greedy+Max-P is not sensitive to ρ because both schemes solve (P) in a centralized manner assuming

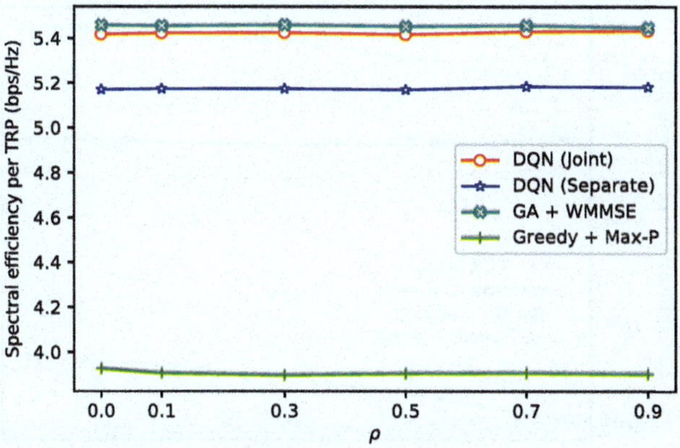

Fig. 5.6 Average spectral efficiency per TRP under different correlation coefficient ρ

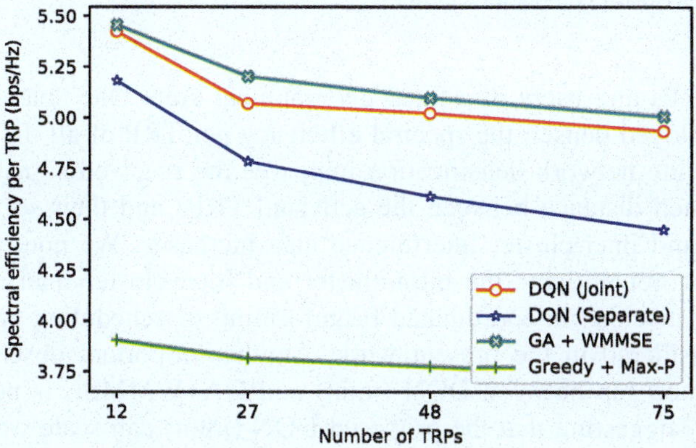

Fig. 5.7 Average spectral efficiency per TRP vs. number of TRPs

global and accurate CSI is available at a central controller. Differently, the proposed DQN is implemented at each TRP with local CSI and limited CSI exchange with neighboring clusters.

5.7.4 Scalability

In the multi-agent DQL, each TRP acts as an agent that takes its own actions based on some local information. To observe how the proposed DQL performs in a network with a large number of TRPs, Fig. 5.7 shows the average spectral efficiency per TRP of the considered algorithms with varied number of TRPs from 12 to 75. To do so, the number of sites is increased from 4 to 25 and each site has three cells. The number of users per cell is set to 10 as before. This results in higher

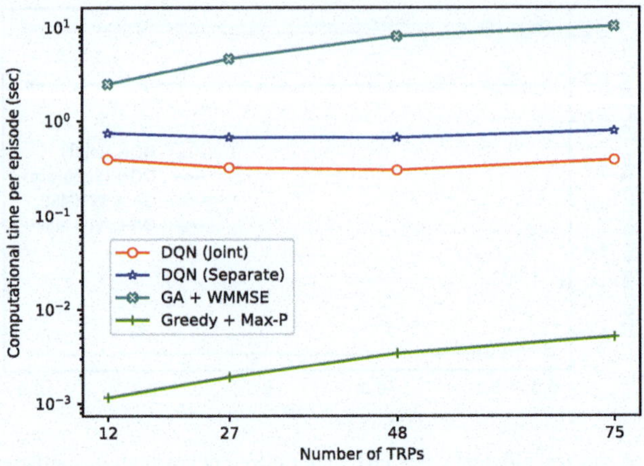

Fig. 5.8 Comparison of computational time

density of TRPs and users in a fixed geographical area. One can see that as the TRPs are deployed denser, the spectral efficiency per TRP of all the schemes gets reduced. Despite network densification improves the received signal strength due to the shortened distance between the activated TRPs and their served users, the intra-cluster and inter-cluster interference also increases. We note that the MRT precoder does not mitigate the intra-cluster and inter-cluster interference, which needs to be canceled via coordinated beamforming or scheduling across different cells not considered in the present work. Another important observation is that the performance gap between DQN (joint) and GA+WMMSE is nearly constant in the figure, suggesting that the proposed DQN (joint) can scale well even in the interference-limited scenario. However, the difference between DQN (separate) and GA+WMMSE increases from 5% to 11% as the number of TRPs increases from 12 to 75. This indicates that with more TRPs, i.e., agents in the network, the actions taken by individual agent deviate more severely from the global optimal.

We also compare the complexity of each algorithm by showing their computational time per episode in Fig. 5.8. All the algorithms are implemented in Python on an Intel CPU at 2.5 GHz (128 GHz RAM) desktop. It is shown that when the number of TRPs increases from 12 to 75, the computational time of the proposed DQN (joint) and DQN (separate) is nearly constant. The computational time of the DQN mainly depends on the number of states. In the proposed DQN, the information exchange within the same cluster is limited to a fixed number of candidate UEs such that the number of states per DQN is kept unchanged. Clearly this results in a certain performance loss because only partial information about the neighboring cells is available to each agent. As discussed in Fig. 5.7, the spectral efficiency loss of DQN (joint) is quite small while that of DQN (separate) is slightly larger. Although GA+WMMSE achieves the highest spectral efficiency, its computation time is about ten times longer than the proposed DQN. The above results allow

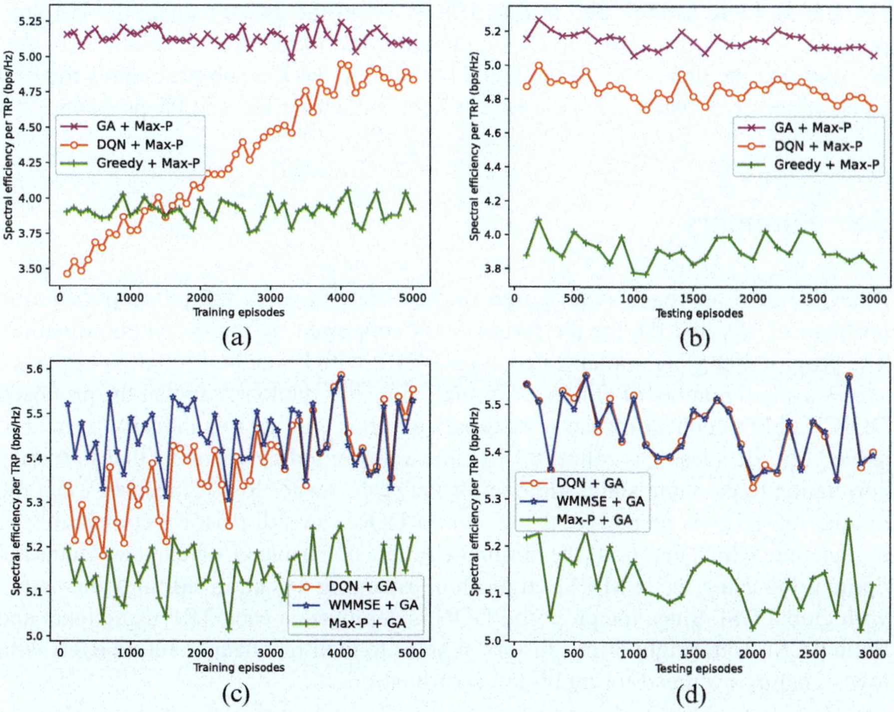

Fig. 5.9 Performance of separate DQN: (**a**) Training results of the UA sub-problem, (**b**) Testing results of the UA sub-problem, (**c**) Training results of the PA sub-problem, and (**d**) Testing results of the PA sub-problem

us to conclude that the proposed DQN with limited information exchange among agents can adapt to network topology changes with desired scalability.

5.7.5 Closer Look at DQN

To better understand how DQN performs in solving the UA and PA sub-problems, we show the results of using DQN to solve the two sub-problems individually in Fig. 5.9 along with those from the benchmark schemes. As before, the SE per TRP is considered as the performance metric. For a fair comparison, all the active TRPs transmit with the maximum power in the considered schemes. The results clearly show that the proposed DQN can efficiently migrate to a better solution during the training period and consistently locate the close-to-optimal solution during the testing period. From Fig. 5.9c, d, it is observed with the UA solution obtained by GA, the proposed DQN performs as good as WMMSE for solving the PA sub-problem. However, the proposed DQN only delivers sub-optimal solution for the

UA sub-problem as indicated in Fig. 5.9a, b. From the above results, the UA sub-problem seems to be harder to DQN comparing to the PA sub-problem. It is thus inferred that the deficiency of the DQN for solving the UA sub-problem is the root cause of the error when using the DQN to solve the joint UA and PA problem.

5.8 Summary

A DQL-based approach is proposed in this work to solve the joint optimization problem of UA and PA for the future RAN supported by multi-cell coordination. The proposed DQN is implemented at each TRP with limited information exchange across a small number of clusters. Simulation results demonstrate that the proposed DQN is able to deliver closed-to-optimal solutions that are dynamically formed to serve UEs. Besides, it is robust to the time-varying fading channels that introduce correlation to the input state. Although some performance loss is incurred when the number of TRPs is increased, the proposed DQN can still perform comparably to the scheme, which first finds the optimal UA decision using GA and then obtains the optimal PA using the WMMSE algorithm, assuming a centralized implementation with global CSI. Since the proposed DQN is deployed at each TRP using local and limited CSI from neighboring clusters, it helps to intelligentize the future RAN with low signaling overhead for multi-cell coordination.

References

1. 3GPP TR 36.819 (V.11.2.0), Coordinated multi-point operation for LTE physical layer aspects (2013)
2. 3GPP TSG RAN WG1 Meeting #86bis, Final report. R1-1608562 (2016)
3. X. Foukas, G. Patounas, A. Elmokashfi, M.K. Marina, Network slicing in 5G: survey and challenges. IEEE Commun. Mag. **55**(5), 94–100 (2017)
4. Q. Shi, M. Razaviyayn, Z.-Q. Luo, C. He, An iteratively weighted mmseapproach to distributed sum-utility maximization for a mimo interfering broadcastchannel. IEEE Trans. Signal Process. **59**(9), 4331–4340 (2011)
5. K. Shen, W. Yu, Fractional programming for communication systems - Part I: power control and beamforming. IEEE Trans. Signal Process. **66**(10), 2616–2630 (2018)
6. Q. Ye, B. Rong, Y. Chen, M. Al-Shalash, C. Caramanis, J.G. Andrews, User association for load balancing in heterogeneous cellular networks. IEEE Trans. Wireless Commun. **12**(6), 2706–2716 (2013)
7. S.G. Kiani, G.E. Oien, D. Gesbert, Maximizing multicell capacity using distributed power allocation and scheduling, in *Proceedings IEEE Wireless Communications and Networking Conference (WCNC), Hong Kong* (2007), pp. 1690–1694
8. T. Zhou, Z. Liu, J. Zhao, C. Li, L. Yang, Joint user association and power control for load balancing in downlink heterogeneous cellular network. IEEE Trans. Veh. Technol. **67**(3), 2582–2593 (2018)
9. S. Bassoy, H. Farooq, M.A. Imran, A. Imran, Coordinated multi-point clustering schemes: a survey. IEEE Commun. Surv. Tuts. **19**(2), 743–764 (2017)

10. 3GPP TSG RAN WG1 Meeting #95, Intra-site vs. inter-site clustering for nc-jt and dps in dense urban. R1-1813612 (2018)
11. 3GPP TSG RAN WG1 Meeting #95, Comaprison of nc-jt performnance with coordinated and independent scheduling in dense urban scenario. R1-1813613 (2018)
12. H. Sun, X. Chen, Q. Shi, M. Hong, X. Fu, N.D. Sidiropoulos, Learning to optimize: training deep neuralnetworks for interference management. IEEE Trans. Signal Process. **66**(20), 5438–5453 (2018)
13. W. Lee, M. Kim, D.-H. Cho, Deep power control: transmit power control scheme based on convolutional neural network. IEEE Commun. Lett. **22**(6), 1276–1279 (2018)
14. L. Sanguinetti, A. Zappone, M. Debbah, Deep learning power allocation in massive MIMO, in *Proceedings of Asilomar Conference on Signals, Systems, and Computers, Pacific Grove* (2018), pp. 1257–1261
15. D. Li, H. Zhang, K. Long, W. Huangfu, J. Dong, A. Nallanathan, User association and power allocation based on Q-learning in ultra dense heterogeneous networks, in *Proceedings of IEEE Global Communications Conference (GLOBECOM), Waikoloa* (2019), pp. 1–5
16. O. Naparstek, K. Cohen, Deep multi-user reinforcement learning for dynamic spectrum access in multichannel wireless networks, in *Proceedings of IEEE Global Communications Conference (Globecom), Singapore* (2017), pp. 1–7
17. K.I. Ahmed, E. Hossain, A deep Q-learning method for downlink power allocation in multi-cell network (2019). arXiv:1904.13032
18. L. Xiao, H. Zhang, Y. Xiao, X. Wan, S. Liu, L.-C. Wang, H.V. Poor, Reinforcement learning-based downlink interference control for ultra-dense small cells. IEEE Trans. Wireless Commun. **19**(1), 423–434 (2020)
19. Y.S. Nasir, D. Guo, Multi-agent deep reinforcement learning for dynamic power allocation in wireless networks. IEEE J. Sel. Areas Commun. **37**(1), 2239–2250 (2019)
20. F. Meng, P. Chen, L. Wu, Power allocation in multi-user cellular networks with deep Q learning approach, in *Proceedings of IEEE International Conferenceon Communications (ICC), Shanghai* (2019), pp. 1–6
21. S. Xu, P. Liu, R. Wang, S.S. Panwar, Realtime scheduling and power allocation using deep neural networks, in *Proceedings of IEEE Wireless Communications and Networking Conference (WCNC), Marrakesh* (2019), pp. 1–5
22. H. Ding, F. Zhao, J. Tian, D. Li, H. Zhang, A deep reinforcement learning for user association and power control in heterogeneous networks. Ad Hoc Netw. **102**(1), 102069 (2019)
23. Z.-Q. Luo, S. Zhang, Dynamic spectrum management: complexity and duality. IEEE J. Sel. Top. Signal Process. **2**(1), 57–73 (2008)
24. H. van Hasselt, A. Guez, D. Silver, Deep reinforcement learning with double Q-learning, in *Proceedings of Thirtieth AAAI Conference on Artificial Intelligence*, vol. 30, no. 1 (2016)

Chapter 6
Routing Algorithms for Heterogeneous Vehicular Networks

Yujie Tang and Wen Wu

6.1 Introduction

Conventionally, ad hoc mode transmissions based on the connections between vehicles and roadside units (RSUs) in vehicular networks are referred to as vehicular ad hoc networks (VANETs), which mainly focus on road safety applications. With the new era of the Internet of Things (IoT), the conventional VANETs have evolved to the Internet of Vehicles (IoV) [1].

IoV is an important component of intelligent transportation systems (ITS) and has emerged as an active research topic in recent years. The main goal of IoV is to alleviate traffic congestion, enhance road safety, improve transportation efficiency, and provide Internet access and entertainment service for both passengers and drivers [2]. In IoV, vehicles and RSUs such as roadside base stations or access points are installed with short range wireless transceivers. Vehicles are connected through vehicle-to-infrastructure (V2I) or vehicle-to-vehicle (V2V) communication links [3]. There is a suite of wireless standards referred to as dedicated short-range communication (DSRC) [4], which includes IEEE 802.11p, IEEE 1609.1/.2/.3/.4 protocol family, and SAE J2735 message set dictionary [5]. DSRC supports both V2I and V2V communications. In IoV, RSUs are fixed infrastructure deployed along the road to provide services for vehicles. The vehicles are equipped with on-board unit (OBU) so that they can communicate with other vehicles and RSUs. In addition to DSRC, the existing Wi-Fi and cellular technology such as long-term evolution (LTE) can also be the options for next-generation vehicular networks.

Y. Tang (✉)
School of Computer Science and Technology, Algoma University, Sault Ste. Marie, ON, Canada
e-mail: yujie.tang@algomau.ca

W. Wu
Pengcheng Laboratory, Shenzhen, P.R. China

L. Cai et al. (eds.), *Broadband Communications, Computing, and Control for Ubiquitous Intelligence*, Wireless Networks,
https://doi.org/10.1007/978-3-030-98064-1_6

IoV has encountered several unique challenging issues. First, the vehicular network topology is highly dynamic due to rush hours and traffic jams. Second, the high mobility characteristic of vehicles results an intermittent connectivity between the vehicles and RSUs or among the vehicles. In other words, communication links in IoV may have frequent outages when they are exchanging information. However, traditional routing schemes such as ad hoc on-demand distance vector (ADOV) [6], optimized link state protocol (OLSR) [7], and dynamic source routing (DSR) [8] are not efficient or fully applicable to IoV. Therefore, in order to address the challenges in IoV, various ad hoc routing protocols have been proposed [9–11]. These protocols adopt hello messages to detect neighbor nodes. The hello messages are used to carry the vehicle's geographic position, velocity, and moving direction information. This kind of information tends to exhaust the communication bandwidth due to periodic beaconing. Moreover, heavy control cost occurs under the rapidly changing network topology. Software-defined network (SDN) has emerged as a solution to control the networks with low communication cost. In the SDN paradigm, it separates the data plane and control plane to simplify the network management and expedite system evolution [12]. There are three advantages of SDN-based IoV. First, the network management and control can be simplified through the SDN controller since it decouples the control and data planes as well as provides elasticity for IoV. Second, the SDN controller can collect the global information to make the optimal routing decision, and switches report vehicle status information to the SDN controller, but they do not need to exchange beacon information with each other periodically. Routing overhead can be reduced. Third, since routing decisions may change due to the failures of some switches, the central SDN controller is aware of the connections among switches, which makes the routing accurate and flexible. Nevertheless, there are still many challenges need to be solved. Vehicles are moving fast and vehicular network topology changes dynamically. Therefore, the decisions made by the SDN controller need to be timely. In addition, even though the SDN controller can make the optimal routing decision based on the global information, it is not easy to achieve.

In order to address the open issues, we introduce a unicast routing protocol for IoV [13] in this chapter. The main goal is to minimize the delay without continuously monitoring vehicle locations since the continuous monitoring may cause a large amount of overhead. The routing decision may not be accurate enough due to the high mobility of vehicles, so mobility prediction would be a better choice to estimate the vehicle connections. The proposed centralized routing scheme with mobility prediction (CRS-MP) can guarantee more timely and reliable transmissions for IoV, in which the SDN controller can intelligently predict vehicle arrivals and therefore vehicle connections.

The remainder of this chapter is organized as follows. Section 6.2 introduces an overview of the routing algorithms in IoV. Specifically, the CRS-MP routing algorithm is presented in Sect. 6.3. Simulation results are demonstrated in Sect. 6.4. Finally, conclusions are given in Sect. 6.5.

6.2 Background

6.2.1 Unicast Routing Algorithms

Through unicast routing algorithms, a source vehicle is able to deliver a message to a specific destination vehicle. Therefore, the main goal of unicast routing in IoV is to transmit data from one vehicle to the other via wireless multi-hops. Unicast routing can be classified into four categories: position-based, topology-based, map-based, and path-based. In position-based routing, packets are forwarded based on locations of vehicles' neighbors and destinations, while in topology-based routing, packets are delivered based on the information of network links. In map-based routing, packets are forwarded according to the forwarding path, which is calculated based on vehicle's location and placement on the map. Path-based routing makes use of the vehicle mobility prediction to calculate the packet delivery delay and find the next hop to forward a packet. In this chapter, the routing algorithm presented in Sect. 6.3 belongs to the unicast routing.

6.2.2 Broadcast Routing Algorithms

In IoV, vehicles can share emergency, traffic, weather, and other information through broadcast transmissions. Broadcast routing algorithms enable the broadcast transmissions that one vehicle delivers a message to other vehicles within its transmission range. However, the broadcast protocols rely on large message dissemination and hence may cause a high communication overhead and message congestion on the network. Therefore, broadcast storm is a big challenge in broadcast transmissions. In order to solve this issue, a two-phase multi-hop broadcast routing algorithm is proposed in [14], which focuses on the urban area. An emergency broadcast routing algorithm is presented by Durresi et al. The algorithm is operated based on geographical locations of the vehicles in a partitioned highway. Sensors are installed in vehicles and they continuously gather important information. If any emergency is detected, it will be broadcasted to other vehicles [15]. In [16], Fang and Luo proposed a two-timer-based broadcast routing algorithm, which is totally distributed and only relies on GPS information, but it has faster packet penetration speed and smaller delay compared with slotted routing algorithm and the traditional flooding-based routing algorithm.

6.2.3 Geocast Routing Algorithms

Geocast routing algorithms are basically location-based multicast routings. In geocast routing, messages can be delivered to a group of vehicles in an IoV based on their geographic locations. Many IoV applications can benefit from it. Celes et al. in [17] presented a spatial information-based approach for routing in IoV, in which user trajectories can be combined with a geocast strategy to improve the data delivery rate in sparse IoV. In [18], the authors proposed a vehicle-mobility-aware and vehicle-contact-aware geocast routing algorithm for urban IoV from the delay-tolerant network perspective to deal with the high mobility and transient connectivity issues.

6.2.4 Related Work in Routing Algorithms

Routing algorithms for IoV have been widely investigated in the literature [19–23]. Most of existing works primarily focus on broadcast routing protocols for IoV. In fact, message broadcast wastes the bandwidth and increases the communication overhead. Therefore, we consider the unicast scenario [9, 10, 24] in which messages are delivered by V2V transmission instead of broadcasting. Security issues are studied in [25–27], which may be integrated in our work, but they are beyond the scope of this work. In addition, the establishment of routing path in IoV needs frequent negotiation since the topology may change very fast. Thus, conventional routing protocols can still result in network congestion, although the monitoring can be done by basic safety message. Nevertheless, in our scheme, the routing-related control messages are sent through LTE, which avoids frequent rebroadcasting and makes the network less congested.

To establish the end-to-end route, transmit packets timely, and reduce the information exchange cost, there are several research works focusing on investigating the routing algorithms based on SDN. Destounis et al. in [28] provided a practical way of keeping the routing cost small within a given reconfiguration constraint. Nevertheless, this SDN-based routing algorithm focuses on the routing policy among the nodes with wired connections. In [29], Duan et al. assumed that the SDN controller monitors the location of vehicles without the update information from vehicles. This kind of assumption reduces the communication cost but increases the complexity of SDN controller. Liu et al. formulated the cooperative data scheduling algorithm in [30], in which the RSU chooses source and destination vehicles and corresponding data for V2V communication. At the same time, the RSU broadcasts a message to vehicles that are informed to tune to the V2I channel. Nevertheless, only one-hop V2V communication is considered in this work. In [31], the SDN controller sets up the routing paths of the periodic warning messages to the destination RSUs based on geographical and topological information. However, when network size increases, it becomes difficult to manage the dynamically changing requests. Thus, it brings some new challenges, though SDN-based IoV improves the network efficiency by increasing throughput, reducing latency, and enhancing reliability of the network.

The most challenging issue for SDN-based IoV is that even though SDN can make decision on optimal routing path based on the global information, the network topology changes very fast in IoV, especially when the vehicle speed is high. Thus, mobility prediction becomes a crucial component in the SDN controller, since the predicted information can help the SDN controller make decision ahead of time. Usually, machine learning can be used for prediction. In [32], Shen et al. proposed a novel adaptive fuzzy logic inference system to predict and estimate the probability information for wireless communication networks, and the prediction is performed using the recursive least square approach. However, ANN is a more powerful approach in machine learning. It is widely used in solving various classification and forecasting problems. In [33], the authors proposed a spectrum detection method by adopting ANN. In [34], ANN is adopted by the authors to predict the traffic flow by developing an optimized structure through layer-by-layer feature granulation with a greedy-layer-wise unsupervised learning algorithm. Nevertheless, the presented work focuses on the vehicle mobility prediction by using ANN. A traffic flow prediction method is proposed in [35] by using a deep learning approach. However, the prediction period is quite long, which is not feasible for IoV due to the high mobility feature in IoV.

Moreover, many research works focus on designing efficient routing algorithms for IoV. Since the vehicle traffic varies during the day, i.e., the IoV can be either sparsely connected (non-rush hour) or fully connected (rush hour), which depends on the time of the day [2]. In majority of existing works, the vehicle arrival is usually modeled as a Poisson process [36, 37]. However, in reality, there are multiple busy periods due to rush hours. Thus, because of the characteristic of non-homogeneous Poisson process (NHPP), it is more appropriate to model vehicle arrivals that exhibit in multiple busy periods. The stochastic monotonicity and related properties of the inter-arrival time of an NHPP have been studied by Kochar [38] and Pellerey et al. [39]. Nevertheless, in NHPP, the arrival rate function is used for the probability and delay analysis, which is very challenging to obtain due to the dynamic traffic conditions and random vehicle arrivals. The next section presents a mobility prediction based on NHPP model.

6.3 Machine Learning-Based Routing Algorithm for IoV with Mobility Prediction

A unicast routing protocol, referred to as CRS-MP, is introduced in this section, in which ANN is employed to learn and predict the vehicle arrival rate. Afterwards, based on the prediction, the RSU/BS integrates a stochastic urban traffic model in the analysis to estimate the average delay and successful transmission probability. Depending on the locations of source vehicle and destination vehicle, whether the SDN controller decides the routing path or the RSU/BS makes the routing decision can be determined.

For each RSU/BS, during the request scheduling process, vehicles in the V2I channel and the V2V channel can transmit simultaneously since V2I and V2V communications are using different spectrum access technologies. The motivation of requests' scheduling is serving the vehicles that are far away from each other and alleviating the burden of the RSU/BS in rush hours as well as improving the vehicle connectivity through the assistance of RSU/BS in non-rush hours. In particular, the main contributions of the proposed routing algorithm are outlined as follows:

- In the CRS-MP routing scheme, the controller does not need to continuously monitor vehicle locations. ANN is implemented by the controller to learn and predict the vehicle arrival rate, which reaps the powerful learning capability of ANN. Through building the statistical traffic model, the RSU/BS can do the traffic mobility estimation based on the vehicle arrival rate function and then make routing decisions and inform the vehicles.
- In the mobility prediction, successful transmission probability and average time delay of not only single-hop and multi-hop V2V transmissions but also V2I communications are analyzed in vehicles based on the NHPP traffic model and the learnt arrival function.
- The routing problem is aiming to minimize the overall vehicular service delay in IoV and is formulated as an integer programming problem that is NP-hard.
- The NP-hard routing problem is solved by performing a bipartite matching algorithm, and a request delivery algorithm is designed to further improve the network performance. Meanwhile, the serving capability of RSU/BS is taken into consideration.

6.3.1 Network Model

We present the network model of a joint V2I and V2V communication system for IoV in urban area. The ubiquitous coverage can be achieved through cellular networks, i.e., BS, as shown in Fig. 6.1. A distributed SDN controller is adopted in this work, which copes with the distributed and heterogeneous nature of modern overlay networks. Each controller is responsible for an SDN domain that refers to cities or communities. There exists a stable TCP/IP connection between distributed controllers and the BS/RSUs, which minimizes the probability of unsuccessful transmission and packet loss. Each SDN controller collects the information from all the BS and RSUs within its charged area, and the BS and the RSUs are considered as the switches that can communicate with each other. The RSUs are located in each intersection since deploying RSUs at intersections can reduce the need for non-line-of-sight transmissions along the road to provide better service coverage. The RSUs serve the vehicles within the RSUs' coverage area, while the BS serves the vehicles that are out of the coverage area of any RSU. The vehicles send their requests as well as the IP addresses of both themselves and their destinations to the RSU/BS. Then, the RSU/BS delivers the information to the controller. Afterwards, the SDN controller converts the IP addresses to the vehicle indexes according to the locations of both source vehicle and destination vehicle and makes the routing decision.

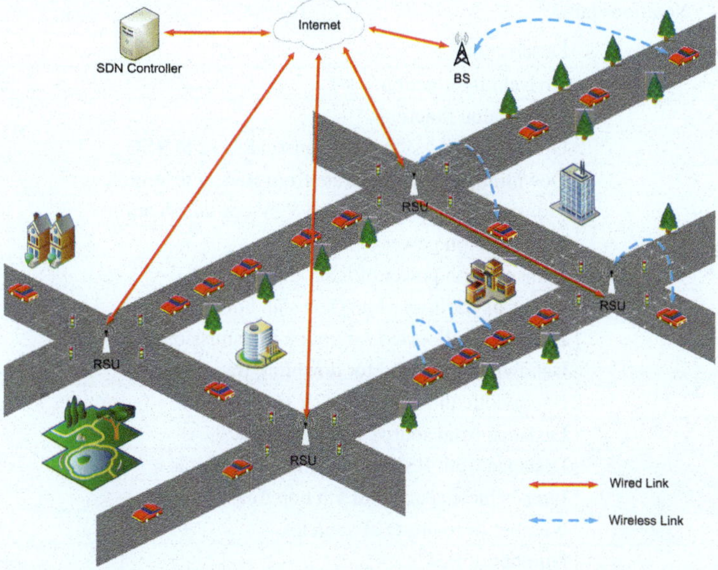

Fig. 6.1 Scenario in urban area

Both IEEE 802.11p and LTE interfaces are installed on each vehicle since this hybrid architecture has been recently adopted to exploit both the low cost of IEEE 802.11p and the wide-range low-latency communication of the cellular networks [40]. DSRC that is based on IEEE 802.11p protocol is used for V2V communications, where a 75 MHz band is allocated with 7 channels, each with 10 MHz bandwidth. Meanwhile, V2I communications operate through LTE, in which multiple vehicles can share the spectrum resources by orthogonal frequency division multiple access. The communications among the BS and RSUs are via wired connections. The neighboring RSUs communicate with each other using different bandwidth to avoid the interference. The control channel is for the standard usage in DSRC such as the safety message broadcasting, while the other 6 channels are used for V2V communications. The routing-related messages are sent through the LTE system. Orthogonal signaling is assumed for LTE in each time slot (one request scheduling period), and each resource block is allocated to each request. Each RSU can serve up to C vehicles, and the transmission range of each RSU is D meters. The rest of the notations used in this chapter are listed in Table 6.1.

6.3.2 Statistical Mobility Model

An NHPP $\mathcal{A}$ counts the number of vehicles that enters the road, and $\mathcal{A}$ is characterized by its arrival rate function $\lambda(t)$, $(t \geq 0)$ which is also referred to

Table 6.1 Notations table

Notation	Description
$\lambda(t)$	Arrival rate function
$\Lambda(t)$	Mean value function
R_m^y	Uplink transmission rate from vehicle m to RSU y
R_n^y	Downlink transmission rate from RSU y to vehicle n
R_{V_k}	Transmission rate of per hop V2V communication
P_V	Transmission power of vehicles
P_y	Transmission power of RSU y to vehicles
N_0	Power of additive white Gaussian noise (AWGN)
W_m	Bandwidth allocated for uplink transmission
W_n	Bandwidth allocated for downlink transmission
W_V	Bandwidth allocated for V2V transmission
d_m	Distance from source vehicle m to RSU y
d_n	Distance from RSU y to destination vehicle n
d_k	Inter-vehicle spacing of kth hop transmission
X	Number of requests of vehicles
Y	Number of RSUs
Z	Size of request packet
H	Number of hops transmitting from vehicle m to vehicle n

as the intensity function, and $\Lambda(t) = \int_0^t \lambda(u)du$ denotes the mean value function, which is the expected number of arrivals on the time interval $(0, t)$. Let $N(x, t)$ be the number of vehicles in the road segment $(0, x]$ at time t, and $n(x, t)$ be the density of vehicles in road segment $(0, x]$ at time t. The expectation of $N(x, t)$ is expressed by:

$$\mathbb{E}[N(x, t)] = \int_0^t \lambda(u)du. \tag{6.1}$$

The relationship between $N(x, t)$ and $n(x, t)$ would be described by:

$$n(x, t) = \frac{\partial N(x, t)}{\partial x}. \tag{6.2}$$

6.3.2.1 Inter-Arrival Time Distribution

Given that the NHPP starts at $T_0 = 0$, the arrival times are denoted by $T_1, T_2, \ldots, T_n$, and the corresponding inter-arrival times are represented by $X_1 = T_1, X_2 = T_2 - T_1, \ldots, X_n = T_n - T_{n-1} (n = 1, 2, \ldots)$. According to the deduction from (3) in [41], the probability density function of T_n, i.e., $f_n(t)$, is

$$f_n(t) = \frac{\lambda(t)e^{-\Lambda(t)}\,[\Lambda(t)]^{n-1}}{(n-1)!}, t > 0, n \geq 1. \tag{6.3}$$

For $t > 0$, the probability density functions of X_1 and X_n, i.e., $g_1(t)$ and $g_n(t)$, are given, respectively, by (see (7) of [41]):

$$g_1(t) = \lambda(t)e^{-\Lambda(t)}, \tag{6.4}$$

$$g_n(t) = \int_0^{+\infty} \lambda(x)\lambda(t+x)e^{-\Lambda(t+x)}\frac{[\Lambda(x)]^{n-2}}{(n-2)!}dx, n \geq 2. \tag{6.5}$$

6.3.2.2 Inter-Vehicle Spacing Distribution

The vehicles are moving at an average speed v, and the inter-vehicle spacing follows the NHPP with intensity function $\lambda_S(s)$. Thus, the inter-vehicle spacing $\phi_n(s)$ can be characterized by the following distribution which is given by:

$$\phi_1(s) = \lambda_S(s)e^{-\Lambda_S(s)}, s > 0, \tag{6.6}$$

$$\phi_n(s) = \int_0^{+\infty} \lambda_S(x)\lambda_S(s+x)e^{-\Lambda_S(s+x)}\frac{[\Lambda_S(x)]^{n-2}}{(n-2)!}dx, \quad s > 0, n \geq 2, \tag{6.7}$$

where the parameter $\lambda_S(s)$ and $\Lambda_S(s)$ can be represented by:

$$\lambda_S(s) = \lambda\left(\frac{s}{v}\right) \tag{6.8}$$

and

$$\Lambda_S(s) = \int_0^s \lambda_S(u)du, \tag{6.9}$$

respectively.

6.3.3 *Channel Model*

The uplink transmission rate R_m^y and the downlink transmission rate R_n^y of RSU y for V2I communication are described by:

$$R_m^y = W_m \cdot \log_2\left(1 + \frac{P_V}{N_0}\delta d_m^{-\gamma} |h|_{m,y}^2\right) \tag{6.10}$$

and

$$R_n^y = W_n \cdot \log_2\left(1 + \frac{P_y}{N_0}\delta d_n^{-\gamma} |h|_{y,n}^2\right), \tag{6.11}$$

respectively. P_V represents the transmission power of the vehicles. P_y is the transmission power of RSU y. When $y = 0$, it refers to the transmission power of BS. W_m denotes the bandwidth allocated for uplink, while W_n is the bandwidth allocated for downlink. N_0 is the power of additive white Gaussian noise (AWGN), and δ is the log-normal shadowing component, with a mean of 0 dB and standard deviation σ_S. d_m and d_n are the distances from the source vehicle m to RSU y and from RSU y to the destination vehicle n, respectively. γ is the path fading exponent, which is typically chosen within [2, 4]. $|h|_{m,y}$ and $|h|_{y,n}$ are the Rayleigh-distributed fading magnitude with $\mathbb{E}[|h|^2] = 1$.

The transmission rate of per hop V2V communication is presented by:

$$R_{V_k} = W_V \cdot \log_2\left(1 + \frac{P_V}{N_0}\delta d_k^{-\gamma} |h|_k^2\right), \tag{6.12}$$

where W_V is the bandwidth allocated for V2V communication, and d_k is the inter-vehicle spacing of kth hop transmission, which is a positive value since backward transmission is considered. $|h|_k$ is the Rayleigh-distributed fading magnitude with $\mathbb{E}[|h|^2] = 1$. d_m, d_n, and d_k can be estimated through the probability density function of inter-vehicle spacing. Based on the estimation of the distance, we can obtain the transmission rates R_m^y, R_n^y, and R_{V_k}.

6.3.4 ANN Model

ANN imitates the neural structure of the human brain and it is a powerful technique. In ANN, knowledge can be obtained by the network through a learning process, and the knowledge can be stored. There are three main benefits of ANN. First, ANN can adapt itself according to the input data without knowing any information of the signal. Second, ANN can approximate any function with arbitrary accuracy. Third, ANN is a nonlinear model and can be applied to most of real-world applications.

As the aforementioned merits, ANN has a natural propensity for storing experiential knowledge and makes it available for use. Therefore, we use ANN to predict the vehicle arrival function and further estimate the successful transmission probability and average delay of multi-hop transmission.

There are many different types of ANN, and we focus on a simple and effective model of ANN, i.e., back-propagation neural network (BPNN), which has three layers: input layer, hidden layer, and output layer. A two-layer neural network is selected in this work. There are 20 neurons in the hidden layer and the number the neurons is chosen according to the empirical value. The transfer function is "transig" in the hidden layer. The other layer is the output layer with one neuron, and the transfer function is "purelin." Function "transig" and "purelin" are the two transfer functions of BPNN. A training function "trainlm," one of the training functions of BPNN, is chosen.

After obtaining the ANN training results, SDN can get the real-time traffic arrival rates. Based on that, SDN estimates the successful transmission probability and average delay of multi-hop transmission of each request. Then, the SDN controller is able to make routing decisions.

6.4 Performance Evaluation

The simulation model is built based on the system architecture described in Sect. 6.2. We consider that all the vehicles drive in the same direction and the arrival rate of vehicles in each lane follows the NHPP. A wide range of vehicle speeds are simulated to evaluate the system performance under different scenarios. In each lane, the vehicle speed is selected by the given specific vehicle arrival rate function. The communication characteristics are simulated based on DSRC for V2V communication, and LTE for V2I and I2I communications. The scheduling period is the same as the maximum waiting time. Each vehicle can submit a request to the SDN controller at any time. The total number of submitted requests varies from 10% to 90% out of the total number of vehicles in each lane. Detailed vehicular system parameters are summarized in Table 6.2.

The proposed ANN model is applied to the data collected from the Coquitlam database as a numerical example. The traffic data are collected every 1 min from individual detectors along the roads. We adopt the data from the vehicle volumes

Table 6.2 Simulation parameter

Parameter	Value
Mobility model	Non-homogeneous Poisson process
Transmission protocol	LTE for V2I & DSRC for V2V
RSU transmission range	500 m
Vehicle transmission range	200 m
Serving capability of RSU	10
Transmission power of RSU	20 dBm
Transmission power of vehicle	10 dBm
Bandwidth	10 MHz
Packet size	5 Mbit
Number of iterations	1000

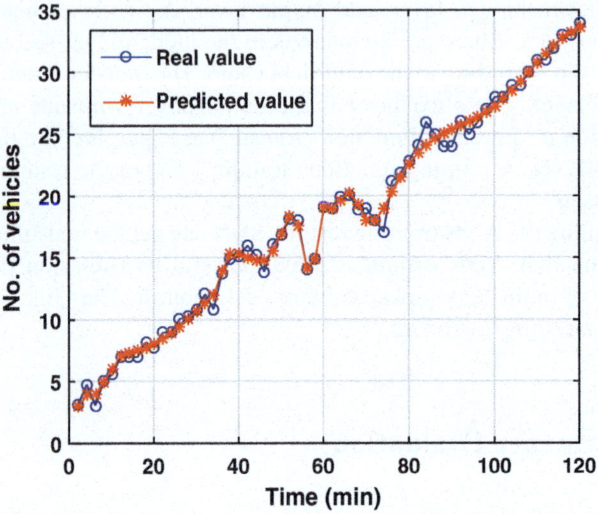

Fig. 6.2 ANN prediction with 20 epochs

in the intersection of Eagleridge Dr. and Lansdowne Dr., Coquitlam, BC, Canada. The traffic flow data collected in the weekdays of the year 2016 are divided into two groups: one for training and the other one for testing. Training parameters are set as follows: training goal is 10^{-1}, training epoch is 20, and learning rate is 0.01. During the training process, the thresholds and weights of each unit in the network are adjusted iteratively in such a way that the error between the actual output and the desired output is reduced. We apply ANN to predict the vehicle arrivals with a setting of 20 epochs and then use the least square method to fit the arrival rate function.

Testing results and fitting function are shown in Figs. 6.2 and 6.3. The vehicle arrival rate is predicted by ANN as shown in Fig. 6.2 and the arrival rate function is fitted as shown in Fig. 6.3. The x-axis represents the time that varies within two hours, and the y-axis denotes the number of vehicle arrivals per minute. Blue circles are the real vehicle arrival rate, and red stars are the predicted vehicle arrival rate. The green line is the function we fit based on the predicted data. The iterations of least square method is set to 20, and the obtained arrival rate function is $\lambda(t) = 1.6415 \times 10^{-13}x^{10} - 5.4755 \times 10^{-11}x^9 + 7.7203 \times 10^{-09}x^8 - 5.9885 \times 10^{-7}x^7 + 2.7844 \times 10^{-5}x^6 - 0.00079247x^5 + 0.013523x^4 - 0.12985x^3 + 0.61408x^2 - 0.49383x + 3.0104$.

Figures 6.4 and 6.5 shows the probability density function (PDF) and cumulative distribution function (CDF) of the inter-vehicle spacing, respectively. n represents the nth inter-vehicle spacing between vehicle n and vehicle $n+1$. In order to simplify the analysis, we set the vehicle arrival function as a linear function $\lambda(t) = a \cdot t + b$, in which $a = 6.25 \times 10^{-5}$, $b = 0.05$. Accordingly, the vehicle arrival rate varies

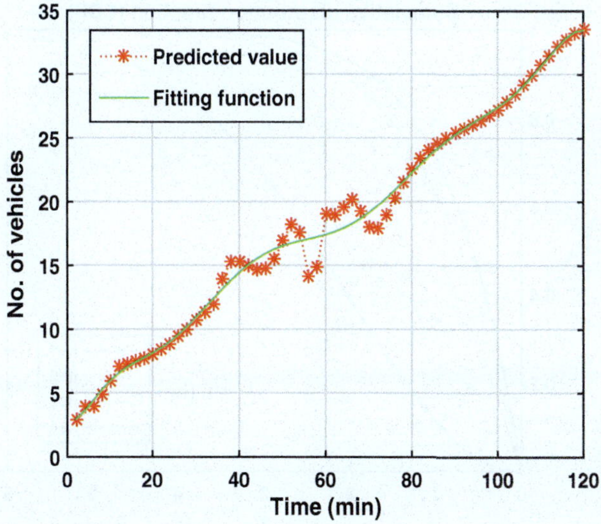

Fig. 6.3 Least square fitting

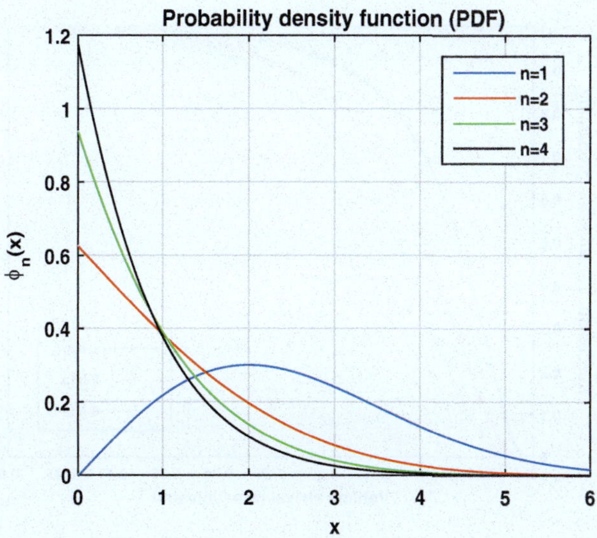

Fig. 6.4 PDF of the inter-vehicle spacing

from 0.05 (vehicle/sec) to 0.5 (vehicle/sec) within 2 h. As shown in the figures, we can see that as n increases, the inter-vehicle spacing becomes smaller.

The successful transmission probability of V2V is shown in Fig. 6.6, which is calculated in the SDN controller based on the ANN training results. It can be seen that successful transmission probability decreases with the increase of the number

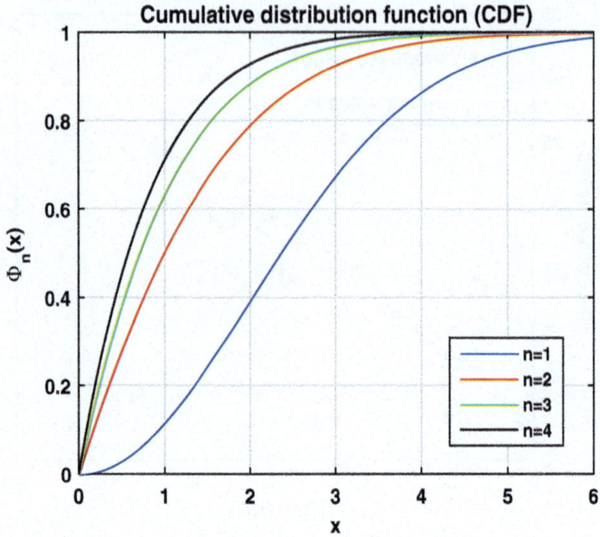

Fig. 6.5 CDF of the inter-vehicle spacing

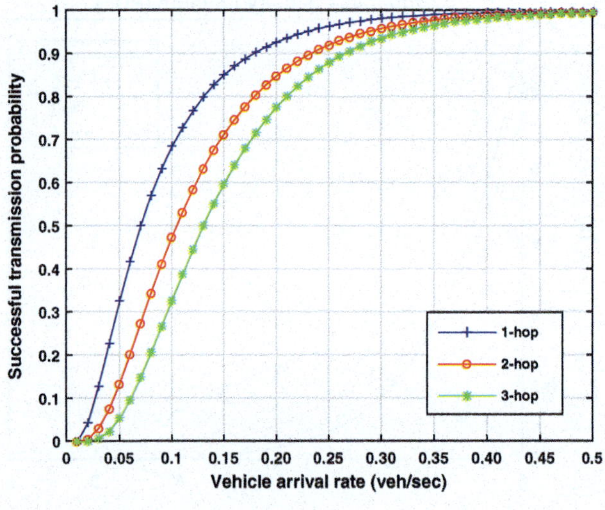

Fig. 6.6 V2V n-hop successful transmission probability

of hops. Moreover, the higher vehicle arrival rate corresponds to higher successful transmission probability.

The average delays of V2I and V2V transmission mode are shown in Figs. 6.7 and 6.8, respectively. In Fig. 6.7, we compare three cases with different maximum waiting time: $T = 1$, $T = 3$, and $T = 5$. A longer waiting time leads to a higher average delay. The V2I average delay decreases to approximately 0 as the vehicle arrival rate reaches 0.15 (veh/sec) for all the three cases. Moreover, when the vehicle

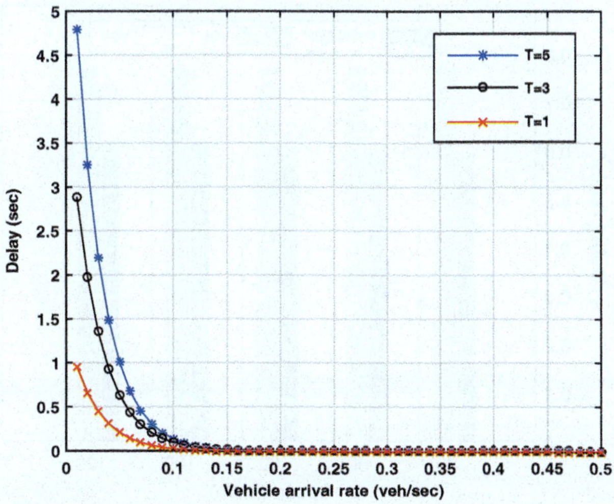

Fig. 6.7 V2I average delay

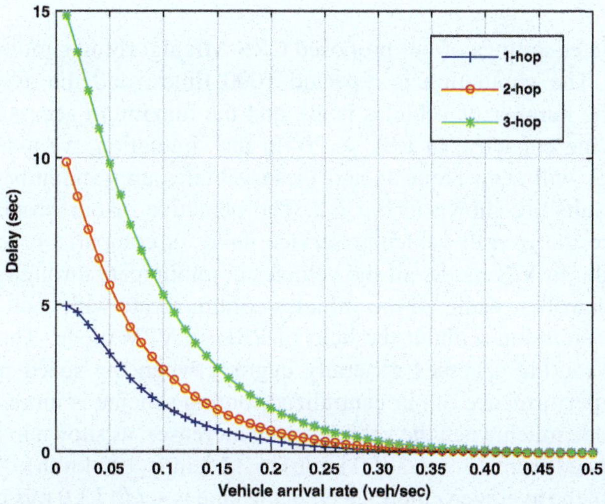

Fig. 6.8 V2V average delay

arrival rate is smaller than 0.1, the maximum waiting time has more impact on the average delay. In this case, the lower the vehicle arrival rate and the larger the maximum waiting time, the higher the average delay. We set the maximum waiting time $T = 5$ in Fig. 6.8. The average delay increases with the increase of the number of hops, and the delay decreases to almost 0 when the vehicle arrival rate is larger than 0.25 (veh/sec), since it is the rush hour when the vehicle density should be high.

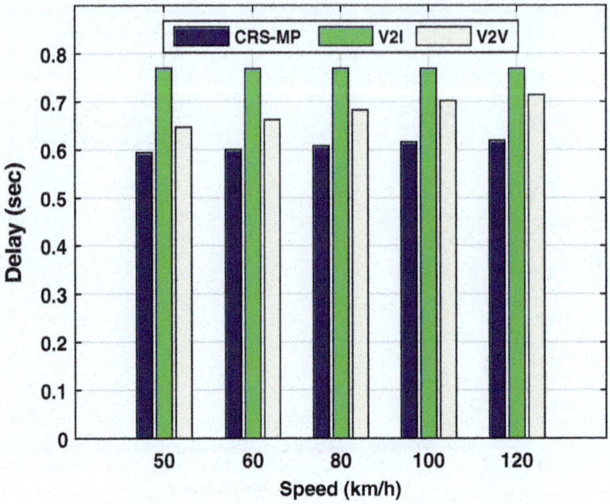

Fig. 6.9 Comparison of the delay among CRS-MP, V2I, and V2V with different speeds

The parameter settings of our proposed CRS-MP algorithm simulation are listed in Table 6.2. The simulation is repeated 1000 times, and the average delay is calculated. The number of vehicles is 40, and the maximum accessing number of vehicles (serving capacity) of RSU is 20. In the simulation, we assume that 60% of the vehicles will send requests and these vehicles are randomly selected. The simulation results are shown in Fig. 6.9. The objective of our proposed CRS-MP is to minimize the overall vehicular service delay in comparison with V2I mode and V2V mode. In V2I mode, all the vehicles communicate through the RSU with the help of controller, while all the vehicles communicate with each other through multi-hop transmission without the help of RSU in V2V mode. The performance of traditional routing schemes generally degrade when the speed increases [42]. However, the performance of our centralized routing scheme is more robust since it does not degrade much when the vehicle speed increases, as shown in Fig. 6.9 where we vary the speed from 13.9 m/s (50 km/h) to 33.3 m/s (120 km/h).

In Fig. 6.10, the average vehicle speed in the lane is set to 13.9 m/s (50 km/h), and the speed limit is 16.7 m/s (60 km/h). The number of vehicles is set to 40, 80, and 120 in each sub-figure, respectively. In each sub-figure, we vary the request fraction with 10, 50, and 90%. The serving capacity of RSU is 100. Figure 6.10 shows that our proposed CRS-MP scheme outperforms the routing scheme without mobility prediction as it achieves lower delay when the number of vehicles increases under different request fraction scenarios. Moreover, the delay of the scheme without mobility prediction increases faster than our proposed routing scheme when the request fraction becomes higher since the serving capacity of RSU is limited. When the number of requesting vehicles is close to the serving capacity of RSU, the subsequent vehicles have to wait to access the RSU. Therefore, the overall vehicular service delay increases faster without mobility prediction.

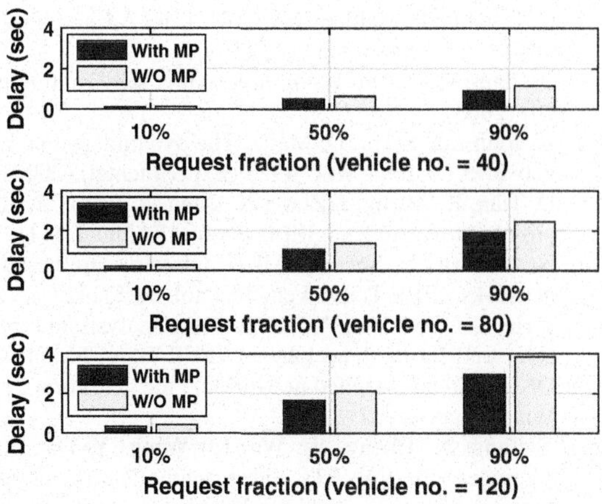

Fig. 6.10 Comparison of the delay of routing scheme with and without mobility prediction with different numbers of vehicles under different request fraction scenarios

6.5 Conclusion

In this chapter, we first presented a brief introduction of routing algorithms in IoV. Then, we discussed different types of routing algorithms adopted in IoV in detail. Afterwards, we demonstrated the proposed centralized routing scheme for end-to-end unicast communication in IoV. The proposed routing scheme has the prediction capability and selects the optimal routing path based on the global information. To adapt to the dynamic changing network topology, the proposed routing scheme can choose either V2I or V2V communication. We have simulated our proposed routing scheme with NHPP and compared it with other routing protocols (pure V2I and pure V2V) in IoV. Simulation results have shown that our proposed CRS-MP scheme outperforms other routing schemes in terms of overall vehicular service delay. Besides, the proposed scheme is more robust when the vehicle speeds vary. For our future work, we will implement practical vehicle arrival model in order to make the mobility prediction more accurate for enhancing the routing performance of IoV.

References

1. X. Shen, R. Fantacci, S. Chen, Internet of vehicles. Proc. IEEE. **108**(2), 242–245 (2020)
2. W. Xu, H. Zhou, N. Cheng, F. Lyu, W. Shi, J. Chen, X. Shen, Internet of vehicles in big data era. IEEE/CAA J. Autom. Sin. **5**(1), 18–35 (2018)
3. N. Cheng, L. Feng, J. Chen, W. Hu, H. Zhou, S. Zhang, X. Shen, Big data driven vehicular networks. IEEE Netw. **32**(6), 160–167 (2018)
4. FCC, Amendment of the commission's rules regarding dedicated short-range communication services in the 5.850–5.925 GHz band. FCC Report and order. 06–110 (2006)
5. Y.L. Morgan, Notes on DSRC & WAVE standards suite: its architecture, design, and characteristics. IEEE Commun. Surv. Tuts. **12**(4), 504–518 (2010)

6. C. Perkins, E. Belding-Royer, S. Das, Ad hoc on-demand distance vector (AODV) routing. RFC 3561 (Experimental) (2003)
7. T. Clausen, P. Jacquet, Optimized link state routing protocol (OLSR). RFC 3626 (Experimental) (2003)
8. D. Johnson, Y. Hu, D. Maltz, The dynamic source routing protocol (DSR) for mobile ad hoc networks for IPv4. RFC 4728 (Experimental) (2007)
9. D. Tian, K. Zheng, J. Zhou, X. Duan, Y. Wang, Z. Sheng, Q. Ni, A microbial inspired routing protocol for VANETs. IEEE Internet Things J. **5**(4), 2293–2303 (2018)
10. N. Alsharif, X. Shen, iCAR-II: infrastructure-based connectivity aware routing in vehicular networks. IEEE Trans. Veh. Technol. **66**(5), 4231–4244 (2017)
11. J. He, L. Cai, J. Pan, P. Cheng, Delay analysis and routing for two-dimensional VANETs using carry-and-forward mechanism. IEEE Trans. Mobile Comput. **16**(7), 1830–1841 (2017)
12. ONF, Software-defined networking: the new norm for networks. Open Networking Foundation White Paper (2012)
13. Y. Tang, N. Cheng, W. Wu, M. Wang, Y. Dai, X. Shen, Delay-minimization routing for heterogeneous VANETs with machine learning based mobility prediction. IEEE Trans. Veh. Technol. **68**(4), 3967–3979 (2019)
14. G. Korkmaz, E. Ekici, F. Özgüner, Ü Özgüner, Urban multi-hop broadcast protocol for inter-vehicle communication systems, in *Proceedings of 1st ACM International Workshop Vehicular Ad Hoc Network* (2004), pp. 76–85
15. M. Durresi, A. Durresi, L. Barolli, Emergency broadcast protocol for inter-vehicle communications, in *Proceddings of 11th International Conference Parallel Distributed Systems* (2005), pp. 402–406
16. S. Fang, T. Luo, A novel two-timer-based broadcast routing algorithm for vehicular ad-hoc networks, in *Proceedings of IEEE International Conference on Green Computing and Communications* (2013), pp. 1518–1522
17. C. Celes, R.B. Braga, C.T. De Oliveira, R.M.C. Andrade, A.A.F. Loureiro, GeoSPIN: an approach for geocast routing based on SPatial INformation in VANETs, in *Proceedings of IEEE 78th VTC Fall* (2013), pp. 1–6
18. L. Zhang, B. Yu, J. Pan, GeoMobCon: a mobility-contact-aware geocast scheme for urban VANETs. IEEE Trans. Veh. Technol. **65**(8), 6715–6730 (2016)
19. A.M. Mezher, M.A. Igartua, Multimedia multimetric map-aware routing protocol to send video-reporting messages over VANETs in smart cities. IEEE Trans. Veh. Technol. **66**(12), 10611–10625 (2017)
20. N. Li, J.-F. Martínez-Ortega, V.H. Díaz, J.A.S. Fernandez, Probability prediction-based reliable and efficient opportunistic routing algorithm for VANETs. IEEE/ACM Trans. Netw. **26**(4), 1933–1947 (2018)
21. M.H. Eiza, T. Owens, Q. Ni, Q. Shi, Situation-aware QoS routing algorithm for vehicular ad hoc networks. IEEE Trans. Veh. Technol. **64**(12), 5520–5535 (2015)
22. M.A. Togou, A. Hafid, L. Khoukhi, SCRP: stable CDS-based routing protocol for urban vehicular ad hoc networks. IEEE Trans. Intell. Transp. Syst. **17**(5), 1298–1307 (2016)
23. M.A. Salkuyeh, B. Abolhassani, An adaptive multipath geographic routing for video transmission in urban VANETs. IEEE Trans. Intell. Transp. Syst. **17**(10), 2822–2831 (2016)
24. D. Lin, J. Kang, A. Squicciarini, Y. Wu, S. Gurung, O. Tonguz, MoZo: a moving zone based routing protocol using pure V2V communication in VANETs. IEEE Trans. Mobile Comput. **16**(5), 1357–1370 (2017)
25. H. Zhu, X. Lin, R. Lu, Y. Fan, X. Shen, SMART: a secure multilayer credit-based incentive scheme for delay-tolerant networks. IEEE Trans. Veh. Technol. **58**(8), 4628–4639 (2009)
26. H. Li, H. Zhu, S. Du, X. Liang, X. Shen, Privacy leakage of location sharing in mobile social networks: attacks and defense. IEEE Trans. Depend. Sec. Comput. **15**(4), 646–660 (2018)
27. H. Zhu, C. Fang, Y. Liu, C. Chen, M. Li, X. Shen, You can jam but you cannot hide: defending against jamming attacks for geo-location database driven spectrum sharing. IEEE J. Sel. Areas Commun. **34**(10), 2723–2737 (2016)

28. A. Destounis, S. Paris, L. Maggi, G.S. Paschos, J. Leguay, Minimum cost SDN routing with reconfiguration frequency constraints. IEEE/ACM Trans. Netw. **26**(4), 1577–1590 (2018)
29. X. Duan, Y. Liu, X. Wang, SDN enabled 5G-VANET: adaptive vehicle clustering and beamformed transmission for aggregated traffic. IEEE Commun. Mag. **55**(7), 120–127 (2017)
30. K. Liu, J.K.Y. Ng, V.C.S. Lee, S.H. Son, I. Stojmenovic, Cooperative data scheduling in hybrid vehicular ad hoc networks: VANET as a software defined network. IEEE/ACM Trans. Netw. **24**(3), 1759–1773 (2016)
31. Y. Liu, C. Chen, S. Chakraborty, A software defined network architecture for geoBroadcast in VANETs, in *Proceedings of IEEE International Conference on Communications* (2015), pp. 6559–6564
32. X. Shen, J.W. Mark, J. Ye, User mobility profile prediction: an adaptive fuzzy inference approach. Wireless Netw. **6**(5), 363–374 (2000)
33. Y. Tang, Q. Zhang W. Lin, Artificial neural network based spectrum sensing method for cognitive radio, in *Proceedings of International Conference on Wireless Communications Networking and Mobile Computing* (2010), pp. 1–4
34. H.-F. Yang, T.S. Dillon, Y.-P. Chen, Optimized structure of the traffic flow forecasting model with a deep learning approach. IEEE Trans. Neural Netw. Learn. Syst. **28**(10), 2371–2381 (2017)
35. Y. Lv, Y. Duan, W. Kang, Z. Li, F. Wang, Traffic flow prediction with big data: a deep learning approach. IEEE Trans. Intell. Transp. Syst. **16**(2), 865–873 (2015)
36. Q. Ye, J. Li, K. Qu, W. Zhuang, X. Shen, X. Li, End-to-end quality of service in 5G networks-examining the effectiveness of a network slicing framework. IEEE Veh. Technol. Mag. **13**(3), 65–74 (2018)
37. L. Zhu, C. Li, Y. Wang, Z. Luo, Z. Liu, B. Li, X. Wang, On stochastic analysis of greedy routing in vehicular networks. IEEE Trans. Intell. Transp. Syst. **16**(6), 3353–3366 (2015)
38. S.-C. Kochar, Some results on interarrival times of nonhomogeneous Poisson processes. Probabil. Eng. Inf. Sci. **10**(1), 75–85 (1996)
39. F. Pellerey, M. Shaked, J. Zinn, Nonhomogeneous Poisson processes and logconcavity. Probabil. Eng. Inf. Sci. **14**(3), 353–373 (2000)
40. S. Ucar, S.C. Ergen, O. Ozkasap, Multihop-cluster-based IEEE 802.11p and LTE hybrid architecture for VANET safety message dissemination. IEEE Trans. Veh. Technol. **65**(4), 2621–2636 (2016)
41. L. Baxter, Reliability applications of the relevation transform. Naval Res. Logist. Quart **29**(2), 323–329 (1982)
42. P. Fazio, F.D. Rango, C. Sottile, A predictive cross-layered interference management in a multichannel MAC with reactive routing in VANET. IEEE Trans. Mobile Comput. **15**(8), 1850–1862 (2016)

Chapter 7
Teaching from Home: Computer and Communication Network Perspectives

Jianping Pan

7.1 Introduction

Starting from early 2020, CoViD-19 has fundamentally changed how teaching and learning are done from K-12 schools to colleges and universities around the world [1]. Many education institutions had to move their in-person teaching online without advance notice [2]. Although there are many online conferencing, lecturing, and meeting (CLM) platforms such as Blackboard Collaborate Ultra, WebEx, Zoom, etc., this sudden and massive move still created a lot of new challenges for teachers and students [3]. Online teaching or distance education is not entirely new, but often supported by professional information technology (IT) staff in education institutions [4]. Teaching from home, on the other hand, is totally new for most instructors who have to deliver their lectures, tutorials, and even labs online. Many teachers and students have noticed considerable degradation of their teaching and learning experience.

Due to the lack of dedicated IT support staff, teaching from home encountered technical challenges in addition to pedagogical ones. Many instructors were caught off guard, even though most of them do have Internet access at home. However, their work-from-home computers and Internet access are not intended for teaching activities, especially synchronous lecturing and online discussion (e.g., office hours). Although Blackboard, WebEx, and Zoom all increased their network and data center capacity and improved their software on short notice, teachers and students still observed unacceptable audio/video quality degradation during prearranged sessions.

J. Pan (✉)
Department of Computer Science, University of Victoria, Victoria, BC, Canada
e-mail: pan@uvic.ca

L. Cai et al. (eds.), *Broadband Communications, Computing, and Control for Ubiquitous Intelligence*, Wireless Networks,
https://doi.org/10.1007/978-3-030-98064-1_7

Upon close examination, many of the issues happened at their home and from it to the Internet, as most home Internet access has been designed and optimized for Email, Web browsing, and video streaming-like applications, i.e., the massive data stream is mainly going from the Internet to the home user.

Online teaching and learning, as its name implies, is a two-way, synchronous and interactive communication process, where one or a few teaching staff interact with a potentially large population of students possibly scattered around the world. Online CLM platforms dealt with this challenge by deploying their cloud meeting platforms all around the world in dedicated data centers, often interconnected by private network links with high quality of service (QoS) guarantees, such as sufficient link bandwidth, limited delay variation, and negligible communication loss. Home Internet access, on the other hand, is likely arranged by individual consumers, constrained by available service providers and plans in certain regions, which usually advertise download data rates much higher than upload ones and can easily become the bottleneck for two-way communications to the Internet. If the lecturer's audio, video, or screen-sharing streams were delayed or lost, it will affect all students regardless of their own locations or network provisioning.

Therefore, the uplink capacity and reliability become the bottleneck of "teaching from home" and are the main focus of this book chapter. Based on the experience since Spring 2020 when we switched to online teaching in the middle of the semester, and the input from professional IT support staff, this book chapter first presents the challenges brought by this new teaching and learning paradigm. Next, it examines the possible technologies and alternatives in home networks and Internet access, leveraging the decades-long advance of computer networking research and education. Further, it proposes a few new approaches and solutions to improving the capability and reliability of wireless fidelity (WiFi) home networks and digital subscriber line (DSL) and cable modem (CM) Internet access, which are commonly used by many instructors at home. The purpose of this book chapter is to create the much needed discussion on these technical issues that have been impeding the successful delivery of online teaching during the pandemic, and it can offer further insights into the future online and distance education paradigm, where "lifetime teaching and learning anywhere" is the ultimate goal, regardless of whether there is another "stay at home" order due to pandemic or other reasons, as well as for home, small- and medium-sized business (SMB) without dedicated IT infrastructure and support staff.

The rest of the book chapter is organized as follows. Section 7.2 scans the literature on related work, and Sect. 7.3 summarizes and compares the existing networking technologies for teaching from home. Section 7.4 proposes feasible approaches to addressing the WiFi interference problem and Internet access reliability problem and makes some recommendations. Further discussion is offered in Sect. 7.5, and Sect. 7.6 concludes the book chapter with future work and directions.

7.2 Related Work

Both home network and Internet access have been well studied and developed in academia and industry, and there is a rich body of the literature on distance education (e.g., pedagogy) and IT technical support in not-for-profit institutions and for-profit organizations [4]. Mature online lecturing, meeting, and conferencing (CLM) tools are readily available at affordable cost, many of which offer free or extended free services during the pandemic, and some have been integrated at least partially with mainstream learning management systems (LMS) [3]. Thus we refer interested readers to each branch of the related work for the status quo and the state-of-the-art.

However, study on "teaching from home" is quite rare and was considered unrealistic pedagogically and technically. Here, "home" refers to the places not where traditional classroom education happens, regardless at K-12, college, or university levels [4]. There have been some attempts on "learning from home" and online learning with various degrees of success and acceptance. Nevertheless, classroom teaching and learning are still the mainstream in normal days, and many hi-tech equipments such as computers, video/data projectors, smart boards, etc., become more and more commonplace. Flipped classroom also happens, where students conduct some, if not all, learning activities in their own time, probably at home, but come to classroom for face-to-face interaction and discussion with instructors and other classmates, supported by many newer LMS systems [5]. Regardless, none of them have gone that far to totally "home," which was set precedent by this pandemic worldwide. SMB such as YouTube broadcasters may encounter similar problems.

With the "stay at home" orders in various forms, teachers and students have to continue their teaching and learning missions entirely online, and for teachers, most likely to instruct from their own home. This is a brand new adventure for many instructors. There are lots of pedagogical challenges, but the focus of this book chapter is on technical ones. Of course, pedagogy is more important, and we try to achieve the same pedagogical goals as classroom teaching, with the assistance of existing technologies, to the maximum possibility first [4–6]. A lot of teachers, students, and some literature have pointed out the long preparation and low efficiency of online teaching and learning, contributed by many factors beyond the scope of this book chapter. Here, we differentiate teaching from home vs. the usual teaching from classroom or office and learning from home and have identified the bottleneck at the instructor's first hop to the Internet, i.e., home networks and Internet access.

The majority of the existing home network and Internet access technologies is designed, engineered, and optimized to deliver massive data from the Internet to home users for Email, Web browsing, and video streaming-like applications. For example, DSL and CM both have more bandwidth allocated to downlink (from the Internet to home) than uplink (vice versa). Even the WiFi access points (AP) in our home and cellular base stations (BS) on the street are engineered to give more opportunities to downlink traffic. These asymmetric links work well until we have the need for broadcasting from home, for teaching or other purposes. There are symmetrically allocated links such as Ethernet, leased circuits, and fiber optics, but they are mostly available in business and backbone settings nowadays, even though the networking research communities and standardization bodies have recognized the need for symmetric links, driven by the previous ups and downs of consumer peer-to-peer (P2P) applications, where the uplink was also a bottleneck. However, with synchronous teaching, meeting, and discussion from home, the bottleneck is severer as the audio and video sources come from ordinary houses. Most CLM platforms allow audio streams to "call in" through telephone systems or bridges, which is very cumbersome and incurs additional cost for education entities.

In this book chapter, we are motivated to make the best out of the existing technologies, to improve the capability and reliability of home network and Internet access. It seems to be a short-term solution but can also shed light into the future of online teaching and learning, for lifetime anywhere, and family Skype video calls.

7.3 Network Technologies Involved

In this section, we first examine the network technologies involved in supporting teaching from home, by host computers, home networks, and Internet access, from the computer and communication network support viewpoint, as illustrated in Fig. 7.1 with recommendations proposed in Sect. 7.4.

7.3.1 Host Computers

Most online CLM tools can run as a standalone application (normally requires download and installation on Windows, Mac OS, and Linux desktop or laptop computers), or an app (lightweight application on portable devices such as iOS and Android tablet computers or smart phones), or even in a Web browser (without additional download and installation and thus operating systems, OS, independent).

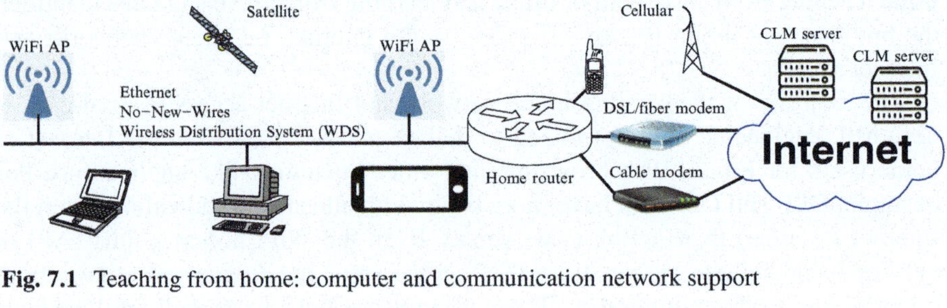

Fig. 7.1 Teaching from home: computer and communication network support

Besides user preferences, here we are concerned about their impact on the computer and communication network support for online teaching.

7.3.1.1 Desktop, Laptop, or Tablet?

The choice of desktop, laptop, or tablet computers for online teaching is mainly device availability and user preferences. Different educational institutions may have different policies to bring institutional equipment home for work or teaching, and some educators have to use their personal devices. Most desktop computers come with Ethernet network interface controller (NIC), for wired network connectivity most common in workplace. At home, Ethernet wall socket may not be available, so alternate wires (see Sect. 7.3.2.2) or wireless (Sect. 7.3.2.3) interfaces and adapters are needed. For laptop computers, most of them come with WiFi interfaces for mobility, but WiFi coverage may vary at home and have high interference from neighbors (see Sect. 7.3.2.3). Some old laptops may have Ethernet NIC embedded, and for newer ones, external Ethernet or additional WiFi adapters via PCMCIA or USB ports are also feasible. Tablet computers are very convenient for annotation during online lecturing, and most of them only have embedded WiFi and some may have cellular Internet capabilities (e.g., through 4G or the emerging 5G mobile communication systems). For tablets and smartphones, external Ethernet interface may be possible through dedicated adapters with micro-USB, Lightning, or USB-C connectors. The form factor further affects the sensitivity of internal antennas, as well as human body (hand and grip gestures) shadowing effect on WiFi signals.

7.3.1.2 Windows, Mac OS, or Linux?

Windows, Mac OS, and Linux, and their tablet and smartphone counterparts, such as iOS and Android, all have the capability of being connected to the Internet through the standard TCP/IP protocol stack. Again, the choice for teaching is mainly personal preferences but dependent on the device availability. From the viewpoint of network support, all these mainstream operating systems come with some network diagnosis tools, such as `ping` for end-host reachability and `traceroute` (or `tracert` on Windows) to discover the routing path. More advanced tools (e.g., `tcpdump` to capture packets and observe protocol interactions) with better user interface (`wireshark`) are also available with additional packages or installation, e.g., Windows or Mac OS `Network` or `Wireless Diagnostics`. Popular network performance testing websites, e.g., `speedtest.net`, further allow users to check their achievable download and upload throughput and ping time to one of the available test servers (often auto-selected by testing websites according to the user location and server availability and load) through any web browser, thus OS independent and convenient. These tools are useful for teachers at home too.

7.3.1.3 Other Necessary Peripherals

Besides host computers running online CLM tools, instructors may choose to use wireless camera (for multi-view), headset (microphone with in or on-ear buds), and in-hand presenters to enrich their presentation. Many of these devices use either Bluetooth, WiFi, or proprietary radio technologies, but often in the same license-free channels as WiFi, which may cause some extra noise and interference. Also many of these devices are powered by batteries and use power-saving techniques extensively to reduce the need of frequent recharging, at the cost of additional delay for audio and video, increasing the mouth-to-ear latency and variation (e.g., voice cutoff or skipping at the beginning of a talk spurt). Whenever possible, wired connectivity (e.g., by USB) of such peripherals to host computer is preferred, especially when the host computer relies on WiFi for Internet access.

7.3.2 Home Networks

As the "last-meter" technology, home network is responsible to interconnect home computers and connect them to the Internet.

7.3.2.1 Ethernet Structured Wiring

Ethernet is the most preferred way of constructing local-area computer networks (LAN) and universally adopted in workplace such as office and commercial buildings. It also becomes common in newly built houses and apartment buildings. Wherever Ethernet is available, it is highly recommended to host computers for reliability and consistency. Even if the host computer does not have an Ethernet interface, various Ethernet adapters are available for different desktop, laptop, and tablet computers and smart phones. However, for most existing houses, Ethernet wiring is not available, and it is very expensive and cumbersome to retrofit for Ethernet structured wiring. Thus, the following options can be considered and are in fact more widely used at home.

7.3.2.2 No-New-Wires Home Backbone

Most existing houses have telephone and television cables wired and sockets installed in some if not all rooms on different floors. Regardless, almost all rooms have power line and outlets for electricity. IP television (IPTV) at the beginning of this century has witnessed the booming of the so-called no-new-wires (NNW) technologies, to transport Ethernet frames over telephone, television, and electricity wires, through an extra adapter connected to computers by wired or wireless Ethernet or USB. Older adapters and technologies only allow networking over a

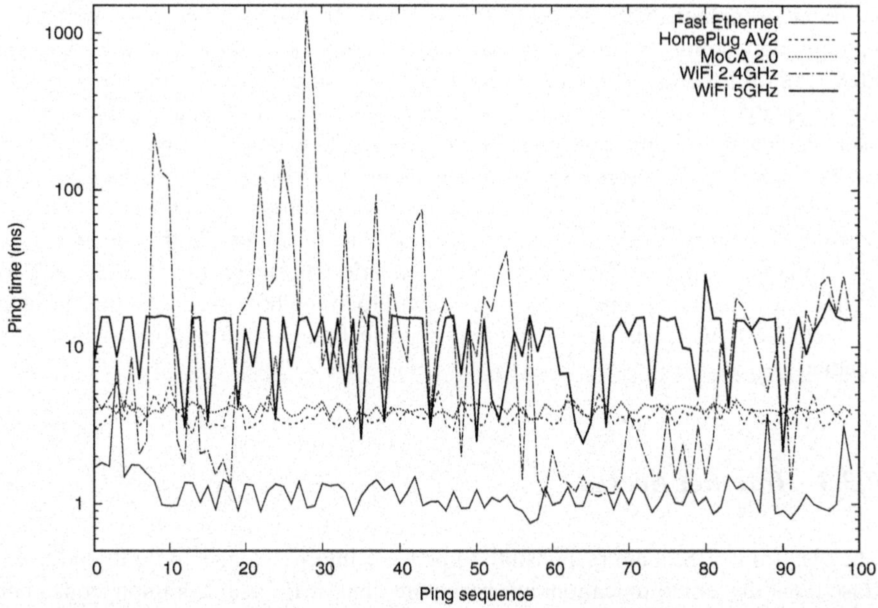

Fig. 7.2 Ping time from the host to home gateway through Ethernet vs. HPPA vs. MoCA vs. WiFi

given type of wires, e.g., HPNA for telephone wires, MoCA for coaxial cables and HPPA (HomePlug) for power lines, and the connectivity is limited, so is the capacity, as each kind of these wires shares their capacity, sometimes even with neighbors. The newer adapters following the G.hn standards can run over different wires, and some even multiple (different kinds of) wires, greatly improving availability and capacity. However, when compared with the switched `Ethernet`, `MoCA` is still the second choice due to the high noise, interference, and collision in the house as shown in Fig. 7.2.

7.3.2.3 Wireless Home Network

WiFi probably is the most common home network technology preferred by many users, especially due to the support for portability and mobility. However, running in 2.4 GHz license-free industrial, scientific, and medical (ISM) and 5 GHz unlicensed wireless channels also means WiFi has to compete with other WiFi and household devices such as cordless phones, microwave ovens, and baby monitors. Particularly, the high-power microwave ovens running in 2.4 GHz frequency bands can easily kill any ongoing WiFi or Bluetooth sessions, as shown in Fig. 7.2 around `ping` #30 for `WiFi 2.4 GHz`, despite various techniques to avoid so. For office buildings, WiFi access points (AP) and channel allocation have been carefully surveyed and arranged, so the interference between nearby APs is minimized. However, in a home environment, WiFi AP is collocated with Internet service provider (ISP)'s modem, depending on the location of point of entry to a house. A single WiFi AP often cannot have an adequate coverage for the entire house, especially when the AP is located at a corner of a house where the modem is located. Even worse, users can easily find many WiFi APs around their house by a simple channel scan, as shown in Fig. 7.4, some even stronger than their own (e.g., `Cable` and `DSL-2.4 GHz` and `5 GHz`). Certain coordination with neighbors is possible but not always feasible. Compared with `Ethernet`, 1-hop `WiFi` has much higher delay (in 64-byte round-trip time by `ping`) and more variation as shown in Fig. 7.2, even for 5 GHz due to heavier propagation loss. We will focus on how to address this problem in Sect. 7.4.1, which is one of the two main technical contributions of this book chapter.

7.3.3 Internet Access

The "last-mile" ISPs are responsible to provide Internet connectivity to end users. Based on the communication infrastructure that ISPs use, common consumer-market Internet access technologies are summarized below and further compared for the purpose of online teaching.

7.3.3.1 Fiber, Cellular, or Satellite?

Fiber optics are the most common communication medium used by the Internet backbone and commercial Internet access networks commonly found in business organizations, education institutions, and government agencies, mainly due to its high capacity and cost, often associated with the need to lay down the fiber optical cable. Fiber to the node, curb, building, and home (FTTN/C/B/H, or FTTx) starts to appear on the consumer market, especially in some countries with emerging economy and highly concentrated population. However, it is still not readily and widely available in many places around the world at consumer level, other than some pilot projects such as Google Fiber. Cellular coverage is almost ubiquitous in urban and suburban areas, but the high cost of data plans in many countries still limits it to an emergency replacement or backup only for home Internet access. Similar concerns are for satellite-based Internet access.

7.3.3.2 Telephone Service Providers

DSL through telephone service providers is one of the two most common home Internet access technologies. Initially designed to carry voice traffic with limited bandwidth and data rate, unshielded twisted pairs (UTP) are the most common wires from telephone companies to customer premises in local loop. Dial-up modem was the first widely adopted Internet access technology, followed by DSL where larger bandwidth is freed over shorter distance through the same UTP wires with limited capacity and susceptible to electromagnetic noise and interference. However, due to the wide availability of dedicated telephone wires to most houses, DSL is still very popular, although some telephone companies are now motivated to bring fiber optics to consumers in selected markets. DSL is less likely affected by neighbors.

7.3.3.3 Television Service Providers

Coaxial cables due to its shield construction and thus much wider bandwidth and better electromagnetic properties were initially used for cable TV broadcasting. With the booming of the Internet, television service providers also upgraded their infrastructure with bidirectional power amplifiers and hybrid fiber-cable (HFC) networks to provide Internet services. Due to the large link bandwidth, cable modem (CM) often can provide higher data rates than their DSL competitors. On the other hand, neighbors do share the same drop cable, and thus the bandwidth and achievable throughput can vary significantly.

As shown in Fig. 7.3, `DSL` has smaller delay and less variation than `Cable` modem, as the latter is indeed affected by neighbors, and `Fiber` has the smallest delay, while `LTE` the highest. Compared with Fig. 7.2, the "last-mile" delay around 10 ms is actually smaller and more stable than the "last-meter" in-home WiFi.

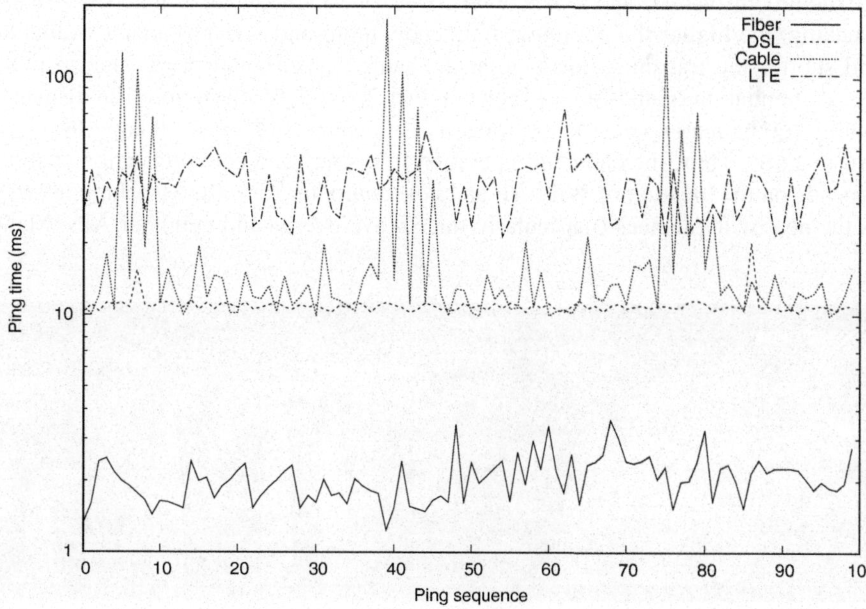

Fig. 7.3 Ping time from the home gateway to first ISP router by Fiber vs. DSL vs. Cable vs. LTE

7.4 Improvement for Online Teaching

Based on the summary and comparison of the existing technologies above, in this section we focus on how to improve WiFi home networks and leverage both DSL and CM ISPs for reliability.

7.4.1 WiFi Interference Avoidance

Many home Internet access issues are actually the problem caused by WiFi networks at home. Service providers often advise their customers to troubleshoot their Internet access problems with a wired Ethernet cable to their so-called modem, AP, or router. A `ping` and `traceroute` can easily identify the additional delay caused by home WiFi networks, due to the poor coverage and severe interference. The following approaches can address these issues with the technologies already existing in most homes.

7.4.1.1 A Better (Al)located WiFi AP

As analyzed above, WiFi home networks have two major issues: coverage and interference. Most DSL or cable modems come with an IEEE 802.11a/b/g/n/ac WiFi AP running in 2.4 GHz and 5 GHz, with 20, 40, or 80 MHz-wide channels. Normally speaking, the higher the operation frequency, more and wider channels available, and shorter the transmission range at the same transmission power due to more signal attenuation (path loss), as shown in Fig. 7.4 with received power in dBm as a quality (Q) indicator, so higher Q for `2.4 GHz` channels (`1` to `14`) than `5 GHz` ones (`36` to `165`). Thus, the choice of operation frequency and communication channel depends on the location of WiFi AP and host computer for online teaching, as well as the nearby appliances (particularly microwave ovens) and neighbor APs. Many

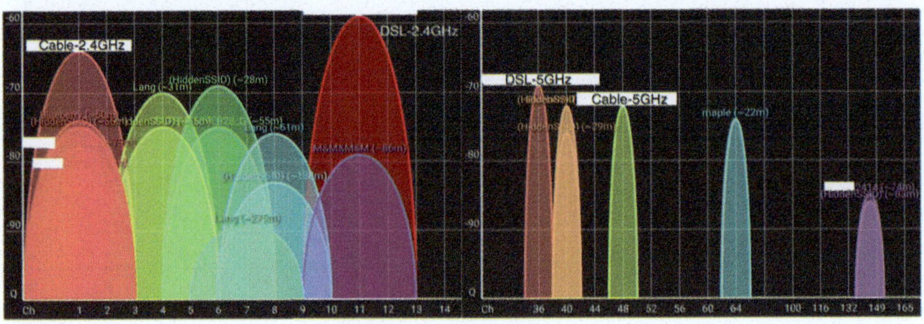

Fig. 7.4 Home WiFi signals in 2.4 GHz and 5 GHz channels

newer APs allow them to "automatically" select a channel based on observation, and some third parties (e.g., `WiFi Analyzer`) offer tools to survey and visualize wireless channels to help consumers choose a less congested channel with stronger signals, e.g., the purposely spaced out `5 GHz` channels in Fig. 7.4. Nevertheless, a single WiFi AP, as the default setting for many users (channel `1`, `11`, `40` and `140`), still suffers the whole-house coverage and interference problems.

7.4.1.2 Wired Interconnected WiFi APs

For some houses, a single WiFi AP is not sufficient to cover the entire house well, especially when the DSL or cable modem is at one corner of the house. To improve the coverage, multiple WiFi APs at different locations can be deployed and interconnected by Ethernet cables if available through the LAN ports of these APs, which is very similar to the setting in workplace. If Ethernet is not available, NNW in Sect. 7.3.2.2 can be used, as shown in Fig. 7.1. With multiple WiFi APs, certain coordination is needed to designate one as the Internet gateway to the outside world with DSL or cable modem, and other APs running in access point mode only, with coordinated addressing and routing if multiple subnets exist. On the other hand, these WiFi APs can run in different channels to minimize the interference among themselves. Instructors can choose the best operation frequency and channel for their host computer. This is often the best home network configuration. Unfortunately, Ethernet is not always available, and NNW can introduce delay variation and security concerns.

7.4.1.3 Wireless Interconnected WiFi APs

On the other hand, when neither Ethernet nor NNW links are available, WiFi APs can be interconnected without wires through wireless distribution system (WDS) [7], which is equivalent to a wired home network backbone. Such approach is often used in cellular systems to interconnect BSs in their wireless backhaul network. Not all DSL or cable modems with integrated AP support WDS in their stock firmware, but many off-the-shelf consumer WiFi APs, especially those powered by `OpenWRT` and `DD-WRT`, can be easily configured to support WDS and have more advanced and flexible configuration. Due to the wireless interconnection, further attention on channel selection is needed to avoid the interference between the home backbone and access networks. By associating to nearby APs, WDS offers a smooth roaming experience, similar to a wired backbone.

7.4.2 *WAN Reliability Augmentation*

Both consumer-grade DSL and cable Internet access services suffer reliability issues, far below what fiber optics can offer in commercial workplace. For instructors to lose connection to the Internet, even briefly or intermittently, is unacceptable for a potentially large group of students during lectures. In the following, we examine and compare DSL- and CM-based Internet access and the possibility to leverage both ISPs when feasible to improve reliability.

7.4.2.1 DSL vs. Cable Modem

As discussed in Sects. 7.3.3.2 and 7.3.3.3, DSL and cable both have their pros and cons. DSL is not affected by neighbors but has limited bandwidth and is more susceptible to noise and interference. CM has more bandwidth but has to share the capacity with neighbors, especially for the uplink. For example, an advertised 25/5 Mbps (for downlink and uplink, respectively) DSL plan only achieves a 3 Mbps uplink, but the ping time from the DSL modem to the first DSL ISP router is lower and more stable due to the dedicated uplink. An advertised 50/5 Mbps CM plan can achieve a 59 Mbps downlink during off-peak hours, but its ping time to the first CM ISP router is a bit higher and highly variable due to the shared capacity, as shown in Fig. 7.3. According to the most CLM platforms, a 500 kbps uplink is sufficient for a standard-definition video stream, which is well accommodated by most DSL and CM links, but delay and loss affect the live video streaming much more.

However, from the DSL and CM ISP networks to CLM data centers, depending on how and where CLM providers deploy their services, the varying bandwidth and delay can cause additional QoS fluctuation, as illustrated in Table 7.1 with `traceroute` to a public enhanced DNS server. In terms of reliability, both DSL and CM can vary by providers and regions, the cable plant, and supporting infrastructures. Consumer-grade ISPs and plans also have routine maintenance and unexpected outage without guaranteed backup and recovery as allowed by their service agreement. Thus, relying on one DSL or CM service provider is often not sufficient for high reliability. Paying higher cost for a business service plan is an option, but in the following we explore other more flexible alternatives.

Table 7.1 Traceroute from the home gateway to 1.1.1.1: Cable vs. DSL ISP

Hop	Cable modem	DSL (router IP, RTT)
1	XX.66.224.1, 10.153 ms	10.31.254.1, 6.553 ms
2	YY.59.161.241, 13.243 ms	* * *
3	YY.163.72.22, 11.705 ms	AAA.11.12.198, 11.644 ms
4	YY.163.68.18, 13.340 ms	BBB.41.104.52, 10.739 ms
5	ZZ.81.81.10, 13.713 ms	**1.1.1.1**, 10.973 ms
6	**1.1.1.1**, 14.765 ms	

Bold indicates the destination (1.1.1.1) reached

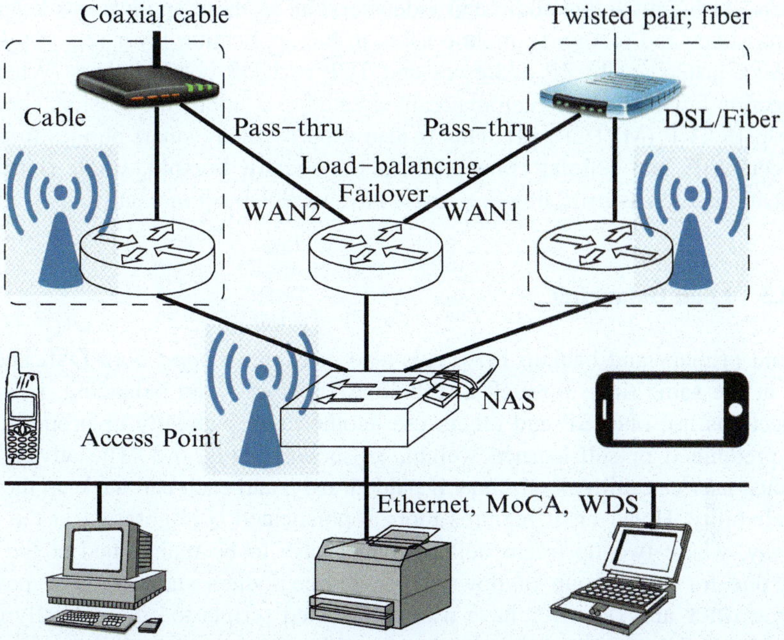

Fig. 7.5 Home network improvement for teaching from home

7.4.2.2 Primary vs. Backup

As shown in Fig. 7.5, we subscribed to two ISPs, one DSL and one CM, which are often available in and competing for the same market. Note that some DSL and CM ISPs wholesale from other major ISPs and then resale to consumers, but here we know these two ISPs are actually independent in terms of their wiring infrastructures and maintenance schedules, to improve reliability. Depending on the service quality and cost of these two ISPs, one can be designated as the primary upstream ISP (e.g., the one offers a flat monthly fee or without data cap) and the other backup (the one charges by the data amount transferred, including cellular or satellite ISPs). To facilitate the automatic switch between the primary and backup upstream ISP, the WiFi AP (or an interconnected group of them) with routing functionalities and connected to both DSL and cable modems shall check the liveliness of the primary ISP, e.g., by `pinging` a known IP address periodically, and then set the default route to the backup ISP when the primary one fails. Depending on the user-defined policy, the home gateway can keep checking the primary ISP periodically and switch back when the primary one becomes available. In this case, there is only one active ISP at any time by default routing. It improves the reliability, unless both fail at the same time, without additional capacity.

Most modern Web-era applications, including Blackboard, WebEx, and Zoom, can sustain the switch of ISPs, and thus the change of the publicly assigned IP address, during an active audio and video session, as these applications keep their session states and recognize mobile users in the application layer (e.g., by HTTP cookies), instead of by IP addresses and TCP or UDP port numbers. When one connection fails, others are automatically created to continue the session, similar to multi-path TCP (MPTCP) [8]. This is also used by many smart phones to switch between WiFi and cellular connections automatically. For old, single-connection applications such as `ssh`, however, users have to reconnect manually.

7.4.2.3 Load Balancing

Beyond primary and backup ISPs, it is also possible to *bond* both DSL and CM ISPs at the same time, through a technique known as load balancing, i.e., some connections use one ISP and others use another, either equally or proportionally to a predefined or self-learned weight, as shown in Fig. 7.5. The advantage is obvious: user can utilize both links if paid already, and each can back up the other for reliability. However, it requires more sophisticated configuration at the home gateway, where two upstream default routes have to be maintained at the same time, one for each group of flows. Open-source routers such as those powered by `OpenWRT` and `DD-WRT` have user-contributed scripts to automatically create virtual LAN (VLAN) for different upstream ISPs, define rules to split traffic, check network connectivity periodically, and fail over to the other link when necessary, under the so-called Dual WAN capability [9]. Most full-blown Linux systems, e.g., Ubuntu, have multi-homing capability, and some low-cost SMB routers, such as TP-Link R470T+, offer multi-WAN capability with very simple and intuitive graphic user interface (GUI)-based configuration. Table 7.2 lists the delay and throughput to `speedtest` servers hosted by Cable and DSL ISP, through Cable and DSL individually, and jointly as bonded. It shows the great advantage of bonding.

However, there are still some subtle issues with load balancing in terms of the "bonding" granularity, i.e., whether the packets from the same session can be distributed over different upstream ISPs. If so, a single application can fully benefit from both ISPs, in terms of both reliability and capacity, but this capability depends on specific applications and whether they or the transport-layer protocol they use can deal with out-of-order packet arrivals through different paths. For most CLM tools, even free but not open source, we cannot guarantee their behavior. Nevertheless, they seem to be able to handle when video and audio streams are carried by different ISPs, similar in concept but different in technology as the call-in feature in most

Table 7.2 Individual and bonded speed test: Cable vs. DSL ISP

Thru	To Cable hosted server	DSL (ping, down/upload)
Cable	13 ms, 59.18/5.28 Mbps	13 ms, 57.43/5.32 Mbps
DSL	11 ms, 24.55/2.84 Mbps	10 ms, 24.25/2.81 Mbps
Bonded	11 ms, 80.99/8.15 Mbps	10 ms, 83.03/8.14 Mbps

Table 7.3 CLM interruption: host vs. Internet link down vs. up

CLM	Ethernet	WiFi	Cable (t: timer)	DSL (down/up)
App	0/0 sec	1/0 sec	t/0 sec	t/t sec
Web	40/0 sec	40/0 sec	3t/0 sec	2t/3t sec

commercial CLM tools. Table 7.3 compares the interruption due to host interface and Internet access down and up events for `App` and `Web`-based CLM platforms. With bonding, load balancing, and liveliness checking, CLM only suffers in the order of the detection timer, which can be as low as 1 sec and much lower than the down-to-up time of DSL (40 sec) and CM (few minutes).

7.4.3 Recommendations on Teaching from Home

Based on the above summary, comparison, and proposal, and the experience in 2020 and 2021, in this section, we make some recommendations on online teaching in 2021 and beyond. First, use a computer with Ethernet connection to home router whenever possible, and choose an ISP with reasonable data rates, especially the uplink one, but more importantly with less delay and variation and fewer packet losses and service outages. When wired Ethernet is not available, consider NNW or improved WiFi with wired or wireless interconnection if needed. When feasible and affordable, consider to have two independent ISPs to guarantee the reliability for teaching from home, especially when large-scale synchronous lecturing is anticipated. If there are other active users at home at the same time, consider allocating them to use a low-priority WiFi channel and ISP when possible to avoid link congestion.

7.5 Further Discussion

Currently, most colleges and universities planned to have online teaching for undergrad or large classes, and possibly in-person teaching for grad or small classes. Teachers may or may not have to teach from home. However, CoViD-19 spikes may return again later 2021 or early 2022 in north hemisphere when another flu season starts, and instructors may have to teach from home again, if a vaccine or proven medicine is not widely available or accepted. Looking beyond the pandemic and Fall 2021, some further thoughts deserve more discussion:

- **Online or offline?** Regardless another pandemic looming in the next few years, the mixed online and offline teaching is likely to stay with us. Online teaching can help us reach more population to further the education mission.

- **Synchronous or asynchronous?** This book chapter mainly addresses the challenges due to synchronous lecturing from home. Another option is asynchronous lecturing where instructors record video lectures in advance. To compensate the lack of interaction during lectures, additional Q&A sessions can be held, where synchronous communication is needed. We believe that both synchronous and asynchronous communications will be a part of our teaching regardless during or after another pandemic or other events.
- **The future of teaching and learning**. Unarguably, CoViD-19 has fundamentally changed the way how education, as well as other sectors of the societies around the world, conducts their business, once forever. It is unlikely we fully go back to the traditional classroom teaching—it is not all necessary, nor sufficient. However, there are still many other pedagogical challenges due to online teaching and learning, e.g., how to conduct labs and evaluate students against expected learning outcomes meaningfully and truthfully.

7.6 Conclusions

In this book chapter, based on our experience in 2020 and 2021 during the CoViD-19 pandemic and the input from professional IT support, we examined the challenges brought by the sudden massive move to online teaching, particularly teaching from home. By comparing existing technologies and alternatives, we proposed and validated some approaches to improving the capability and reliability of home networks and Internet access, specifically for synchronous lecturing from home to a large student population. The purpose of this book chapter is to create some much needed discussion on this topic, even after the first few waves of CoViD-19. Insights obtained can also be applied to other scenarios such as SMB and "broadcast yourself" from home or even family video calls. After addressing these technical issues, we hope the community can be better equipped to focus on other more challenging issues in pedagogy for enriched teaching and learning experience.

References

1. WHO, Coronavirus Disease (CoViD-19) Pandemic, http://who.int/covid-19
2. UNESCO, CoViD-19 Response: Education, http://en.unesco.org/covid19
3. Wiki, http://en.wikipedia.org/wiki/Comparison_of_web_conferencing_software
4. M. Simonson et al., *Teaching and Learning at a Distance*, 7th edn. (IAP, Charlotte, 2019)
5. C. Reidsema et al. (ed.), *The Flipped Classroom* (Springer, Berlin, 2017)
6. J. Boettcher et al., *The Online Teaching Survival Guide*, 2nd ed. (Wiley, San Francisco, 2016)
7. Wiki, WDS, http://en.wikipedia.org/wiki/Wireless_distribution_system
8. IC Team, Multi-path TCP, http://www.multipath-tcp.org/
9. dd-wrt, Dual WAN, http://wiki.dd-wrt.com/wiki/index.php/Category:Dual_WAN

Part II
Caching, Computing, and Control for Ubiquitous Intelligence

Chapter 8
State Transition Field: A New Framework for Mobile Dynamic Caching

Jie Gao, Mushu Li, Xinhua Ling, Lian Zhao, and Xuemin (Sherman) Shen

8.1 Introduction

Due to the upsurge in the number of devices connected to the Internet and the demand for multimedia services, the role of content caching in wireless networks becomes prominent [1–3]. The recent trend of pushing contents and services closer to users as well as the emergence of virtual and augmented reality further underline the importance of caching. Accordingly, the modeling and analysis of caching have gained a lot of research attention [4, 5], especially the joint study of communication, computation, and caching to deploy services close to mobile users [6, 7].

Depending on the considered scenario, the main benefit of caching can be decreasing the content delivery latency [8], alleviating congestion over the backhaul [9], reducing energy consumption [10], or a combination of the above [11]. While the objectives can be different, the caching performance is largely centered around

J. Gao (✉)
School of Information Technology, Carleton University, Ottawa, Canada
e-mail: JieGao6@cunet.carleton.ca

M. Li · X. (Sherman) Shen
Department of Electrical and Computer Engineering, University of Waterloo, Waterloo, ON, Canada
e-mail: m475li@uwaterloo.ca; sshen@uwaterloo.ca

X. Ling
XLNTec Inc., Waterloo, ON, Canada
e-mail: xinhua@xlntec.com

L. Zhao
Department of Electrical, Computer and Biomedical Engineering, Toronto Metropolitan University, Toronto, ON, Canada
e-mail: l5zhao@ryerson.ca

L. Cai et al. (eds.), *Broadband Communications, Computing, and Control for Ubiquitous Intelligence*, Wireless Networks,
https://doi.org/10.1007/978-3-030-98064-1_8

one measurement, i.e., the cache hit ratio. Since a cache can only accommodate a limited amount of contents, the cache hit ratio is determined by how the cached contents are selected and how they are updated. The problem of selecting the contents to be cached is referred to as *content placement*. In static caching, the cached contents will not change once the content placement problem is solved. In contrast, the cached content can be updated, e.g., as content request and download status changes, in dynamic caching. The problem of selecting new content to cache while replacing existing content is referred to as *content replacement*.

Content placement is important in proactive caching, when contents are downloaded in advance. For example, an edge node can cache contents in advance during off-peak hours to reduce peak-hour network traffic load [12]. The key to proactive caching is adapting to unknown content popularity or network environment, usually leading to a Markov decision problem [13] or a learning problem [14]. In dynamic caching, cached content may be evicted and replaced by new content whenever a cache miss occurs, which leads to a dynamic process that updates cached contents on the fly [15].

The guiding rule for updating the contents is referred to as a *cache replacement scheme*, which has a significant impact on the performance of dynamic caching. With heuristic principles such as the least-frequently-used (LFU) or least-recently-used (LRU), the least popular content is replaced when new content is accepted in classic dynamic caching. Dynamic probabilistic caching originated from computer networks, and the idea is to cache a user-requested content with a certain probability [16, 17]. Because of the dynamic cache-and-replace process, dynamic probabilistic caching may adapt to time-varying content popularity without knowing the popularity. Moreover, dynamic probabilistic caching could achieve fair and efficient content placement in a network with low redundancy in a distributed setting [16, 18]. However, it is difficult to establish a bound on the performance of probabilistic caching, which is typically evaluated numerically.

In this chapter, we introduce the state transition field (STF) theory and the design of dynamic probabilistic caching based on the average content popularity. The STF theory can unify the analysis of different replacement schemes, characterize their features, and intuitively illustrate their differences [19, 20]. The proposed dynamic probabilistic caching exploits content popularity information while making content replacement decisions to achieve high cache hit ratio with reduced number of replacements [21]. In Sects. 8.2 and 8.3, we demonstrate that a replacement scheme corresponds to a unique state transition matrix, which in turn generates a unique STF. The STF determines the expected change of the dynamic cache state distribution just like an electromagnetic field determines the movement of a charged particle placed in it. We introduce the STF theory in the cases of both time-invariant and time-varying content popularity and provide examples to illustrate the STF. Then, in Sect. 8.4, we study caching from another perspective: given the average content popularity, how to design a replacement scheme through determining the state transition matrix. We formulate an equivalent problem of designing the state transition probability matrix of a Markov chain and develop a method to solve the problem. Section 8.5 demonstrates the STF in various scenarios and the probabilistic

Table 8.1 List of symbols

N_c	The number of all contents
N_s	The number of all cache states
L	The cache size limit
C	The set of all contents, i.e., $\{1, \ldots, N_c\}$
$\mathcal{S}$	The set of all cache states, i.e., $\{1, \ldots, N_s\}$
$\mathcal{S}_l$	The set of all cache states that cache content l
$\mathbf{s}_k$	The kth cache state vector
C_k	The set of contents cached in state k
$\mathbf{C}_s$	The cache state matrix, i.e., $[\mathbf{s}_1, \ldots, \mathbf{s}_{N_s}]$
$\mathcal{H}_k$	The set of all neighbors of state k
$\mathcal{H}_{k,l}$	The set of all content-l neighbors of state k
φ_l	The request probability of content l
$\boldsymbol{\varphi}$	The content request probability vector, i.e., $[\varphi_1, \ldots, \varphi_{N_c}]^T$
$\phi_{l,q,k}$	The conditional probability that content l replaces content q given that cache is in state k and content l is requested
$\boldsymbol{\Theta}$	The state transition probability matrix
$\boldsymbol{\Theta}_l$	The conditional state transition probability matrix given that content l is requested
$\boldsymbol{\Theta}(m, k)$	The probability of transitioning from state k to state m
$\boldsymbol{\Theta}_l(m, k)$	The probability of transitioning from state k to state m given that content l is requested
$\eta_k^{(n)}$	The SCP for state k in the duration from the nth to the $(n+1)$th replacement
$\boldsymbol{\eta}^{(n)}$	The SCP vector in the duration from the nth to the $(n+1)$th replacement, i.e., $[\eta_1^{(n)}, \ldots, \eta_{N_s}^{(n)}]^T$
$\lambda_l^{(n)}$	The CCP for content l in the duration from the nth to the $(n+1)$th replacement
$\boldsymbol{\lambda}^{(n)}$	The CCP vector in the duration from the nth to the $(n+1)$th replacement, i.e., $[\lambda_1^{(n)}, \ldots, \lambda_{N_c}^{(n)}]^T$
$\gamma^{(n)}$	The instantaneous cache hit probability at the nth request
$\mathbf{u}(\boldsymbol{\eta})$	The state transition field at $\boldsymbol{\eta}$
$\mathbf{u}_l(\boldsymbol{\eta})$	The content-l state transition field at $\boldsymbol{\eta}$
$u_{m,l}(\boldsymbol{\eta})$	The mth element of the state transition field at $\boldsymbol{\eta}$
$u_{m,l}(\boldsymbol{\eta})$	The mth element of the content-l state transition field at $\boldsymbol{\eta}$

content replacement policy using numerical results. The list of symbols used in this chapter is given in Table 8.1.

8.2 State Transition Field

Consider the scenario of N_c contents and a cache with size L. The set of all contents is denoted by C. Without loss of generality, we assume that all contents are of identical unit size. The caching scenario can be generic, such as caching at a cellular base station, a Wi-Fi access point, a road side unit for vehicles, or a mobile device for device-to-device caching.

The fundamental assumption in this section is that the requested contents represented by integer random variables are independent and identically distributed. This follows from the widely used independent reference model (IRM), a simplification of the actual request process that can be accurate with a large number of requesting users [22] or within a short time frame [23]. As the requested content follows a distribution that is time invariant, the probability of content $l \in C$ being requested can be denoted by φ_l. The probabilities $\{\varphi_l\}_{\forall l}$ are organized into the request probability vector $\boldsymbol{\varphi}$ and referred to as the content popularity.

8.2.1 Content Request and Replacement

If content l is requested but not in the cache, it will be downloaded and, depending on the replacement scheme, may replace one cached content. It is assumed that the download and replacement can be completed before the next content request arrives at the cache.

The timeline of the considered dynamic caching is illustrated in Fig. 8.1. For simplicity of notation, we put a replacement point after each request regardless of whether a replacement actually happens or not. If a replacement occurs following the nth request, it is completed by the nth replacement point.

8.2.2 Cache State

Cache state is introduced to describe the combination of cached contents. There are $N_s = \binom{N_c}{L}$ different possible combinations of cached contents, corresponding to N_s caching states.[1] The set of all cache states is denoted by $\mathcal{S}$, and the set of contents

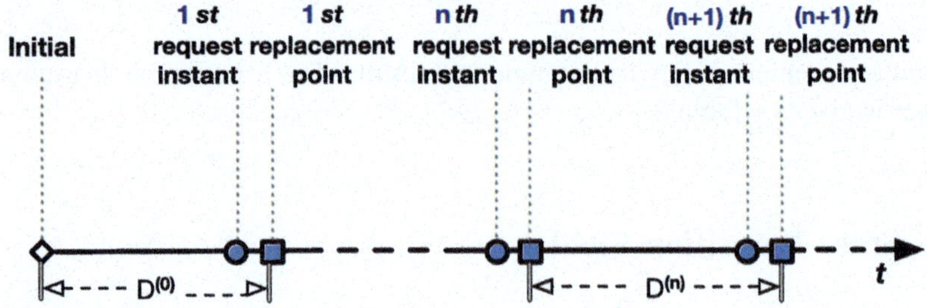

Fig. 8.1 Illustration of the timeline model. $\mathrm{D}^{(n)}$ represents the duration between the nth and the $(n+1)$th replacement points

[1] Here we ignore the cases when the cache is not full.

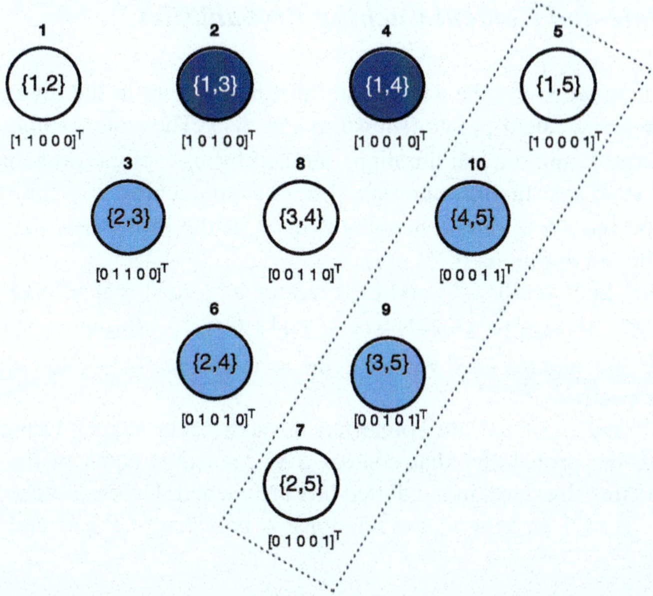

Fig. 8.2 An illustration of states with $N_c = 5$ and $L = 2$. Each circle represents a state. The number above a circle represents the state ID, and the set inside a circle represents the set of cached contents in that state. For example, contents 2 and 5 are cached in state 7

cached in state k is denoted by C_k. The cache state vector for state k is defined as a $N_c \times 1$ vector with elements determined as follows:

$$\mathbf{s}_k(l) = \begin{cases} 1, & \text{if } l \in C_k, \\ 0, & \text{if } l \notin C_k, \end{cases} \quad \forall l \in C, \forall k \in \mathcal{S}, \tag{8.1}$$

where the lth element of vector $\mathbf{s}_k$ corresponds to the lth content. An example with $N_c = 5$ and $L = 2$ is illustrated in Fig. 8.2. In this example, there are $\binom{5}{2} = 10$ states. Each circle in the figure represents a state, while the number above the circle represents the state ID. The set given in the circle of state k is the set of cached contents in state k, i.e., C_k, and the vector beneath state k is $\mathbf{s}_k$. For example, state 7 caches contents $\{2, 5\}$ and is represented by the cache state vector $\mathbf{s}_7 = [0\,1\,0\,0\,1]^{\mathrm{T}}$, where $\cdot^{\mathrm{T}}$ stands for transpose.

A state is a neighbor of state k if its cached contents differ from those cached in state k by just one element. The set of neighbors of state k is denoted as $\mathcal{H}_k$. For any content $l \notin C_k$, a content-l neighbor of state k is a neighboring state that caches l. The set of content-l neighboring states of state k is denoted as $\mathcal{H}_{k,l}$. Using Fig. 8.2 and state 8 as an example, $\mathcal{H}_8$ is the set of all colored states, and $\mathcal{H}_{8,1}$ is the two states with navy blue color.

8.2.3 State and Content Caching Probabilities

The cached contents and the cache state remain constant in the duration between consecutive replacement points (shown in Fig. 8.1). The state caching probability (SCP) for state k and the nth duration, denoted by $\eta_k^{(n)}$, is the probability that the cache is in state k in the nth duration. The content caching probability (CCP) for content l and the nth duration, denoted by $\lambda_l^{(n)}$, is the probability that content l is cached in the nth duration.

Define the SCP vector $\boldsymbol{\eta}^{(n)}$ and CCP vector $\boldsymbol{\lambda}^{(n)}$ such that $\boldsymbol{\eta}^{(n)}(k) = \eta_k^{(n)}$ and $\boldsymbol{\lambda}^{(n)}(l) = \lambda_l^{(n)}$. Evidently, $\mathbf{1}^{\mathrm{T}}\boldsymbol{\eta}^{(n)} = 1$ and $\mathbf{1}^{\mathrm{T}}\boldsymbol{\lambda}^{(n)} = L$. Based on the timeline in Fig. 8.1, the SCP and the CCP vectors at the instant of the $(n+1)$th request are $\boldsymbol{\eta}^{(n)}$ and $\boldsymbol{\lambda}^{(n)}$, respectively.[2]

The SCP and the CCP are connected through cache states. Using Fig. 8.2 as an example, the probability that content 5 is cached is equal to the sum of the probabilities that the states in the dotted box are cached. Define a cache state matrix $\mathbf{C}_{\mathrm{s}} = [\mathbf{s}_1, \ldots, \mathbf{s}_{N_{\mathrm{s}}}]$. In general, the relationship between SCP $\boldsymbol{\eta}^{(n)}$ and CCP $\boldsymbol{\lambda}^{(n)}$ is given by:

$$\boldsymbol{\lambda}^{(n)} = \mathbf{C}_{\mathrm{s}}\boldsymbol{\eta}^{(n)}. \tag{8.2}$$

Given that content l is being requested at the $(n+1)$th request, the conditional instantaneous cache hit probability is $\lambda_l^{(n)}$. The instantaneous cache hit probability at the $(n+1)$th request, denoted by $\gamma^{(n+1)}$, is given by:

$$\gamma^{(n+1)} = \boldsymbol{\varphi}^{\mathrm{T}}\boldsymbol{\lambda}^{(n)}. \tag{8.3}$$

If the cache is at state k while content $l \notin C_k$ is requested, the cache downloads content l and decides whether to replace a cached content with content l. In the general model, the probability of replacing content q with content l when the cache is at state k is denoted by $\phi_{l,q,k}$, for any $q \in C_k$ and $l \notin C_k$. For each state, there are $L(N_{\mathrm{c}} - L)$ possible replacements.

8.2.4 General Cache State Transition Model

A content replacement triggers a cache state transition. For neighboring states k and m that satisfy $m \in \mathcal{H}_{k,l}$ and $k \in \mathcal{H}_{m,q}$, replacing content q with l triggers a transition from state k to state m. The conditional cache state transition probabilities given that content l is requested can be organized into the following matrix $\boldsymbol{\Theta}_l$:

[2] We use lower-case bold letters for vectors, upper-case bold letters for matrices, and calligraphic letters for sets. The superscript $(\cdot)^{(n)}$ is used on letters related to the nth request or replacement. Greek letters are used to represent various probabilities.

$$\boldsymbol{\Theta}_l(m,k)=\begin{cases}1, & \text{if } k=m \text{ and } l\in C_k,\\ 1-\sum\limits_{m'\in\mathcal{H}_{k,l}}\phi_{l,e(k,m'),k}, & \text{if } k=m \text{ and } l\notin C_k,\\ \phi_{l,e(k,m),k}, & \text{if } m\in\mathcal{H}_{k,l},\\ 0, & \text{otherwise,}\end{cases} \tag{8.4}$$

where $e(k,m)$ denotes the unique content that is cached by state k but not state m given that $k \in \mathcal{H}_m$. Accordingly, the overall cache state transition probability matrix in the general case is given by:

$$\boldsymbol{\Theta} = \sum_{l\in C} \varphi_l \boldsymbol{\Theta}_l. \tag{8.5}$$

From the definition of the SCP vector $\boldsymbol{\eta}^{(n)}$ and state transition probability matrix $\boldsymbol{\Theta}$, it can be seen that:

$$\boldsymbol{\eta}^{(n)} = \boldsymbol{\Theta}\boldsymbol{\eta}^{(n-1)}. \tag{8.6}$$

The above model can be extended to the scenario in which each content request (and replacement) involves multiple contents. In such a case, assuming that each request is for a block of B contents ($B < L$), there are $N_\mathrm{B} = \binom{N_\mathrm{c}}{B}$ different blocks. Then, Eq. (8.5) can be extended as follows:

$$\boldsymbol{\Theta}^\mathrm{B} = \sum_{b=1}^{N_\mathrm{B}} \varphi_b^\mathrm{B} \boldsymbol{\Theta}_b^\mathrm{B}, \tag{8.7}$$

where φ_b^B is the probability that the bth block is requested, and $\boldsymbol{\Theta}_b^\mathrm{B}$ is the conditional cache state transition probabilities given that block b is requested. The size of $\boldsymbol{\Theta}_b^\mathrm{B}$ remains $N_\mathrm{S} \times N_\mathrm{S}$. However, for any given state, e.g., state k, the set of its neighbors $\mathcal{H}_k$ will contain more states under block replacement, and the set of its content-l neighbors $\mathcal{H}_{k,l}$ will be replaced by a set of block-b neighbors. The extension is straightforward, and the details are omitted here.

8.2.5 *State Transition Field*

Denote the general SCP without specifying any time instant as $\boldsymbol{\eta}$. Consider $\boldsymbol{\eta}$ as a point in the N_S-dimensional vector space. Driving by the requests and replacements, $\boldsymbol{\eta}$ varies in the following domain:

$$\mathcal{D}=\left\{(\eta_1,\ldots,\eta_{N_\mathrm{S}})\,\middle|\, 0\le\eta_k\le 1, \forall k\in\mathcal{S};\ \sum_k \eta_k = 1\right\}. \tag{8.8}$$

The expected "movement" of $\boldsymbol{\eta}$ in $\mathcal{D}$ after the nth replacement point, assuming a replacement actually happens, is characterized by $\boldsymbol{\eta}^{(n)} - \boldsymbol{\eta}^{(n-1)}$. This difference, in turn, is determined by three factors:

- The current position of $\boldsymbol{\eta}$ in $\mathcal{D}$, i.e., the value of $\boldsymbol{\eta}^{(n-1)}$
- The content popularity $\boldsymbol{\varphi}$
- The state transition probability matrix $\boldsymbol{\Theta}$,

while $\boldsymbol{\Theta}$ is determined by the replacement scheme and generally dependent on $\boldsymbol{\varphi}$ (such dependence is shown in Eq. (8.5)).

Define the STF at the point $\boldsymbol{\eta}^{(n-1)}$ using the aforementioned difference:

$$\mathbf{u}(\boldsymbol{\eta}^{(n-1)}) = \boldsymbol{\eta}^{(n)} - \boldsymbol{\eta}^{(n-1)}. \tag{8.9}$$

Substituting Eq. (8.6) into Eq. (8.9), it follows that:

$$\mathbf{u}(\boldsymbol{\eta}^{(n-1)}) = \boldsymbol{\Theta}\boldsymbol{\eta}^{(n-1)} - \boldsymbol{\eta}^{(n-1)}. \tag{8.10}$$

The STF is a vector field defined over the domain $\mathcal{D}$. It can be seen that understanding the STF can provide insight into the design and performance analysis of replacement schemes. Similar to a magnetic or electric field, the STF can vary in direction and strength at different points in the domain (although the STF exists mathematically but not physically). Under the IRM, the STF does not change over time, as $\boldsymbol{\varphi}$ and $\boldsymbol{\Theta}$ are both constant.

In the definition Eq. (8.9), the $\boldsymbol{\eta}^{(n-1)}$ in the parentheses specifies a point in the domain $\mathcal{D}$. If the STF is known at all points in $\mathcal{D}$, a path can be identified from any initial point. As illustrated in Fig. 8.3, the end point of a path gives the steady state of the replacement scheme, while the number of steps reflects the time for the underlying Markov chain to attain its stationary state from the initial point. Different replacement schemes yield different STFs, and the impact is conveyed through $\boldsymbol{\Theta}$. Therefore, the STF is a complete characterization of replacement schemes.

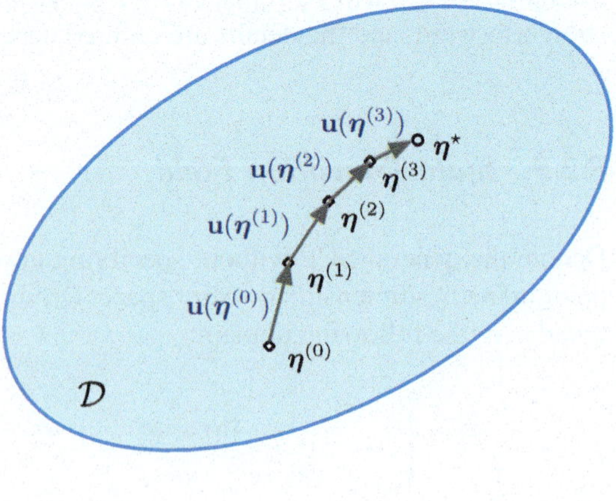

Fig. 8.3 An illustration of STF at four points, i.e., $\boldsymbol{\eta}^{(0)}$ to $\boldsymbol{\eta}^{(3)}$. The end point $\boldsymbol{\eta}^{\star}$ represents the steady state, at which the STF diminishes to an all-zero vector

The STF can be decomposed. Define:

$$\mathbf{u}_l(\boldsymbol{\eta}^{(n-1)}) = \boldsymbol{\Theta}_l \boldsymbol{\eta}^{(n-1)} - \boldsymbol{\eta}^{(n-1)}. \tag{8.11}$$

It follows that:

$$\sum_{l \in C} \varphi_l \mathbf{u}_l(\boldsymbol{\eta}^{(n-1)}) = \sum_{l \in C} \varphi_l \boldsymbol{\Theta}_l \boldsymbol{\eta}^{(n-1)} - \boldsymbol{\eta}^{(n-1)} = \mathbf{u}(\boldsymbol{\eta}^{(n-1)}), \tag{8.12}$$

where the last step uses Eq. (8.5). Accordingly, $\mathbf{u}_l(\boldsymbol{\eta}^{(n-1)})$ can be considered as the content-specific STF that represents the "movement" of $\boldsymbol{\eta}$ from the point $\boldsymbol{\eta}^{(n-1)}$ after content l is requested. The superposition of all content-specific STFs, weighted by the corresponding content popularity, yields the overall STF.

It is not difficult to see that the following equalities hold:

$$\mathbf{1}^{\mathrm{T}} \mathbf{u}_l(\boldsymbol{\eta}^{(n-1)}) = 0, \ \forall l \in C, \ \forall \boldsymbol{\eta}^{(n-1)} \in \mathcal{D}, \tag{8.13}$$

$$\mathbf{1}^{\mathrm{T}} \mathbf{u}(\boldsymbol{\eta}^{(n-1)}) = 0, \ \forall \boldsymbol{\eta}^{(n-1)} \in \mathcal{D}. \tag{8.14}$$

8.2.6 Discussions on the Steady State and the Convergence

In this subsection, we discuss the benefits of using the proposed STF to analyze replacement schemes in practice. First, we use an example to show how the STF can characterize the property of the stationary states. Then, we use another example to show how the STF can be used to compare the convergence rates of replacement schemes.

At the steady state, the overall STF must be equal to $\mathbf{0}$ regardless of the replacement scheme. However, this does not mean that no replacement happens after the steady state is achieved. Instead, contents can still be evicted from or accepted into the cache, while the probabilities of the two events must be equal for any content at the steady state. Therefore, it is not difficult to see that there can be more frequent replacements at the steady state of one replacement scheme than that of another. This frequency of replacement at a steady state can be analyzed by decomposing the STF into content-specific STF using Eq. (8.12), as illustrated in Fig. 8.4. In the illustrated cases, we assume the same content request probabilities, while the content-specific STFs in Fig. 8.4a have much smaller norms than those in Fig. 8.4b. Correspondingly, there can be less frequent replacements at the steady state $\boldsymbol{\eta}^{\star}$ in Fig. 8.4a than at the steady state $\tilde{\boldsymbol{\eta}}^{\star}$ in Fig. 8.4b.

In the case when each replacement incurs a cost or when cache wear-out is a concern, characterizing the frequency of replacement can be of interest. Based on the above discussion, the weighted sum of the norm of content-specific STF can be used as a metric for comparing the frequency of content replacement at the steady state of different replacement schemes. For example, a metric can be calculated as follows:

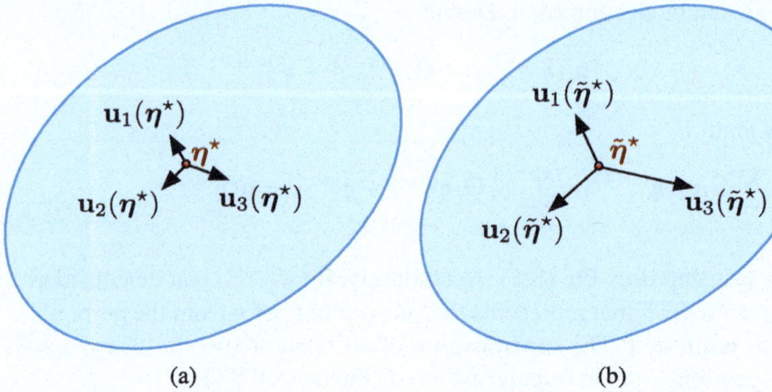

Fig. 8.4 Illustration of decomposing the STF at the steady state. **(a)** Small $\|\mathbf{u}_l(\boldsymbol{\eta}^{\star})\|$. **(b)** Large $\|\mathbf{u}_l(\tilde{\boldsymbol{\eta}}^{\star})\|$

$$M(\boldsymbol{\eta}^{\star}) = \sum_{l \in C} \varphi_l \|\mathbf{u}_l(\boldsymbol{\eta}^{\star})\|, \tag{8.15}$$

where the weights are the content request probabilities.

The rate of convergence depends on the second largest eigenvalue of the transition probability matrix $\boldsymbol{\Theta}$. Since STF is a derivative of state transition probability matrix, it does not provide a new characterization of the rate of convergence in theory. However, we can use STF to develop a metric for comparing the convergence rate of different replacement schemes in practice.

For example, we can generate sample points in the state transition region, as illustrated using hollow circles in Fig. 8.5. Hypothetically, if the STF at every point of the state transition region points toward the steady state $\boldsymbol{\eta}^{\star}$, then the rate of convergence is determined by the strength (norm) of the STF. In practice, the STF at the sample points generally does not point straight toward the steady state. Nevertheless, we can project the STF at a sample point onto the connection line between that sample point and the steady state. This is illustrated with two example sample points, i.e., $\boldsymbol{\eta}_a$ and $\boldsymbol{\eta}_b$, in Fig. 8.5. In this figure, the red solid circle represents the steady state $\boldsymbol{\eta}^{\star}$. The black arrows at sample points $\boldsymbol{\eta}_a$ and $\boldsymbol{\eta}_b$ represent the STF $\mathbf{u}(\boldsymbol{\eta}_a)$ and $\mathbf{u}(\boldsymbol{\eta}_b)$, respectively. The two dashed lines connect $\boldsymbol{\eta}_a$ and $\boldsymbol{\eta}_b$ with the steady state $\boldsymbol{\eta}^{\star}$. The two blue arrows on the dashed lines represent the projection of $\mathbf{u}(\boldsymbol{\eta}_a)$ and $\mathbf{u}(\boldsymbol{\eta}_b)$, respectively. The norm of the projection, aggregated over all sample points, can provide a metric for characterizing the convergence rates of replacement schemes. The accuracy of this approach depends on the number and locations of the chosen sample points.

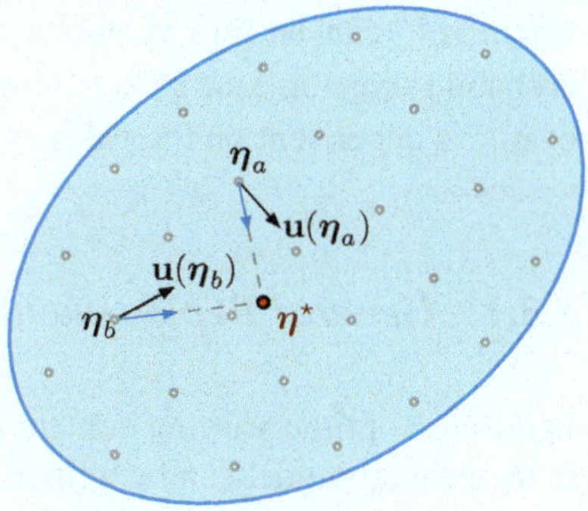

Fig. 8.5 Illustration of sampling the STF for characterizing the convergence rate

8.3 State Transition Field with Time-Varying Content Popularity

As the content popularity becomes time varying, some symbols become dependent on the request instant. We refer to such symbols as request-dependent symbols, which include the following three:

Content Request Probabilities The probability of content l being requested at request instant n, denoted by $\varphi_l^{(n)}$, and the overall content popularity at the nth content request, denoted by $\boldsymbol{\varphi}^{(n)}$.

Instantaneous Cache Hit Probability The instantaneous cache hit probability at the $(n+1)$th request, denoted by $\gamma^{(n+1)}$, is given by:

$$\gamma^{(n+1)} = \left(\boldsymbol{\varphi}^{(n+1)}\right)^{\mathrm{T}} \boldsymbol{\lambda}^{(n)}, \tag{8.16}$$

where $(\cdot)^{\mathrm{T}}$ represents transpose, and $\boldsymbol{\lambda}^{(n)}$ is the CCP vector after the nth round of request and replacement.

Station Transition Matrices The conditional state transition matrix and the state transition matrix are generally time dependent and thus denoted by $\boldsymbol{\Theta}_l^{(n)}$ and $\boldsymbol{\Theta}^{(n)}$, respectively, under time-varying content popularity.

The relation between state and content caching probabilities, i.e., (8.2), from Sect. 8.2 still applies in the case of time-varying content popularity, where $\boldsymbol{\eta}^{(n)}$ is the SCP vector after the nth round of request and replacement. It follows from (8.2) that:

$$\boldsymbol{\eta}^{(n)} = \mathbf{C}_{\mathrm{s}}^{\mathrm{T}} \left(\mathbf{C}_{\mathrm{s}}\mathbf{C}_{\mathrm{s}}^{\mathrm{T}}\right)^{-1} \boldsymbol{\lambda}^{(n)} + \mathbf{n}_C^{(n)}, \tag{8.17}$$

where $\mathbf{n}_C^{(n)}$ can be any vector in the null space of $\mathbf{C}_s$ that renders $\boldsymbol{\eta}^{(n)}$ a valid probability vector, i.e., $\boldsymbol{\eta}^{(n)} \succeq 0$, $\boldsymbol{\eta}^{(n)} \preceq 1$, and $\mathbf{1}^{\mathrm{T}}\boldsymbol{\eta}^{(n)} = 1$. Therefore, the value of $\mathbf{n}_C^{(n)}$ is dependent on the value of $\boldsymbol{\lambda}^{(n)}$.

8.3.1 General Replacement Model

In the case of time-varying content popularity, the state transition probability matrix in the general model can be written as:

$$\boldsymbol{\Theta}^{(n)} = \sum_{l \in C} \varphi_l^{(n)} \boldsymbol{\Theta}_l^{(n)}, \tag{8.18}$$

where C is the set of all contents, and the conditional cache state transition probability matrix given that content $l \notin C_k$ is requested, $\boldsymbol{\Theta}_l^{(n)}$, is given by:

$$\boldsymbol{\Theta}_l^{(n)}(m,k) = \begin{cases} 1, & \text{if } k = m \text{ and } l \in C_k, \\ 1 - \sum\limits_{m \in \mathcal{H}_{k,l}} \phi_{l,e(k,m),k}, & \text{if } k = m \text{ and } l \notin C_k, \\ \phi_{l,e(k,m),k}, & \text{if } m \in \mathcal{H}_{k,l}, \\ 0, & \text{otherwise,} \end{cases} \tag{8.19}$$

where $\phi_{l,q,k}$ denotes the probability of replacing content q with content l given that the cache is at state k and content l is requested. Unlike the case with time-invariant content popularity, the conditional cache state transition probability matrix $\boldsymbol{\Theta}_l^{(n)}$ can be implicitly request dependent as a result of $\phi_{l,q,k}$ being request dependent.

The relationship between the SCP vector $\boldsymbol{\eta}^{(n)}$ and state transition probability matrix $\boldsymbol{\Theta}$ in (8.6) can be rewritten as follows in the case of time-varying content probabilities:

$$\boldsymbol{\eta}^{(n)} = \boldsymbol{\Theta}^{(n)} \boldsymbol{\eta}^{(n-1)}. \tag{8.20}$$

Based on Eqs. (8.2) and (8.20), the resulting CCP vector after the nth request and replacement can be given by:

$$\boldsymbol{\lambda}^{(n)} = \mathbf{C}_s \sum_{l \in C} \varphi_l^{(n)} \boldsymbol{\Theta}_l^{(n)} \boldsymbol{\eta}^{(n-1)}. \tag{8.21}$$

Using Eq. (8.17), it follows that:

$$\boldsymbol{\lambda}^{(n)} = \left(\mathbf{C}_s \sum_{l \in C} \varphi_l^{(n)} \boldsymbol{\Theta}_l^{(n)} \mathbf{C}_s^{\mathrm{T}} \left(\mathbf{C}_s \mathbf{C}_s^{\mathrm{T}} \right)^{-1} \right) \boldsymbol{\lambda}^{(n-1)} + \mathbf{C}_s \sum_{l \in C} \varphi_l^{(n)} \boldsymbol{\Theta}_l^{(n)} \mathbf{n}_C^{(n-1)}. \tag{8.22}$$

It can be seen that the mapping from $\boldsymbol{\lambda}^{(n-1)}$ to $\boldsymbol{\lambda}^{(n)}$ is complicated. Specifically, unlike the mapping between two consecutive SCP vectors, which can be simply written as $\boldsymbol{\eta}^{(n)} = \boldsymbol{\Theta}^{(n)}\boldsymbol{\eta}^{(n-1)}$, the mapping between consecutive CCP vectors cannot be written in a linear form due to the second item in Eq. (8.22), i.e., $\mathbf{C}_s \sum_{l\in C} \varphi_l^{(n)} \boldsymbol{\Theta}_l^{(n)} \mathbf{n}_C^{(n-1)}$. Moreover, despite that Eq. (8.22) seems to have an affine form, the mapping from $\boldsymbol{\lambda}^{(n-1)}$ to $\boldsymbol{\lambda}^{(n)}$ is not affine. This is implicitly conveyed through the variable $\mathbf{n}_C^{(n-1)}$ since the value of $\mathbf{n}_C^{(n-1)}$ depends on $\boldsymbol{\lambda}^{(n-1)}$ and the dependence is nonlinear.

8.3.2 *Instantaneous STF: The General Case*

Under time-varying content popularity, the state transition probability matrix is $\boldsymbol{\Theta}^{(n)}$ when the SCP is $\boldsymbol{\eta}^{(n-1)}$. Therefore, the STF at the instant of the nth request and the point $\boldsymbol{\eta}^{(n-1)}$ is given by:

$$\boldsymbol{u}^{(n)}(\boldsymbol{\eta}^{(n-1)}) = \boldsymbol{\Theta}^{(n)}\boldsymbol{\eta}^{(n-1)} - \boldsymbol{\eta}^{(n-1)}. \tag{8.23}$$

The superscript (n) in $\boldsymbol{u}^{(n)}(\cdot)$ reflects the fact that the STF is no longer static but time varying. The direction and strength of the instantaneous STF depend on both $\boldsymbol{\eta}$, the location in the state transition domain, and n, the request instant. The value of the instantaneous STF $\boldsymbol{u}^{(n)}(\boldsymbol{\eta}^{(n-1)})$ represents the change in the SCP after the nth round of request and replacement. The effect of a replacement scheme on the dynamic SCP over a sequence of requests can be decomposed into the summation over the instantaneous STFs:

$$\boldsymbol{\eta}^{(n+N-1)} - \boldsymbol{\eta}^{(n-1)} = \sum_{t=0}^{N-1} \left(\boldsymbol{\eta}^{(n+t)} - \boldsymbol{\eta}^{(n+t-1)}\right) = \sum_{t=0}^{N-1} \boldsymbol{u}^{(n+t)}(\boldsymbol{\eta}^{(n+t-1)}), \tag{8.24}$$

for any $n \geq 1$ and $N \geq 1$.

Similarly, other metrics can also be studied through instantaneous STFs, e.g., the average cache hit probability. Using instantaneous STFs from the first till the nth request, the average cache hit probability over the n requests can be given by:

$$\gamma_{\text{avg}} = \frac{1}{n}\sum_{t=2}^{n} \left(\boldsymbol{\varphi}^{(t)}\right)^{\mathrm{T}} \mathbf{C}_s \left(\sum_{t'=0}^{t-2} \boldsymbol{u}^{(t'+1)}\right) + \boldsymbol{\varphi}_{\text{avg}}^{\mathrm{T}} \mathbf{C}_s \boldsymbol{\eta}^{(0)}, \tag{8.25}$$

where $\boldsymbol{u}^{(t'+1)}$ is the abbreviation for $\boldsymbol{u}^{(t'+1)}(\boldsymbol{\eta}^{(t')})$, and

$$\boldsymbol{\varphi}_{\text{avg}} = \frac{1}{n}\sum_{t=1}^{n}\boldsymbol{\varphi}^{(t)} \tag{8.26}$$

is the average content popularity over the n requests. The above two equations show that the average cache hit probability over an arbitrary number of requests, starting from any initial SCP $\boldsymbol{\eta}^{(0)}$, can be obtained from instantaneous STFs, instantaneous content request probabilities, and the initial point $\boldsymbol{\eta}^{(0)}$. The inner summation over t' in Eq. (8.25) represents the effect of historical requests and replacements on the instantaneous cache hit probability at the tth request. The decomposition in Eq. (8.24) and the result in Eq. (8.25) demonstrate the importance in analyzing the instantaneous STF under different replacement schemes. If the instantaneous content request probabilities $\boldsymbol{\varphi}^{(t)}, t \in \{1, \ldots, n\}$ can be obtained, the instantaneous STF of a replacement scheme at any point in the state transition region can be calculated using Eqs. (8.18), (8.19), and (8.23). For evaluating and comparing different cache replacement schemes, we can substitute the specific STF of the replacement schemes for $\boldsymbol{u}^{(1)}, \ldots, \boldsymbol{u}^{(t-1)}$ in Eq. (8.25).

The instantaneous STF can also be decomposed into content-specific STF. Define the lth component of $\mathbf{u}^{(n)}(\boldsymbol{\eta}^{(n-1)})$ as:

$$\mathbf{u}_l^{(n)} = \boldsymbol{\Theta}_l^{(n)}\boldsymbol{\eta}^{(n-1)} - \boldsymbol{\eta}^{(n-1)}. \tag{8.27}$$

It can be seen that:

$$\begin{aligned}\mathbf{u}^{(n)}(\boldsymbol{\eta}^{(n-1)}) &= \boldsymbol{\Theta}^{(n)}\boldsymbol{\eta}^{(n-1)} - \boldsymbol{\eta}^{(n-1)} \\ &= \sum_{l\in C}\varphi_l^{(n)}\left(\boldsymbol{\Theta}_l\boldsymbol{\eta}^{(n-1)} - \boldsymbol{\eta}^{(n-1)}\right) \\ &= \sum_{l\in C}\varphi_l^{(n)}\mathbf{u}_l^{(n)}.\end{aligned} \tag{8.28}$$

8.3.3 Impact of STF on Instantaneous Cache Hit Probability

When the content popularity varies over time, a replacement scheme may not lead to any steady state. As a result, the analysis of steady states and rate of convergence is not applicable. Instead, the impact of a replacement scheme on the instantaneous cache hit probability at the next request is investigated.

A replacement after the nth request affects the cache hit probability at the $(n+1)$th request. Consider the time instant right after the nth request and replacement so that $\boldsymbol{u}^{(n)}(\cdot)$ is the current STF and the $(n+1)$th request is the next request. The effect of a replacement scheme can be conveyed through the difference between the

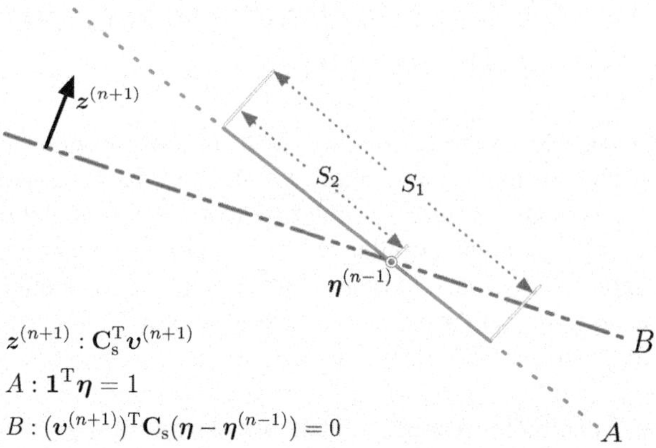

Fig. 8.6 Illustration of the relation between instantaneous cache hit probability, $\boldsymbol{\eta}^{(n)}$, and $\boldsymbol{\varphi}^{(n+1)}$. Area S_1 is the area that $\boldsymbol{\eta}^{(n+1)}$ may fall in, i.e., the intersection of hyperplane A and the subspace $\boldsymbol{\eta}^{(n+1)} \succeq 0$. If $\boldsymbol{\eta}^{(n+1)}$ falls in area S_2, then $(z^{(n+1)})^T \boldsymbol{\eta}^{(n+1)} \geq (z^{(n+1)})^T \boldsymbol{\eta}^{(n)}$

cache hit probability at the $(n+1)$th request with and without a replacement (based on the chosen scheme) after the nth request. This difference is given by:

$$d_{\gamma}^{(n+1)} = \left(\boldsymbol{\varphi}^{(n+1)}\right)^{\mathrm{T}} \mathbf{C}_{\mathrm{s}} \left(\boldsymbol{\eta}^{(n)} - \boldsymbol{\eta}^{(n-1)}\right)$$
$$= \left(\boldsymbol{\varphi}^{(n+1)}\right)^{\mathrm{T}} \mathbf{C}_{\mathrm{s}} \boldsymbol{u}^{(n)}(\boldsymbol{\eta}^{(n-1)}). \tag{8.29}$$

The above result shows that the cache hit ratio at the $(n+1)$th request depends on the content popularity at the $(n+1)$th request, i.e., $\boldsymbol{\varphi}^{(n+1)}$, the STF at the nth request, i.e., $\boldsymbol{u}^{(n)}(\cdot)$, and the SCP at the $(n-1)$th request, i.e., $\boldsymbol{\eta}^{(n-1)}$. Among these three factors, $\boldsymbol{\eta}^{(n-1)}$ reflects the accumulative effect of the previous $n-1$ rounds of request and replacement, $\boldsymbol{u}^{(n)}(\cdot)$ represents the current STF, and $\boldsymbol{\varphi}^{(n+1)}$ represents the content popularity at the next request. Equation (8.29) shows the complication due to time-varying content popularity: $\boldsymbol{\varphi}^{(n+1)}$ and $\boldsymbol{u}^{(n)}(\cdot)$ would reduce to $\boldsymbol{\varphi}$ and $\boldsymbol{u}(\cdot)$, respectively, if the content popularity becomes time invariant.

Some general observations can be made:

1. Define $z^{(n+1)} = \mathbf{C}_{\mathrm{s}}^{\mathrm{T}} \boldsymbol{\varphi}^{(n+1)}$. Then $z^{(n+1)}$ is the state cache hit probability vector at the $(n+1)$th request. Depending on $\boldsymbol{\eta}^{(n-1)}$, $\boldsymbol{\varphi}^{(n)}$, and $\boldsymbol{\Theta}^{(n)}$, $\boldsymbol{\eta}^{(n+1)}$ may fall at any point in the areas S_1 in Fig. 8.6. The replacement after the nth request improves the instantaneous cache hit probability at the $(n+1)$th request if the replacement drives the SCP into the area S_2 in Fig. 8.6.
2. $d_{\gamma}^{(n+1)}$ is small, regardless of $\boldsymbol{\varphi}^{(n+1)}$, when $\boldsymbol{\eta}^{(n-1)}$ is close to the steady state corresponding to $\boldsymbol{\varphi}^{(n)}$ (i.e., the steady state if the content popularity is constant and remains equal to $\boldsymbol{\varphi}^{(n)}$).
3. In the trivial case when $\boldsymbol{\varphi}^{(n+1)}$ approaches $1/N_{\mathrm{c}} \cdot \mathbf{1}$, where N_{c} is the number of contents, the hyperplane $(\boldsymbol{\varphi}^{(n+1)})^T \mathbf{C}_{\mathrm{s}}(\boldsymbol{\eta} - \boldsymbol{\eta}^{(n)}) = 0$ coincides with the hyperplane $\mathbf{1}^{\mathrm{T}} \boldsymbol{\eta} = 1$. In such a case, $d_{\gamma}^{(n+1)}$ becomes zero for any replacement scheme.

8.4 Dynamic Probabilistic Caching with Time-Varying Content Popularity

In classic dynamic caching such as LRU, replacements are usually determined solely by the content request statistics. Such replacements provide adaptivity to time-varying content popularity without any *a priori* information. However, the average content popularity $\{\varphi_k\}_{k\in\mathcal{F}}$ may be available, e.g., through prediction, and the optimal average content caching probabilities $\{p_k^\star\}_{\forall k}$ could be found accordingly. In such a case, using classic dynamic caching while neglecting the average content popularity cannot lead to optimal performance since it ignores useful information. Therefore, we aim to design dynamic probabilistic caching that can exploit the average content popularity $\{\varphi_k\}_{k\in\mathcal{F}}$ to adapt to the content popularity. A logical requirement is that the resulting caching probability should converge to the optimal content placement policy $\boldsymbol{\eta}^\star$ based on the average content popularity in the case of time-invariant content popularity. This section addresses two questions regarding the design of such dynamic caching: (a) what should be the probability of accepting a new content into the cache? and (b) which existing content should be replaced, and with what probability, if the new content is accepted.

8.4.1 The Content Replacement Markov Chain

The content replacement process at the target cache can be modeled using a Markov chain. Our focus here is the design of content replacement that can converge to a target stationary point in the case when the content popularity is time invariant and adapt to the variations in the case when the content popularity is time varying.

In the content replacement Markov chain, a state transition may happen when a requested content is not in the cache. For state m, denote the set of states that state m can transit into by replacing one cached content with content k and the entire set of states that state m can transit into by replacing one cached content with any other content as $\mathcal{H}_m^k$ and $\mathcal{H}_m$, respectively. It is not difficult to see that state m and any state in $\mathcal{H}_m$ must differ in one and only one cached content. Let c be the cache size. The following results hold:

$$\mathcal{H}_m = \bigcup_{k\in\mathcal{F}\setminus\mathcal{G}^m} \mathcal{H}_m^k, \quad |\mathcal{H}_m^k| = c, \quad |\mathcal{H}_m| = c(N_\mathrm{f} - c). \tag{8.30}$$

Consider a pair of neighbor states m and $m' \in \mathcal{H}_m^k$. The question that arises is: with what probability should the cache transition into state m' given that content k is requested while the cache is currently in state m? Denote this conditional probability as $\tau_{m',m}$. Then, the design of dynamic probabilistic caching boils down to finding $\tau_{m',m}$ for every m and $m' \in \mathcal{H}_m^k$ where $k \in \mathcal{F}\setminus\mathcal{G}^m$.

The overall state transition probability matrix of the content replacement Markov chain is determined by two sets of probabilities: the content request probabilities and $\{\tau_{m',m}\}$. Since the instantaneous content request probabilities are unknown, we can only exploit the average content request probabilities $\{\varphi_k\}_{k\in\mathcal{F}}$. Therefore, the design of dynamic content replacement uses $\{\varphi_k\}_{k\in\mathcal{F}}$ instead of the instantaneous content request probabilities. A discussion on the effect of time-varying content popularity will be given later.

Denote the overall state transition probability matrix by $\boldsymbol{\Theta}$. The element at the mth column and the m'th row of $\boldsymbol{\Theta}$ represents the probability that state m transitions into state m'. Then, $\boldsymbol{\Theta}$ can be written as the summation of content-specific conditional state transition matrices as follows:

$$\boldsymbol{\Theta} = \sum_{k\in\mathcal{F}} \varphi_k \boldsymbol{\Theta}_k, \tag{8.31}$$

where $\boldsymbol{\Theta}_k$ is the conditional state transition probability matrix given that content k is being requested. It can be seen that $\boldsymbol{\Theta}$ and $\boldsymbol{\Theta}_k, \forall k$ have many zero elements because $\boldsymbol{\Theta}(m, m')$ and $\boldsymbol{\Theta}(m', m)$ are both zero if $m' \notin \mathcal{H}_m$, and $\boldsymbol{\Theta}_k(m, m')$ and $\boldsymbol{\Theta}_k(m', m)$ are both zero if $m' \notin \mathcal{H}_m^k$. Specifically, based on the conditions in Eq. (8.30), each column of $\boldsymbol{\Theta}$ only has $c(N_\mathrm{f} - c) + 1$ nonzero elements with one on and the rest off the main diagonal. For a column in $\boldsymbol{\Theta}_k$, two cases are possible. If state m caches content k, then the diagonal element $\boldsymbol{\Theta}_k(m, m)$ is the only nonzero element in the mth column. Otherwise, the mth column has c nonzero elements off the main diagonal.

In order to highlight the state transition, the content request probability φ_k at the target cache is alternatively denoted by $\varphi_{m',m}$ if $m' \in \mathcal{H}_m^k$, i.e., if content k must be requested for the cache to transition from state m into state m'. Denote the mth element on the main diagonal of $\boldsymbol{\Theta}$ by $\alpha_{m,m}$. Figure 8.7 illustrates the state transitions. In the illustrated example, $N_\mathrm{f} = 5$ and $c = 2$, and therefore $c(N_\mathrm{f} - c) + 1 = 7$. As a result, each state can transition into itself and six other states.

Given the above notations, the overall transition probability matrix $\boldsymbol{\Theta}$ corresponding to the content replacement Markov chain for the target cache can be written as:

$$\boldsymbol{\Theta} = \begin{array}{c} \\ 1 \\ \vdots \\ m \\ \vdots \\ n \end{array} \!\! \begin{array}{c} \begin{array}{ccccc} 1 & \cdots & m & \cdots & n \end{array} \\ \left[\begin{array}{ccccc} \alpha_{1,1} & \cdots & \varphi_{1,m}\tau_{1,m} & \cdots & \varphi_{1,n}\tau_{1,n} \\ \vdots & \ddots & \vdots & & \vdots \\ \varphi_{m,1}\tau_{m,1} & & \alpha_{m,m} & & \varphi_{m,n}\tau_{m,n} \\ \vdots & & \vdots & \ddots & \vdots \\ \varphi_{n,1}\tau_{n,1} & \cdots & \varphi_{n,m}\tau_{n,m} & \cdots & \alpha_{n,n} \end{array} \right] \end{array}, \tag{8.32}$$

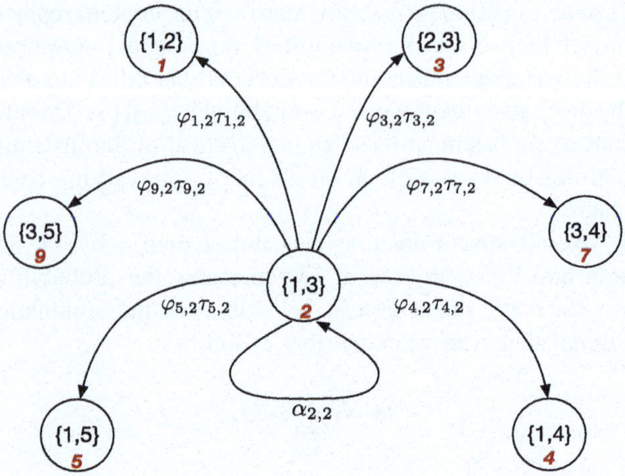

Fig. 8.7 An illustration of the variables in the state transition using the example of state 2, where $N_f = 5$ and $c = 2$. Each circle represents a state, the bracketed numbers in the center represent the cached contents in that state, and the italic number at the bottom represents the state ID

where $\varphi_{m',m} \in [0, 1]$ and $\tau_{m',m} \in [0, 1], \forall m' \in \mathcal{H}_m, \forall m$. The diagonal element $\alpha_{m,m}$ in (8.32) can be found as follows:

$$\alpha_{m,m} = \sum_{k \in \mathcal{G}^m} \varphi_k + \sum_{m' \in \mathcal{H}_m} \varphi_{m',m}(1 - \tau_{m',m}). \tag{8.33}$$

The first item in Eq. (8.33) represents the probability that a content currently in the cache is requested, while the second item represents the probability that a content not in the cache is requested (and downloaded) but not accepted into the cached (i.e., no replacement occurs).

Given $\boldsymbol{\eta}^\star$, the design of dynamic probabilistic caching so that the content caching probabilities converge to the optimal content caching probability vector $\mathbf{p}^\star$ in the case of time-invariant content popularity is equivalent to finding the transition matrix $\boldsymbol{\Theta}$ such that

$$\boldsymbol{\Theta}\boldsymbol{\eta}^\star = \boldsymbol{\eta}^\star. \tag{8.34}$$

If the elements of $\boldsymbol{\Theta}$ could be arbitrarily chosen in the range of [0, 1], the problem can be solved with existing methods, e.g., the Metropolis–Hastings Algorithm [24]. However, as can be seen from Eq. (8.32), there are additional constraints on the elements of $\boldsymbol{\Theta}$. First, each off-diagonal element is a product of two items, and the (m', m)th element should be bounded by $\varphi_{m',m}$. Second, the summation of multiple elements in the same column should also be bounded, i.e.,

$$\sum_{\forall m' \in \mathcal{H}_m^k} \varphi_{m',m}\tau_{m',m} \le \varphi_k, \forall k \in \mathcal{G}^m, \forall m. \tag{8.35}$$

Consequently, the Metropolis–Hastings Algorithm cannot be applied to the considered problem. In the next section, an approach is proposed to construct an irreducible and ergodic Markov chain by designing the matrix $\boldsymbol{\Theta}$ for the target cache so that $\boldsymbol{\eta}^\star$ is the unique steady state.

8.4.2 Generating the State Transition Matrix $\boldsymbol{\Theta}$

Without loss of generality, it is assumed that the elements of $\boldsymbol{\eta}^\star$ are non-zero and arranged in a non-increasing order, i.e., $\eta^q \geq \eta^l$ if $q < l$. An extension to the case when η^j is zero for some j is straightforward.

When a content not in the cache is requested, a replacement may or may not happen. In order to control this factor in our design, we introduce parameters $\{\omega_{k,m',m}\}$ to represent the upper limit on state transition probabilities. Specifically, for any given m and $m' \in \mathcal{H}_m^k$, parameter $\omega_{k,m',m}$ represents the upper limit that state m transits into state m' given that content k is requested. As a result, the following condition must be satisfied:

$$\sum_{m' \in \mathcal{H}_m^k} \omega_{k,m',m} \leq 1, \forall m, \forall k. \tag{8.36}$$

If strict equality holds in the above condition, then content k is always accepted into the cache when the cache state is m. Otherwise, content k may not be accepted into the cache even after it is requested and downloaded.

The next step is to determine which state transitions could happen and with what probabilities. Recall that the elements of $\boldsymbol{\eta}^\star$ are arranged in a non-increasing order. Note that the adjacent states, e.g., state m and state $m+1$, may not be neighbor states using such an order. For any given m, define the functions $V(m)$ for $m \in \{2, \ldots, n\}$ and $X(m)$ for $m \in \{1, \ldots, n-1\}$, both mapping from a state to one of its neighbor states as follows:

$$V(m) = \arg\min_{\hat{m} \in \mathcal{H}_m} \{\eta^{\hat{m}} | \eta^{\hat{m}} \geq \eta^m\}, \tag{8.37a}$$

$$X(m) = \arg\max_{\check{m} \in \mathcal{H}_m} \{\eta^{\check{m}} | \eta^{\check{m}} \leq \eta^m\}. \tag{8.37b}$$

An illustration of $V(\cdot)$ and $X(\cdot)$ when $N_\mathrm{f} = 5$ and $c = 2$ is given in Fig. 8.8a. Using $V(m)$ and $X(m)$ on state m, two cases are possible:

- $X(m) = m+1$ and $V(m+1) = m$: states m and $m+1$ are both adjacent and neighbor states (e.g., states 1 and 2 in Fig. 8.8a).
- $X(m) \neq m+1$ and $V(m+1) \neq m$: states m and $m+1$ are adjacent but not neighbor states (e.g., states 3 and 4 in Fig. 8.8a).

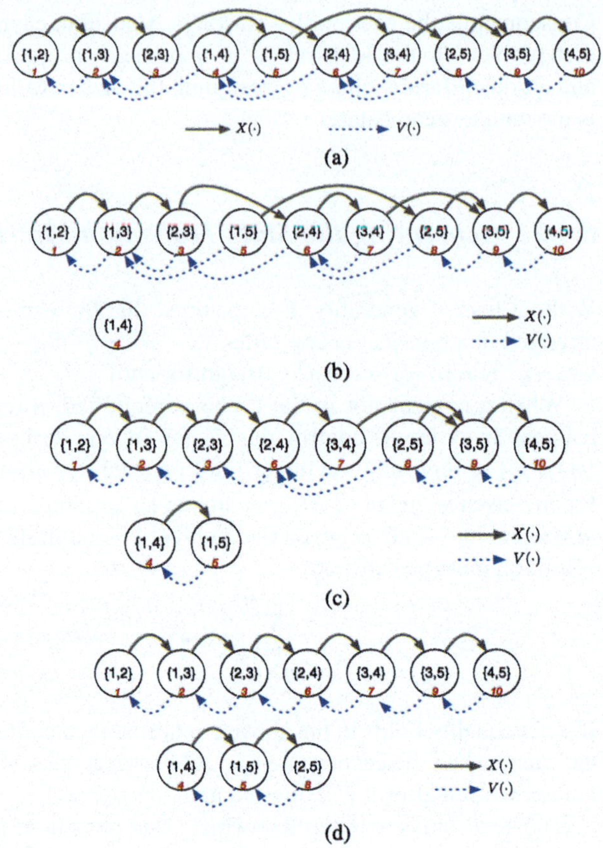

Fig. 8.8 An illustration of using $V(m)$ and $X(m)$ to group states into ordered sequences. The states satisfy $\eta_1^\star \geq \eta_2^\star \geq \cdots \geq \eta_{10}^\star$. (**a**) Step 1: original input sequence. (**b**) Step 2: removing a state from the original sequence. (**c**) Step 3: removing 2nd state from the original sequence. (**d**) Step 4: the final ordered sequences.

To facilitate the design of state transitions, we organize the states into several sequences so that: (1) $\eta_m^\star$ in each sequence is sorted in a non-increasing order; and (2) adjacent states in the same sequence are always neighbors. Denote the original sequence of all states by S_0. Using the procedure in Algorithm 1 repeatedly (by setting the output L as the input sequence S_0 of the next run until the output L is empty), the above-mentioned ordered sequences can be obtained. The procedure is illustrated in four steps in Fig. 8.8.

The connection points identified in Step 3 and Step 5 of Algorithm 1 are the points where the next sequence of states may connect with the current sequence in the Markov chain. We refer to the points where the first state and the last state of the next sequence can connect to as branch and merge points, respectively. Determining one branch point and one merge point for each sequence is not difficult and neglected here. Denote the lth sequence of states by S_l and the length of S_l by K_l. Denote the kth state, the branch point, and the merge point of sequence S_l by $\mathsf{S}_l(k)$, $B(l)$, and $M(l)$, respectively. This is illustrated in Fig. 8.9a.

Algorithm 1 Generating ordered sequence of neighbor states

Input: S_0
Output: S, L
Initialization: $\mathsf{S} = \mathsf{S}_0$; L set to an empty sequence.
for State $m = 2$ to n **do**
  if $X(V(m)) \neq m$ **then**
    Mark $V(m)$ as a potential connection point;
    Remove m from sequence S;
    Add m to the end of sequence L;
  else if $V(X(m)) \neq m$ **then**
    Mark $X(m)$ as a potential connection point;
  end if
end for
return S, L

Given the ordered sequences, the next step is to determine the state transition probabilities. An illustration of this procedure is given in Fig. 8.9a, b, while the details are skipped here. Interested readers are referred to [21] for more information.

The generated Markov chain after the procedure shown in Fig. 8.9a, b satisfies $\boldsymbol{\Theta}\eta^\star = \eta^\star$ but most of the off-diagonal elements in $\boldsymbol{\Theta}$ are 0. As a result, the mixing time can be long. In order to reduce the mixing time, we use a refinement procedure to connect more states based on the fact that the mixing time of the Markov chain is determined by the second largest eigenvalue of the transition matrix [25]. The generated Markov chain after the refinement procedure is illustrated in Fig. 8.9c. Readers are referred to [21] for details.

8.4.3 Discussion on Scalability

The proposed design involves finding the transition probabilities for each cache state, and the overall number of cache states, i.e., $\binom{N_\mathrm{f}}{c}$, can be prohibitively large in practice. Nevertheless, we can limit the number of states to be considered. Next, we provide some methods for reducing the number of states when the number of contents is large.

On the Content Level Although the number of contents can be large, the number of "popular" contents that are worth caching can be small. The study in [26] shows that a large portion of YouTube videos (>70%) are requested only once from an edge network. It follows that a significant portion of contents will be assigned with a caching probability of zero. Denote the number of all contents that are assigned with a positive caching probability as N_p. Then, the number of cache states to be considered decreases from $\binom{N_\mathrm{f}}{c}$ to $\binom{N_\mathrm{p}}{c}$, which is a significant reduction when N_p is much smaller than N_f. Moreover, the "very popular" contents that are assigned with a caching probability of 1 also reduces the number of cache states to be considered.

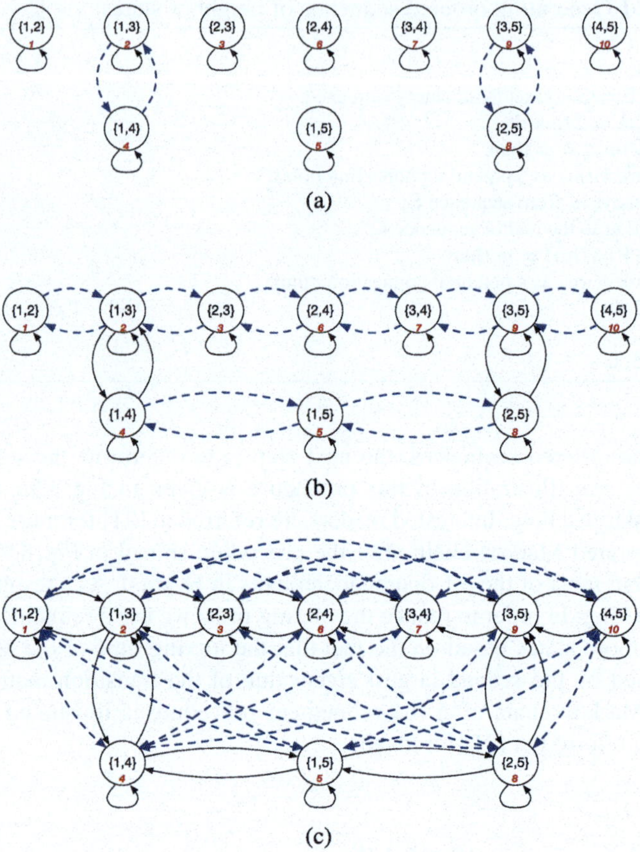

Fig. 8.9 An illustration on generating and refining the underlying Markov chain by updating $\boldsymbol{\Theta}$. (**a**) Ordered state sequence S_1 consists of the seven states in the first row, and S_2 consists of the three states in the second row. The states 2 and 9 are the branch and merge points of S_2, respectively. (**b**) The dashed transition links are created by the update on $\boldsymbol{\Theta}$ in the state transition probability generation procedure. (**c**) The Markov chain created after the refinement procedure. To reduce the number of connections, bi-directional links are used

If N_e contents are assigned a caching probability in (0, 1), then the number of cache states decreases to $\binom{N_e}{c}$.

On the State Level We could limit the considered states to a small number of states with large overall cached content request probabilities. This will significantly reduce the number of states to be considered and the resulting dimension of the state transition matrix. For example, if $N_e = 100$ and $c = 20$, there are more than 10^{20} states. However, we can consider the top 1000 states that cache the most popular contents only. By setting a proper cut-off threshold, the proposed dynamic caching can still yield satisfactory performance. Interested readers are referred to [21], in which there is an example with 10,000 contents but we only consider 30 states.

8.5 Numerical Results

In this section, we demonstrate the numerical results related to the STF and the dynamic caching introduced in Sects. 8.2–8.4.

8.5.1 *State Transition Field with Time-Invariant Content Popularity*

We first demonstrate the STFs obtained from analysis, followed by the STFs obtained from simulations. In general, STF can have high dimensions. We limit most of our demonstration to the case of three dimensions, as it is easy to visualize three-dimensional fields. A three-dimensional subspace in a high-dimensional STF is also illustrated.

Figure 8.10a demonstrates a three-dimensional STF of random replacement (RR), in which a randomly selected content is replaced once a new content is downloaded. In this figure, $N_c = 3$, $L = 2$, and therefore there are only three cache states (i.e., $C_1 = \{1, 2\}, C_2 = \{1, 3\}, C_3 = \{2, 3\}$). The x, y, and z axes correspond to the SCP for the cache states 1, 2, and 3, respectively. The triangular area is the state transition domain $\mathcal{D}$, the square marker represents the center of the triangle, and the circle represents the steady-state SCP $\eta^\star$ in this example. The STF at a point in $\mathcal{D}$ is represented by an arrow originating from that point, while the strength and direction of the STF are shown by the length of the arrow and the direction of the arrowhead, respectively. The straight lines in the x–y plane show the contour of the cache hit probability for the SCP. Note that, in this example, the content popularity φ and the parameter ϕ are predetermined, while different φ and ϕ would create different STFs.

Figure 8.10b demonstrates part of a high-dimensional STF over the surface of an ellipsoid in a three-dimensional subspace. In this example, $N_c = 30$, $L = 3$, and there are 4060 cache states. Three mutually neighbor cache states are selected, corresponding to the three-dimensional subspace in the figure. The STF over the surface of an ellipsoid in this subspace is demonstrated as an example. The x, y, and z axes correspond to the SCP for the three selected cache states. Unlike the case in Fig. 8.10a, the SCPs in Fig. 8.10b are small and do not sum up to 1 since there are many other states. Figure 8.10b serves as an example of high-dimensional STF.

Figure 8.11 shows the STF of RR generated from simulations. The settings on ϕ and φ in Fig. 8.11 are exactly the same as those in Fig. 8.10a. For each point in the STF, M realizations of states are generated based on the corresponding SCP. For each realization, R content requests are generated based on the content popularity. Each data point (i.e., each arrow) in Fig. 8.11a, b is obtained from averaging the state transitions following the $M \times R$ requests. In Fig. 8.11a, M and R are both set to 100. It can be seen that the STF is not accurate, especially in the area close to the steady state, due to insufficient samples. In addition, the arrows point to a

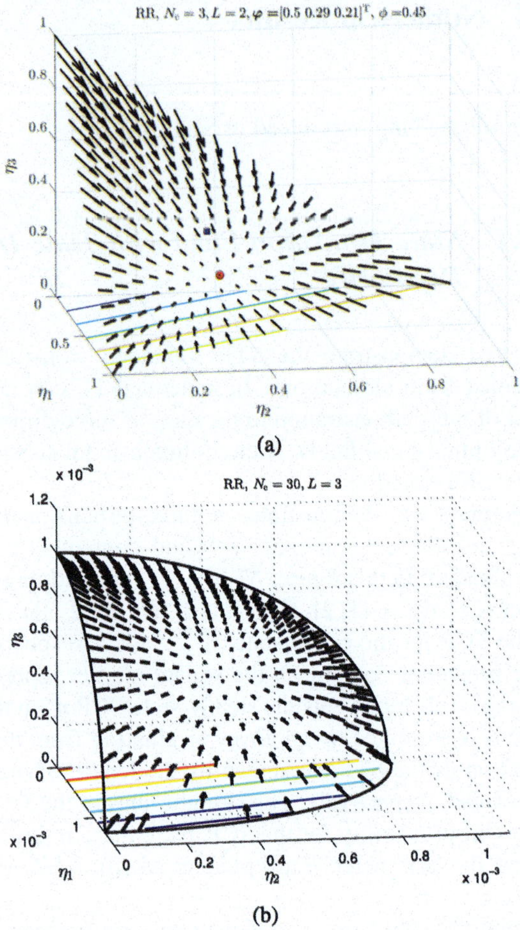

Fig. 8.10 Analytical STF of RR in 3-D. (**a**) Analytical STF of RR, $\phi = 0.45$, $\boldsymbol{\varphi} = [0.5, 0.29, 0.21]^{\mathrm{T}}$. (**b**) Analytical STF of RR, $N_{\mathrm{c}} = 30$, $L = 3$, in a 3-D subspace

steady state slightly deviated from the true steady state in Fig. 8.10a. In Fig. 8.11b, M and R are both increased to 1000. It can be seen that the resulting STF generated based on simulation in Fig. 8.11b becomes an exact match for the analytical STF in Fig. 8.10a.

8.5.2 *State Transition Field with Time-Varying Content Popularity*

Figure 8.12 demonstrates the instantaneous STF under time-varying content popularity using RR as an example. The content popularity at the nth and $(n+1)$th requests is $\boldsymbol{\varphi}^{(n)} = [0.46, 0.30, 0.24]^{\mathrm{T}}$ and $\boldsymbol{\varphi}^{(n+1)} = [0.4, 0.35, 0.25]^{\mathrm{T}}$, respectively. The solid circle with red filling shows where the steady state would be if the

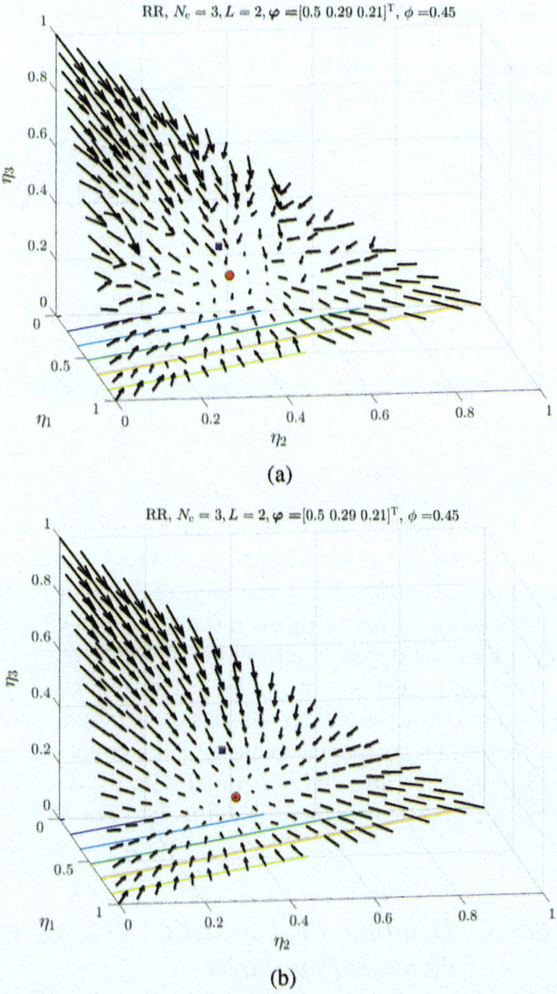

Fig. 8.11 The STF of RR from simulations. (**a**) STF of RR from simulation, $M = 100$, $R = 100$. (**b**) STF of RR from simulation, $M = 1000$, $R = 1000$

content popularity were fixed and equal to $\boldsymbol{\varphi}^{(n)}$. The hollow circle shows where the stationary state would be if the content popularity were fixed and equal to $\boldsymbol{\varphi}^{(n+1)}$. The black triangular area with solid edges represents the state transition domain. The black arrows demonstrate the direction and strength of the STF at the instant of the nth request and the corresponding locations in the state transition domain. The colored straight lines in the x–y plane show the contour of the cache hit probability in the state transition domain. The solid straight line from the origin $(0, 0, 0)$ to the diamond marker in the STF is specified by the vector $\mathbf{C}_s\boldsymbol{\varphi}^{(n+1)}$. Denote the SCP vector $\boldsymbol{\eta}$ at the diamond marker as $\bar{\boldsymbol{\eta}}^{(n)}$. The dashed triangle in blue represents the intersection of the plane $(\boldsymbol{\varphi}^{(n+1)})^T\mathbf{C}_s(\boldsymbol{\eta} - \bar{\boldsymbol{\eta}}^{(n)}) = 0$ with the three planes $\eta_1 = 0$, $\eta_2 = 0$, and $\eta_3 = 0$. The dotted line represents the intersection of the plane $(\boldsymbol{\varphi}^{(n+1)})^T\mathbf{C}_s(\boldsymbol{\eta} - \bar{\boldsymbol{\eta}}^{(n)}) = 0$ with the state transition domain.

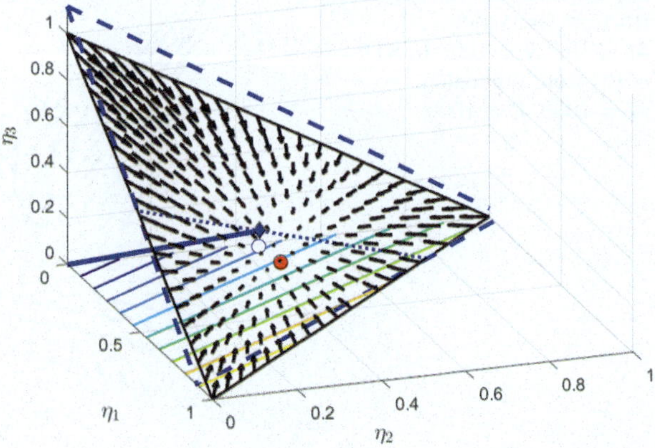

Fig. 8.12 An instantaneous STF of RR and its impact on the instantaneous cache hit probability at the next request

From Fig. 8.12, the effect of the nth replacement, given the replacement scheme of RR and the above change of content popularity from $\boldsymbol{\varphi}^{(n)}$ to $\boldsymbol{\varphi}^{(n+1)}$, can be observed. Specifically, given any SCP, i.e., a point in the state transition domain, if the arrow representing the instantaneous STF at that point can be scaled such that it crosses the dotted line from below to above, the nth replacement yields a smaller cache hit probability at the $(n+1)$th request compared with no replacement. By contrast, if the arrow can be scaled such that it crosses the dotted line from above to below, the nth replacement yields a larger cache hit probability at the $(n+1)$th request. If the arrow is in parallel with the dotted line, the nth replacement has no impact on the cache hit probability at the $(n+1)$th request.

8.5.3 Dynamic Probabilistic Caching with Time-Varying Content Popularity

We first demonstrate the convergence speed of the underlying Markov chain corresponding to the designed $\boldsymbol{\Theta}$. In this illustrative example, 5 contents and a cache with size 2 is considered. Thus, there are 10 pure strategies and the mixed caching strategy is a probability vector with 10 elements. The transition probability matrix $\boldsymbol{\Theta}$ is first generated and then refined as mentioned in Sect. 8.4. For the purpose of illustrating the convergence performance of the proposed $\boldsymbol{\Theta}$, a constant instantaneous content popularity based on Zipf distribution is used in this example. The convergence speed of the underlying Markov chains of $\boldsymbol{\Theta}$ before and after the refinement procedure is shown in the top and bottom subplots of Fig. 8.13, respectively. In each figure, 10,000 tests with randomly generated initial $\boldsymbol{\eta}_0$ are conducted.

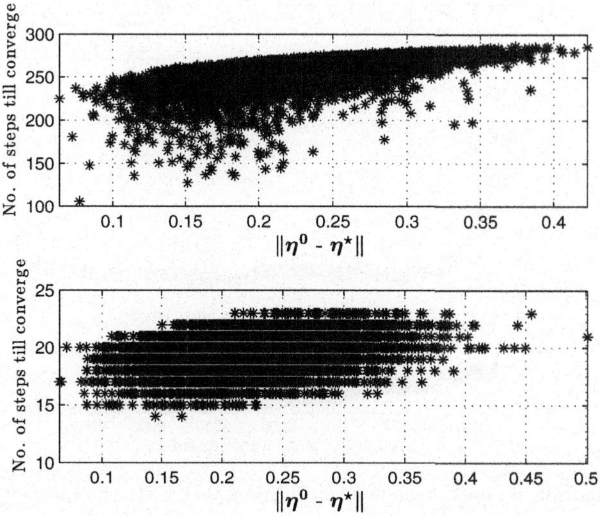

Fig. 8.13 Demonstration of the convergence speed of $\boldsymbol{\Theta}$ before and after the refinement procedure (10,000 random initial states used)

Three observations can be made from Fig. 8.13. First, $\boldsymbol{\eta}$ always converges (in distribution) to the target $\boldsymbol{\eta}^{\star}$ with the designed replacement policy represented by $\boldsymbol{\Theta}$ in the 20,000 tests regardless of the initial caching strategy. Second, the convergence speed is shorter on average when the $\boldsymbol{\eta}^{0}$ is closer to $\boldsymbol{\eta}^{\star}$ and vice versa. Third, the refinement of $\boldsymbol{\Theta}$ significantly reduces the convergence speed, i.e., by a factor of 10.

Next, we demonstrate the comparison of the convergence (in distribution) of replacement policies. In this illustrative example, 15 contents and a cache with size 8 is considered. Thus, there are 6435 cache states, and the state caching probability vector has 6435 elements. The convergence performance of three replacement policies under constant content popularity is compared: the proposed policy corresponding to the designed $\boldsymbol{\Theta}$, LRU, and LFU. The content requests are randomly generated and follows a Zipf distribution. The convergence is represented through the square norm of the difference between the current caching strategy $\boldsymbol{\eta}$ and the target caching strategy $\boldsymbol{\eta}^{\star}$ versus the number of content requests since the beginning of the simulations. The comparison of the convergence performance is shown in Fig. 8.14. It can be seen from this figure that the proposed replacement policy can converge to $\boldsymbol{\eta}^{\star}$ in distribution and thereby implement a given set of caching probabilities when the instantaneous content popularity is a constant. By contrast, LRU or LFU cannot converge to $\boldsymbol{\eta}^{\star}$ under the same condition.

In the last example, we show the benefit of probabilistic content replacement. In this illustrative example, the benefit of dynamic probabilistic content replacement with the designed replacement policy $\{\tau_{m,m'}\}$ is demonstrated with 23 contents and a cache of size 2. A time duration divided into 50 sessions is considered, and 2×10^6 content requests are generated in total. The 23 contents are equally

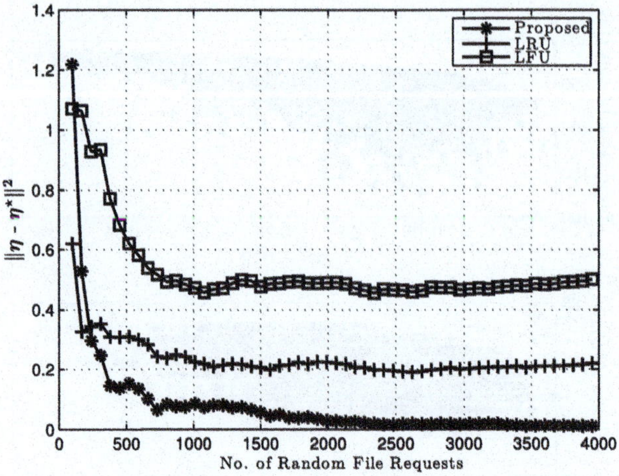

Fig. 8.14 Comparison of replacement policies: the proposed, LRU, and LFU

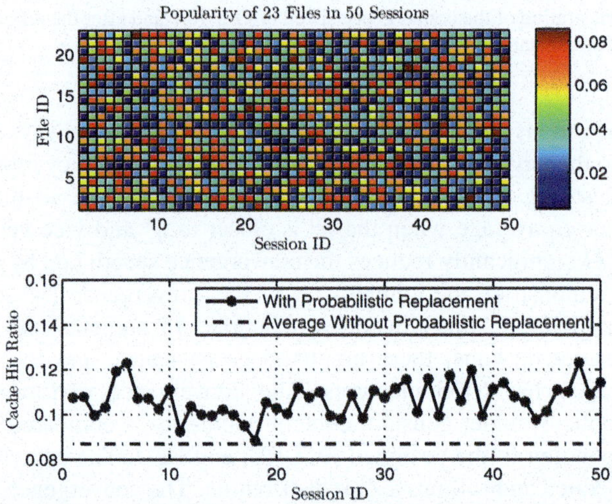

Fig. 8.15 Cache hit ratio with probabilistic replacement when content popularity varies over time—the case of random fluctuation

popular overall, but the popularity of each content varies in each session. Two different cases of variations in content popularity (in terms of content request probability) are considered: random fluctuation and smooth change, as shown in the top subplots of Figs. 8.15 and 8.16, respectively. The corresponding cache hit ratio by using dynamic probabilistic content replacement is given in the bottom subplots of Figs. 8.15 and 8.16, respectively. As the overall popularity is the same for each content, caching any two contents without replacement would lead to a cache hit ratio of 2/23, or 0.087 approximately. It can be seen from Figs. 8.15 and 8.16 that

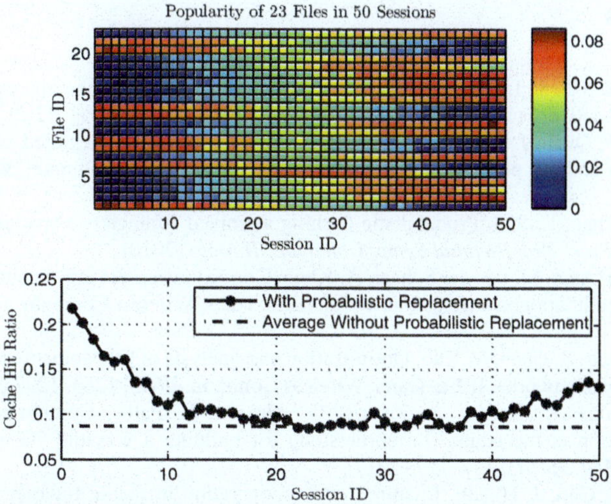

Fig. 8.16 Cache hit ratio with probabilistic replacement when content popularity varies over time—the case of smooth change

the cache hit ratio is improved by using probabilistic content replacement in either case. The overall cache hit ratio is 0.1063 and 0.1085, equivalent to an increase of 22% and 25%, in the cases of Figs. 8.15 and 8.16, respectively. The figures demonstrate that designed probabilistic content replacement based on the average content popularity can improve cache hit ratio by adapting to the varying content popularity when the instantaneous content popularity changes over time.

8.6 Summary

In this chapter, the theory of STF has been introduced as an intuitive yet rigorous new framework for studying dynamic probabilistic caching in the vector space. The investigation has been targeted at revealing insights regarding the relations among content popularity, the knowledge of content popularity, the resulting STFs, and the performance of caching schemes. It has been shown that the STF can be used to model and analyze caching schemes in the cases of time-invariant and time-varying content popularity and that the design of replacement schemes is essentially the manipulation of the STF. The observation inspires our dynamic probabilistic caching design, which can improve the cache hit ratio while significantly reducing the frequency of content replacement. While this chapter has focused on a single cache to illustrate the basic ideas, both the STF theory and the dynamic probabilistic caching design can be extended to handle the case of multiple caches, multi-level caches, and networked caches.

References

1. X. Wang, M. Chen, T. Taleb, A. Ksentini, V.C. Leung, Cache in the air: exploiting content caching and delivery techniques for 5G systems. IEEE Commun. Mag. **52**(2), 131–139 (2014)
2. S. Zhang, W. Quan, J. Li, W. Shi, P. Yang, X. Shen, Air-ground integrated vehicular network slicing with content pushing and caching. IEEE J. Sel. Areas Commun. **36**(9), 2114–2127 (2018)
3. J. Gao, L. Zhao, L. Sun, Probabilistic caching as mixed strategies in spatially-coupled edge caching, in *Proc. 29th Biennial Symp. Commun., Toronto* (2018)
4. S. Müller, O. Atan, M. van der Schaar, A. Klein, Context-aware proactive content caching with service differentiation in wireless networks. IEEE Trans. Wireless Commun. **16**(2), 1024–1036 (2017)
5. K. Li, C. Yang, Z. Chen, M. Tao, Optimization and analysis of probabilistic caching in N -tier heterogeneous networks. IEEE Trans. Wireless Commun. **17**(2), 1283–1297 (2018)
6. E.K. Markakis, K. Karras, A. Sideris, G. Alexiou, E. Pallis, Computing, caching, and communication at the edge: The cornerstone for building a versatile 5G ecosystem. IEEE Commun. Mag. **55**(11), 152–157 (2017)
7. M. Tang, L. Gao, J. Huang, Enabling edge cooperation in tactile Internet via 3C resource sharing. IEEE J. Sel. Areas Commun. **36**(11), 2444–2454 (2018)
8. S. Zhang, P. He, K. Suto, P. Yang, L. Zhao, X. Shen, Cooperative edge caching in user-centric clustered mobile networks. IEEE Trans. Mobile Comput. **17**(8), 1791–1805 (2018)
9. M. Emara, H. Elsawy, S. Sorour, S. Al-Ghadhban, M.S. Alouini, T.Y. Al-Naffouri, Optimal caching in 5G networks with opportunistic spectrum access. IEEE Trans. Wireless Commun. **17**(7), 4447–4461 (2018)
10. T.X. Vu, S. Chatzinotas, B. Ottersten, T.Q. Duong, Energy minimization for cache-assisted content delivery networks with wireless backhaul. IEEE Wireless Commun. Lett. **7**(3), 332–335 (2018)
11. G. Lee, I. Jang, S. Pack, X. Shen, FW-DAS: fast wireless data access scheme in mobile networks. IEEE Trans. Wireless Commun. **13**(8), 4260–4272 (2014)
12. E. Bastug, M. Bennis, M. Debbah, Living on the edge: the role of proactive caching in 5G wireless networks. IEEE Commun. Mag. **52**(8), 82–89 (2014)
13. J. Qiao, Y. He, X. Shen, Proactive caching for mobile video streaming in millimeter wave 5G networks. IEEE Trans. Wireless Commun. **15**(10), 7187–7198 (2016)
14. S.O. Somuyiwa, A. György, D. Gündüz, A reinforcement-learning approach to proactive caching in wireless networks. IEEE J. Sel. Areas Commun. **36**(6), 1331–1344 (2018)
15. R. Pedarsani, M.A. Maddah-Ali, U. Niesen, Online coded caching. IEEE/ACM Trans. Netw. **24**(2), 836–845 (2016)
16. S. Tarnoi, K. Suksomboon, W. Kumwilaisak, Y. Ji, Performance of probabilistic caching and cache replacement policies for content-centric networks, in *Proc. 39th IEEE Conf. Local Computer Networks, Edmonton* (2014), pp. 99–106
17. W. Bao, D. Yuan, K. Shi, W. Ju, A.Y. Zomaya, Ins and outs: optimal caching and re-caching policies in mobile networks, in *Proc. the 18th ACM Mobihoc, New York* (2018), pp. 41–50
18. I. Psaras, W.K. Chai, G. Pavlou, In-network cache management and resource allocation for information-centric networks. IEEE Trans. Parallel Distrib. Syst. **25**(11), 2920–2931 (2014)
19. J. Gao, L. Zhao, X. Shen, The study of dynamic caching via state transition field—the case of time-invariant popularity. IEEE Trans. Wireless Commun. **18**(12), 5924–5937
20. J. Gao, L. Zhao, X. Shen, The study of dynamic caching via state transition field—the case of time-varying popularity. IEEE Trans. Wireless Commun. **18**(12), 5938–5951 (2019)
21. J. Gao, S. Zhang, L. Zhao, X. Shen, The design of dynamic probabilistic caching with time-varying content popularity. IEEE Trans. Mobile Comput. **20**(4), 1672–1684 (2021)
22. S. Tarnoi, V. Suppakitpaisarn, W. Kumwilaisak, Y. Ji, Performance analysis of probabilistic caching scheme using Markov chains, in *Proc. 40th IEEE Conf. Local Computer Networks, Clearwater Beach* (2015), pp. 46–54

23. G.S. Paschos, G. Iosifidis, M. Tao, D. Towsley, G. Caire, The role of caching in future communication systems and networks. IEEE J. Sel. Areas Commun. **36**(6), 1111–1125 (2018)
24. C. Robert, G. Casella, *Monte Carlo Statistical Methods* (Springer Science & Business Media, New York, 2013)
25. S. Boyd, P. Diaconis, L. Xiao, Fastest mixing Markov chain on a graph. SIAM Rev. **46**(4), 667–689 (2004)
26. N. Carlsson, D. Eager, Ephemeral content popularity at the edge and implications for on-demand caching. IEEE Trans. Parallel Distrib. Syst. **28**(6), 1621–1634 (2017)

Chapter 9
Deep Reinforcement Learning for Mobile Edge Computing Systems

Ming Tang and Vincent W. S. Wong

9.1 Introduction

Recently, humans use mobile devices to accomplish many computational intensive tasks, such as artificial intelligence, distributed data analysis, virtual reality, and augmented reality. Despite the fact that mobile devices have become increasingly powerful, they may not be capable of processing all their tasks locally and meeting the delay requirements of the tasks. Mobile edge computing (MEC) [1], also known as multi-access edge computing [2] and fog computing [3], is emerging as a promising architecture. In MEC systems, edge nodes equipped with processing and storage resources are deployed close to the mobile devices. Thus, mobile devices can offload their computational intensive tasks to the edge nodes for processing. When compared with cloud computing systems [4], MEC systems can provide mobile devices with a faster response and hence a low task latency. Mao et al. in [1], Qiu et al. in [5], and Ranaweera et al. in [6] provided comprehensive surveys in the area of MEC.

The environment in MEC systems may involve time-varying and complex system dynamics, such as time-varying task arrivals, device mobility, wireless channel variation, and the interaction among mobile devices. Meanwhile, the operators of the MEC systems, mobile devices, and edge nodes may not be aware of these system dynamics a priori. Such unknown system dynamics impose challenges on addressing the deployment, management, and scheduling problems in MEC systems. On the other hand, conventional network optimization approaches (e.g., online optimization [7], game-theoretic approach [8]) always rely on the modeling

M. Tang · V. W. S. Wong (✉)
Department of Electrical and Computer Engineering, The University of British Columbia, Vancouver, BC, Canada
e-mail: mingt@ece.ubc.ca; vincentw@ece.ubc.ca

L. Cai et al. (eds.), *Broadband Communications, Computing, and Control for Ubiquitous Intelligence*, Wireless Networks,
https://doi.org/10.1007/978-3-030-98064-1_9

of the environment in MEC systems. When the modeling of the environment does not match the practical systems, these conventional approaches may fail to provide a satisfactory performance. In addition, it may be challenging for these approaches to address time-varying environment in MEC systems.

Deep reinforcement learning (DRL) [9] is a promising technique to address the unknown and complex system dynamics in MEC systems [10]. With DRL, an agent (e.g., a mobile device, an edge node, or a network operator in MEC systems) can learn to make decisions (e.g., in terms of deployment, management, and scheduling) by interacting with the environment. During the learning process, the agent continuously gathers its experience from the interaction with the environment and gradually learns the decision-making policy that optimizes its objective. In comparison to conventional reinforcement learning (RL) approaches [11], DRL employs deep learning techniques to tackle the curse of dimensionality issue. Due to the strong capability of deep learning for analyzing and abstracting data, DRL is capable of handling complex systems with large state spaces [9].

In this chapter, we aim at providing an overview on how DRL techniques can benefit the MEC systems. In the rest of this chapter, we first present an overview of DRL in Sect. 9.2. In Sect. 9.3, we demonstrate the application of DRL techniques in MEC systems with a case study, which focuses on the task offloading problem. Finally, in Sect. 9.4, we outline several challenges and future research directions. For notation, let $\mathbb{Z}_{++}$ denote the set of positive integers.

9.2 Overview of Deep Reinforcement Learning

In this section, we first introduce the general DRL problem formulation. Then, we present the main idea for obtaining the optimal policy using DRL algorithms. Finally, we summarize some existing DRL algorithms.

9.2.1 DRL Problem Formulation

In general, a DRL problem can be formulated as a discrete time stochastic control process [9]. In this problem, an agent interacts with the environment. Suppose there are a set of time slots $\mathcal{T}$. At the beginning of time slot $t \in \mathcal{T}$, given the state of the environment $s(t)$, the agent gathers an observation $o(t)$. Based on the observation, the agent chooses an action $a(t)$. After that, the agent obtains a reward $r(t)$. The environment then transits to the next state $s(t+1)$, and the agent gathers the next observation $o(t+1)$. Through such an interaction with the environment, the agent aims at learning an optimal policy that maximizes the expected long-term reward.

For the sake of simplicity, let us consider a fully observable system. In this system, the agent can observe the actual state of the environment, i.e., $o(t) = s(t)$ for all $t \in \mathcal{T}$. Meanwhile, suppose the transition of the state follows Markovian stochastic control processes. The scenario with partially observable system and non-Markovian environment can be found in [9, Section 10]. Let $\mathcal{S}$ and $\mathcal{A}$ denote the state and action spaces. We denote π as the policy of the agent. Note that there are two types of policy: deterministic policy $\pi(s) : \mathcal{S} \rightarrow \mathcal{A}$ and stochastic policy $\pi(s, a) : \mathcal{S} \times \mathcal{A} \rightarrow [0, 1]$, where $\pi(s, a)$ is the probability that action a is selected given current state s. Given any state $s \in \mathcal{S}$, let $V^{\pi}(s) : \mathcal{S} \rightarrow \mathbb{R}$, also called the value function, denote the expected long-term reward in state s under policy π. For example, under the scenario with an infinite time horizon $\mathcal{T} = \{1, 2, \ldots\}$, the value function $V^{\pi}(s) \triangleq \mathbb{E}[\sum_{i=0}^{\infty} \gamma^i r(t+i) \mid s(t) = s, \pi]$. That is, given any $s(t) = s$, the value function $V^{\pi}(s)$ is equal to the expected cumulative future discounted reward, where $\gamma \in (0, 1]$ is the discount factor. The value function under various scenarios can be found in [11, Section 3.5]. The objective of the agent is to find a policy π^* that maximizes $V^{\pi}(s)$. On the other hand, as an alternative to the value function, Q-value function $Q^{\pi}(s, a) : \mathcal{S} \times \mathcal{A} \rightarrow \mathbb{R}$ is sometimes considered, which is the expected long-term reward of choosing action $a \in \mathcal{A}$ in state s under policy π.

9.2.2 Determine the Optimal Policy with Deep Learning

To find the optimal policy, in conventional RL approaches, the agent estimates one or multiple of the following components during the learning process [9, Section 3.2]:

(a) Value function $V^{\pi}(s)$ or Q-value function $Q^{\pi}(s, a)$
(b) Policy $\pi(s)$ or $\pi(s, a)$
(c) The model of the environment, i.e., the transition function (from current state to the next state) and the reward function (from state and action to the reward)

For the RL approaches requiring the estimation of components (a) and (b), they are called model-free algorithms. When component (c) is considered, the associated algorithm is called a model-based algorithm. Furthermore, the algorithms that estimate (a) and (b) are called value-based and policy-based algorithms, respectively. With the estimation of those components (a), (b), or (c), the agent can obtain the optimal policy accordingly. For example, in a model-free value-based Q-learning algorithm, the agent estimates the Q-value function and exploits a policy of choosing the action with the maximum Q-value given each state. Due to the definition of Q-value function, such a policy can maximize the expected long-term reward.

On the other hand, most real-world problems are very complex, e.g., the state and action space can be high-dimensional and continuous. Thus, estimating the value function, policy, and model can be challenging and requires a huge amount of computational resource and memory. To address this issue, in DRL algorithms, deep learning techniques are used for estimating the value function, policy, and model. In particular, deep learning essentially relies on neural networks to estimate a mapping from some input to some output, i.e., $f : \mathcal{X} \rightarrow \mathcal{Y}$. The mapping is characterized by the parameters of the neural network, denoted by θ, and can be represented by

$y = f(x \mid \theta)$. DRL employs neural networks to estimate certain mappings, such as value function, policy, and model. During the learning process in DRL, based on the gathered experience (i.e., state, action, next state, reward), the agent gradually updates the parameters of the neural network (i.e., θ) using deep learning techniques in order to make the neural network achieve an accurate approximation of the actual mapping (e.g., the actual value function, policy, and model). Since neural networks are capable of addressing complicated mappings with large input and output spaces, they provide DRL the capability of addressing complex real-world environment.

9.2.3 Existing DRL Algorithms

For value-based DRL algorithms, deep Q-learning (DQL) [12] aims at estimating the Q-value function using neural networks. In addition to DQL, double deep Q-network (DQN) [13] can handle the issue of overestimation in DQL. Meanwhile, dueling DQN [14] offers a more accurate estimation of the Q-value function by separately learning the value resulting from the state and action. Instead of estimating the expected reward in the value function as in DQL, distributional DQN [15] aims at estimating the distribution of the cumulative reward under each state. Such a distribution characterizes the randomness of the reward in the system.

Policy gradient algorithms belong to policy-based DRL algorithms and are commonly used. These algorithms directly learn an optimal policy that maximizes the expected long-term reward using gradient ascent. In the area of DRL, deep deterministic policy gradient (DDPG) [16] learns the representation of a deterministic policy, where this approach is applicable for continuous action space. Distributed distributional DDPG [17] is a variant of DDPG that can be run in a distributional fashion. Asynchronous advantage actor-critic (A3C) [18] exploits the approximation of both policy and value function using neural networks. Built upon A3C, actor-critic with experience replay [19] exploits the experience replay, which can decrease the data correlation and increase the sample efficiency. Soft actor-critic [20] encourages policy exploration by maximizing the entropy of the policy and the expected long-term reward simultaneously. Twin delayed deep deterministic [21] exploits double DQN to address the overestimation issue in actor-critic methods. In addition, trust region optimization [22] introduces the idea of trust region to bound the change in policy to guarantee monotonic reward improvement. As a variant of trust region optimization, proximal policy optimization [23] introduces a penalty term in the objective function to alleviate the change in policy, and it is easy for implementation. In comparison to those value-based DRL methods, policy gradient algorithms are capable of handling continuous action space and stochastic policy.

For model-based DRL algorithms, the agent either knows the model of the environment a priori or uses the gathered experience to predict the model. The model will be used for planning, i.e., taking the model as an input, the agent finds the optimal policy for the interaction with the environment. For discrete action space, lookahead search can be used by building a decision tree and exploring the potential trajectories in the tree [9, Section 6.1]. Monte Carlo tree search [24] is a typical family of approaches to lookahead search. Such methods have been incorporated with deep learning for addressing real-world complex problem, e.g., the game of Go [25]. For continuous action space, trajectory optimization can be used. For a function that is differentiable, the agent can optimize the policy along trajectories using gradient ascent. Plaat et al. in [26] and Franccois et al. in [9, Section 6] provided comprehensive surveys for model-based DRL algorithms. In comparison to model-free DRL algorithms, model-based DRL algorithms are more sample efficient. That is, with the learned model, model-based algorithms can converge with fewer samples (or gathered experience) than model-free algorithms.

9.3 Case Study: Deep Q-Learning for Task Offloading in MEC

Due to the huge potential of addressing complex and dynamics systems, DRL has been applied in various problems in MEC systems, such as task offloading, mobility management, and edge maintenance. Luong et al. in [27, Section IV] and Wang et al. in [10, Section VIII] surveyed the existing works using DRL in MEC systems.

In this section, we present a case study on the task offloading problem in MEC systems, which is based on our earlier work [28]. In particular, in MEC systems, edge nodes may have limited amount of processing capacity. The tasks offloaded by different mobile devices at an edge node will share the processing capacity of the edge node. Thus, from the perspective of a mobile device, if many other mobile devices choose to offload to a particular edge node, then this mobile device may choose not to offload to the same edge node in order to reduce the task delay. We refer to the number of concurrent mobile devices offloading to an edge node as the *load level* of the edge node. Such a load level depends on the task arrivals and offloading decisions of all mobile devices. Thus, it is time varying and unknown to the mobile devices a priori. This makes it challenging for a mobile device to make the task offloading decision (i.e., whether to offload or not, and if yes, which edge node to choose). On the other hand, although the challenge can be mitigated by letting a centralized entity make a decision for the mobile device, such a centralized decision making may require global information of the system and incur a high signaling overhead.

To address the unknown load level dynamics, we propose a distributed DQL-based task offloading algorithm. The proposed algorithm is a model-free value-based approach that enables each mobile device to make its offloading decision without knowing the task models and offloading decisions of other mobile devices.

In the following subsections, we first present the system model and task offloading problem, respectively. Then, we propose the DQL-based algorithm for MEC systems. Finally, we evaluate the performance of our proposed algorithm.

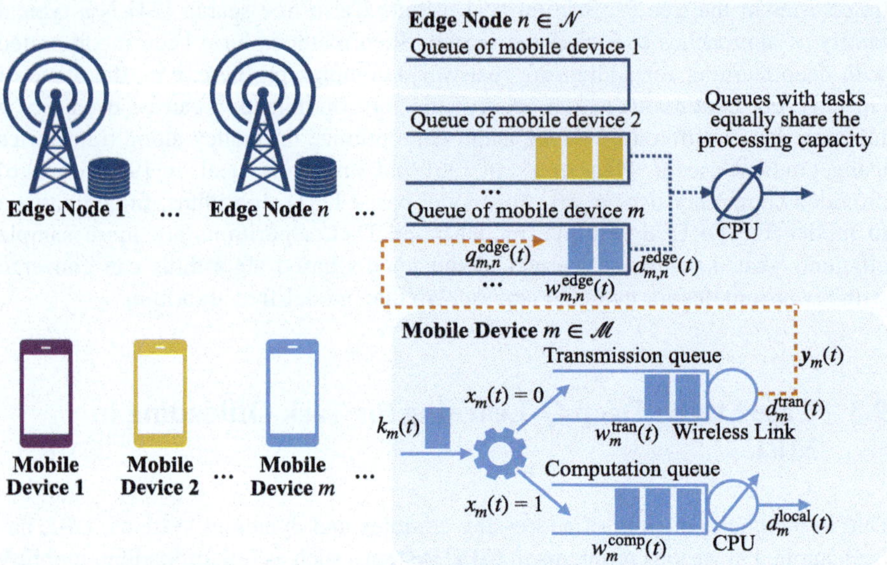

Fig. 9.1 An illustration of an MEC system with edge nodes and mobile devices

9.3.1 System Model

We consider an MEC system that has a set of mobile devices $\mathcal{M} = \{1, 2, \ldots, M\}$ and a set of edge nodes $\mathcal{N} = \{1, 2, \ldots, N\}$. We consider a time-slotted system with a set of time slots $\mathcal{T} = \{1, 2, \ldots, T\}$. Let Δ (in seconds) denote the duration of each time slot. An illustration of an MEC system is shown in Fig. 9.1.

In the following, we first present the task model and task offloading decisions. Then, we discuss the local processing model and edge node offloading model.

9.3.1.1 Task Model

At the beginning of time slot $t \in \mathcal{T}$, mobile device $m \in \mathcal{M}$ either may have a new computational task to be processed or does not have any new task arrival. As in some existing works (e.g., [29]), we assume that the mobile device has a new task arrival at time slot t with a certain probability. If mobile device m has a new task, then we refer to this task using an index $k_m(t) \in \mathbb{Z}_{++}$. For presentation simplicity, if mobile device m does not have any new task, we set $k_m(t) = 0$.

Let $\lambda_m(t)$ (in bits) denote the size of task $k_m(t)$. For presentation simplicity, we set $\lambda_m(t) = 0$ if $k_m(t) = 0$. We set the size of task $k_m(t)$ to be from a discrete set $\Lambda \triangleq \{\lambda_1, \lambda_2, \ldots, \lambda_{|\Lambda|}\}$, where $|\Lambda|$ denotes the cardinality of set Λ. We consider a setting where any task of mobile device m has a deadline τ_m (in time slots). That is, task $k_m(t)$ will be dropped if it has not been completely processed within τ_m

time slots. Moreover, as in some existing works (e.g., [30, 31]), we assume that the number of CPU cycles required for processing a task is proportional to the size of the task. Some examples satisfying this assumption include file compression, video segment encoding and decoding, and object detection in video streaming. Let ρ_m (in CPU cycles per bit) denote the processing density of any task of mobile device m. Thus, $\rho_m \lambda_m(t)$ CPU cycles are required for processing a task of mobile device m with size $\lambda_m(t)$.

9.3.1.2 Task Offloading Decision

If mobile device $m \in \mathcal{M}$ has a new task at the beginning of time slot $t \in \mathcal{T}$, then it needs to decide *whether to process task* $k_m(t)$ *locally or offload it to an edge node*. We use binary variable $x_m(t) \in \{0, 1\}$ to denote this decision. We set $x_m(t) = 1$ if the task is processed locally and set $x_m(t) = 0$ if the task is offloaded to an edge node.

If mobile device m decides to offload task $k_m(t)$ to an edge node, then it needs to decide *which edge node to choose*. We use binary variable $y_{m,n}(t) \in \{0, 1\}$ to denote whether mobile device m chooses edge node $n \in \mathcal{N}$ to offload task $k_m(t)$ or not. We set $y_{m,n}(t) = 1$ if mobile device m chooses edge node n and set $y_{m,n}(t) = 0$ otherwise. Note that exactly one edge node can be chosen to offload task $k_m(t)$, i.e.,

$$\sum_{n \in \mathcal{N}} y_{m,n}(t) = \mathbb{1}(x_m(t) = 0), \ m \in \mathcal{M}, t \in \mathcal{T}, \tag{9.1}$$

where indicator function $\mathbb{1}(x_m(t) = 0) = 1$ if $x_m(t) = 0$, and $\mathbb{1}(x_m(t) = 0) = 0$ otherwise. Let vector $\boldsymbol{y}_m(t) = (y_{m,n}(t), n \in \mathcal{N})$.

9.3.1.3 Local Processing Model

If mobile device m decides to process task $k_m(t)$ locally (i.e., $x_m(t) = 1$), then it will place task $k_m(t)$ in the computation queue for local processing. The tasks in the computation queue are processed in a first-in first-out (FIFO) manner. We use f_m^{device} (in CPU cycles) to denote the processing capacity of mobile device $m \in \mathcal{M}$. We assume that f_m^{device} does not change across time slots. Meanwhile, we assume that if a task has been processed in a time slot, then the processing of the next task in the computation queue will start at the beginning of the next time slot.

For mobile device $m \in \mathcal{M}$, at the beginning of time slot $t \in \mathcal{T}$, let $w_m^{\text{comp}}(t)$ (in time slots) denote the remaining number of time slots until all the tasks placed in the computation queue before time slot t have been either processed or dropped. Note that if task $k_m(t)$ is to be placed in the computation queue, then $w_m^{\text{comp}}(t)$ corresponds to the number of time slots that task $k_m(t)$ will wait in the computation queue for processing, i.e., the queuing delay of task $k_m(t)$ at the computation queue.

Here, we use notation w for the short form of "wait." The expression of $w_m^{\text{comp}}(t)$ is derived as follows. For mobile device $m \in \mathcal{M}$, $w_m^{\text{comp}}(t) = 0$ for $t = 1$, and

$$w_m^{\text{comp}}(t) = \min\left\{\left[w_m^{\text{comp}}(t-1) + \left\lceil \frac{\lambda_m(t-1)x_m(t-1)}{f^{\text{device}}\Delta/\rho_m} \right\rceil - 1\right]^+, \tau_m - 1\right\},$$
$$t \in \mathcal{T} \setminus \{1\}, \qquad (9.2)$$

where $\lceil \cdot \rceil$ is the ceiling function, and operator $[z]^+ = \max\{z, 0\}$. Specifically, for $t \in \mathcal{T} \setminus \{1\}$, if task $k_m(t-1)$ was placed in the computation queue, then given $w_m^{\text{comp}}(t-1)$, the first term in the min operator corresponds to the remaining number of time slots until task $k_m(t-1)$ has been processed since the beginning of time slot t. The second term corresponds to the remaining number of time slots until task $k_m(t-1)$ has been dropped. On the other hand, if either $\lambda_m(t-1) = 0$ or $x_m(t-1) = 0$, then we have $w_m^{\text{comp}}(t) = [w_m^{\text{comp}}(t-1) - 1]^+$. The second term in the min operator is canceled out, because $w_m^{\text{comp}}(t) < \tau_m$ holds for $t \in \mathcal{T}$. In this case, no task was placed in the computation queue in time slot $t-1$. Thus, from time slot $t-1$ to t, the associated remaining number of time slots is decremented by one.

Suppose task $k_m(t)$ is processed locally (i.e., $x_m(t) = 1$). Let $d_m^{\text{local}}(t)$ denote the corresponding delay for local processing, i.e., the number of time slots required to process task $k_m(t)$. The expression of $d_m^{\text{local}}(t)$ is derived as follows:

$$d_m^{\text{local}}(t) = \left\lceil \frac{\lambda_m(t)}{f^{\text{device}}\Delta/\rho_m} \right\rceil, m \in \mathcal{M}, t \in \{t' \mid t' \in \mathcal{T}, k_m(t') \neq 0, x_m(t') = 1\}. \qquad (9.3)$$

Let $\text{Delay}_m(t)$ denote the delay of task $k_m(t)$. Recall that the tasks in the computation queue are processed in an FIFO manner. If task $k_m(t)$ is processed locally, then the delay of task $k_m(t)$ can be derived as follows:

$$\text{Delay}_m(t) = \min\left\{w_m^{\text{comp}}(t) + d_m^{\text{local}}(t), \tau_m\right\},$$
$$m \in \mathcal{M}, t \in \{t' \mid t' \in \mathcal{T}, k_m(t') \neq 0, x_m(t') = 1\}. \qquad (9.4)$$

Specifically, the delay of task $k_m(t)$ is equal to its queuing delay at the computation queue, i.e., $w_m^{\text{comp}}(t)$, plus the processing delay, i.e., $d_m^{\text{local}}(t)$. Note that if the delay exceeds τ_m time slots, then the task will be dropped immediately. Without loss of generality, if task $k_m(t)$ has been dropped, then we set the value of $\text{Delay}_m(t)$ to be τ_m.[1]

---

[1] This setting is for the simplicity of mathematical presentation. For any task $k_m(t)$ that has been dropped, the value of $\text{Delay}_m(t)$ will not be taken into account in our proposed algorithm according

9.3.1.4 Edge Node Offloading Model

If mobile device $m \in \mathcal{M}$ decides to offload task $k_m(t)$ to an edge node (i.e., $x_m(t) = 0$), then it places task $k_m(t)$ in the transmission queue. Tasks in the transmission queue are sent in an FIFO manner. After task $k_m(t)$ has been sent to the chosen edge node based on decision $\boldsymbol{y}_m(t)$, it will be processed by the edge node.

Transmission to an Edge Node The tasks in the transmission queue will be forwarded to the chosen edge node using a wireless network interface. We assume that the mobile devices transmit using orthogonal channels. Thus, there is no interference between mobile devices. Let $|h_{m,n}|^2$ denote the channel gain from mobile device $m \in \mathcal{M}$ to edge node $n \in \mathcal{N}$. Let P denote the transmission power of the mobile device. Thus, the transmission rate from mobile device m to edge node n can be computed as follows:

$$r_{m,n}^{\text{tran}} = W \log_2 \left(1 + \frac{|h_{m,n}|^2 P}{\sigma^2}\right), \quad m \in \mathcal{M}, n \in \mathcal{N}, \tag{9.5}$$

where W denotes the bandwidth allocated to the channel, and σ^2 denotes the received noise power at the edge node. We assume that the transmission rate $r_{m,n}^{\text{tran}}$ does not change across time slots. If a task has been sent in a time slot, then the next task in the transmission queue will be sent at the beginning of the next time slot.

For mobile device $m \in \mathcal{M}$, at the beginning of time slot $t \in \mathcal{T}$, let $w_m^{\text{tran}}(t)$ (in time slots) denote the remaining number of time slots until all the tasks placed in the transmission queue before time slot t have been either processed or dropped. If task $k_m(t)$ is to be offloaded to an edge node, then $w_m^{\text{tran}}(t)$ also corresponds to the number of time slots that task $k_m(t)$ will wait in the transmission queue, i.e., the queuing delay of task $k_m(t)$ at the transmission queue. The expression of $w_m^{\text{tran}}(t)$ is given as follows. For mobile device $m \in \mathcal{M}$, $w_m^{\text{tran}}(t) = 0$ for $t = 1$, and

$$w_m^{\text{tran}}(t) = \min\left\{\left[w_m^{\text{tran}}(t-1) + \left\lceil \sum_{n \in \mathcal{N}} \frac{\lambda_m(t-1) y_{m,n}(t-1)}{r_{m,n}^{\text{tran}} \Delta} \right\rceil - 1\right]^+, \tau_m - 1\right\},$$
$$t \in \mathcal{T} \setminus \{1\}. \tag{9.6}$$

The interpretation of $w_m^{\text{tran}}(t)$ is similar to that of $w_m^{\text{comp}}(t)$. If there is no task arrival in time slot $t-1$ (i.e., $\lambda_m(t-1) = 0$), then $w_m^{\text{tran}}(t) = [w_m^{\text{tran}}(t-1)-1]^+$. Meanwhile, if task $k_m(t-1)$ was placed in the computation queue (i.e., $x_m(t-1) = 1$), then $y_{m,n}(t-1) = 0$ for all $n \in \mathcal{N}$ according to (9.1), and hence $w_m^{\text{tran}}(t) = [w_m^{\text{tran}}(t-1) - 1]^+$.

to Sects. 9.3.2 and 9.3.3. Meanwhile, the delay of a dropped task is not accounted when we evaluate the average delay of the tasks with our proposed algorithm and benchmark methods in Sect. 9.3.4.

If task $k_m(t)$ is offloaded to an edge node, then the number of time slots required to send task $k_m(t)$ to the edge node, denoted by $d_m^{\text{tran}}(t)$, is computed as follows:

$$d_m^{\text{tran}}(t) = \left\lceil \sum_{n \in \mathcal{N}} \frac{\lambda_m(t) y_{m,n}(t)}{r_{m,n}^{\text{tran}} \Delta} \right\rceil, m \in \mathcal{M}, t \in \{t' \mid t' \in \mathcal{T}, k_m(t') \neq 0, x_m(t') = 0\}. \tag{9.7}$$

Processing at an Edge Node At any edge node $n \in \mathcal{N}$, the tasks from different mobile devices are placed in different queues. We refer to the queue that stores the tasks of mobile device $m \in \mathcal{M}$ as the queue of mobile device m. We assume that when a task has been sent to an edge node in a time slot, the edge node places the task into the associated queue at the beginning of the next time slot.

At edge node $n \in \mathcal{N}$, let $q_{m,n}^{\text{edge}}(t)$ (in bits) denote the occupancy of the queue of mobile device $m \in \mathcal{N}$ at the end of time slot $t \in \mathcal{T}$. Let $k_{m,n}^{\text{edge}}(t)$ denote the index of the task placed in the queue of mobile device m at the beginning of time slot t. Specifically, if task $k_m(t')$ is offloaded to edge node n in time slot $t-1$ (i.e., $t' + w_m^{\text{tran}}(t') + d_m^{\text{tran}}(t') - 1 = t - 1$ and $y_{m,n}(t') = 1$), then $k_{m,n}^{\text{edge}}(t) = k_m(t')$. If there does not exist such a task, then we set $k_{m,n}^{\text{edge}}(t) = 0$. We denote $\lambda_{m,n}^{\text{edge}}(t)$ (in bits) as the size of task $k_{m,n}^{\text{edge}}(t)$. If $k_{m,n}^{\text{edge}}(t) = 0$, then we set $\lambda_{m,n}^{\text{edge}}(t) = 0$. In time slot t, we refer to the queue of mobile device m as an *active queue* if either the queue is non-empty or there exists a new task arrival at the queue. Thus, the set of active queues at edge node n in time slot t, denoted by $\mathcal{B}_n(t)$, is defined as

$$\mathcal{B}_n(t) = \left\{ m \;\middle|\; q_{m,n}^{\text{edge}}(t-1) > 0 \text{ or } \lambda_{m,n}^{\text{edge}}(t) > 0, m \in \mathcal{M} \right\}. \tag{9.8}$$

Let $B_n(t)$ denote the number of active queues, i.e., $B_n(t) = |\mathcal{B}_n(t)|$.

Let f_n^{edge} (in CPU cycles per second) denote the processing capacity of edge node n. Within each time slot $t \in \mathcal{T}$, the active queues in set $\mathcal{B}_n(t)$ equally share the processing capacity of edge node $n \in \mathcal{N}$. This is the generalized processor sharing (GPS) model with equal processing capacity sharing [32]. Note that the number of active queues, i.e., $B_n(t)$, varies across time slots and is unknown to the mobile devices and edge nodes a priori. This corresponds to the unknown load level dynamics at the edge nodes and leads to the associated uncertain processing delay.

At the beginning of time slot t, let $w_{m,n}^{\text{edge}}(t)$ (in time slots) denote the remaining number of time slots until all the tasks placed in the queue of mobile device m at edge node n before time slot t have been either processed or dropped. Due to the unknown load level dynamics at the edge nodes, the mobile devices and edge nodes are unaware of the value of $w_{m,n}^{\text{edge}}(t)$ before all those tasks have been either processed or dropped. Let $d_{m,n}^{\text{edge}}(t)$ denote the number of time slots required to process task $k_{m,n}^{\text{edge}}(t)$. For presentation simplicity, we set $d_{m,n}^{\text{edge}}(t) = 0$ if $k_{m,n}^{\text{edge}}(t) = 0$, and $w_{m,n}^{\text{edge}}(1) = 0$. The values of $w_{m,n}^{\text{edge}}(t)$ and $d_{m,n}^{\text{edge}}(t)$ satisfy the following constraints:

$$w_{m,n}^{\text{edge}}(t) = \min\left\{\left[w_{m,n}^{\text{edge}}(t-1) + d_{m,n}^{\text{edge}}(t-1) - 1\right]^{+}, \tau_m - 1\right\},$$
$$m \in \mathcal{M}, n \in \mathcal{N}, t \in \mathcal{T} \setminus \{1\}, \tag{9.9}$$

$$\sum_{t'=t+w_{m,n}^{\text{edge}}(t)}^{t+w_{m,n}^{\text{edge}}(t)+d_{m,n}^{\text{edge}}(t)-1} \frac{\mathbb{1}\left(m \in \mathcal{B}_n(t')\right) f_n^{\text{edge}} \Delta}{\rho_m B_n(t')} \geq \lambda_{m,n}^{\text{edge}}(t),\ m \in \mathcal{M}, n \in \mathcal{N}, t \in \mathcal{T}, \tag{9.10}$$

$$\sum_{t'=t+w_{m,n}^{\text{edge}}(t)}^{t+w_{m,n}^{\text{edge}}(t)+d_{m,n}^{\text{edge}}(t)-2} \frac{\mathbb{1}\left(m \in \mathcal{B}_n(t')\right) f_n^{\text{edge}} \Delta}{\rho_m B_n(t')} < \lambda_{m,n}^{\text{edge}}(t),\ m \in \mathcal{M}, n \in \mathcal{N}, t \in \mathcal{T}. \tag{9.11}$$

The intuition of (9.9) is similar to those for $w_m^{\text{comp}}(t)$ in (9.2) and $w_m^{\text{tran}}(t)$ in (9.6). In inequality (9.10), $t + w_{m,n}^{\text{edge}}(t)$ and $t + w_{m,n}^{\text{edge}}(t) + d_{m,n}^{\text{edge}}(t) - 1$ correspond to the time slots when the processing of task $k_{m,n}^{\text{edge}}(t)$ starts and ends, respectively. Thus, inequalities (9.10) and (9.11) ensure that the processing of task $k_{m,n}^{\text{edge}}(t)$ can be accomplished by time slot $t + w_{m,n}^{\text{edge}}(t) + d_{m,n}^{\text{edge}}(t) - 1$ and cannot be accomplished by $t + w_{m,n}^{\text{edge}}(t) + d_{m,n}^{\text{edge}}(t) - 2$.

For a task $k_m(t)$ that was offloaded to edge node n, it is placed in the associated queue of edge node n at the beginning of time slot $\phi_m(t) \triangleq t + w_m^{\text{tran}}(t) + d_m^{\text{tran}}(t)$. Thus, its queuing delay at edge node n is $w_{m,n}^{\text{edge}}(\phi_m(t))$, and its processing time is $d_{m,n}^{\text{edge}}(\phi_m(t))$. Thus, the delay of task $k_m(t)$ is derived as follows:

$$\text{Delay}_m(t) = \min\left\{w_m^{\text{comp}}(t) + d_m^{\text{local}}(t) + \sum_{n \in \mathcal{N}} y_{m,n}(t)\left(w_{m,n}^{\text{edge}}(\phi_m(t)) + d_{m,n}^{\text{edge}}(\phi_m(t))\right), \tau_m\right\},$$
$$m \in \mathcal{M}, t \in \{t' \mid t' \in \mathcal{T}, k_m(t) \neq 0, x_m(t) = 0\}. \tag{9.12}$$

Note that the above expressions (9.2)–(9.4), (9.6), (9.7), and (9.9)–(9.12) are used for presenting our system model. In practical systems, each mobile device can directly observe the delay of the tasks $\text{Delay}_m(t)$ after those tasks have either been processed or dropped.

9.3.2 Task Offloading Problem

We consider a fully observable system, where the mobile devices can observe the actual values of the state (e.g., queue information, task size). In particular, at the beginning of time slot $t \in \mathcal{T}$, mobile device $m \in \mathcal{M}$ observes its state. If mobile device m has a new task to be processed, then it will choose an action for the task, i.e., whether to offload the task or not, and which edge node to offload the task to. The state and action will result in a cost. We define the cost as the delay of the task if the task has been processed, and define it as a penalty if the task has been dropped. The objective is to find an optimal policy, i.e., a mapping from state to action, that minimizes the expected long-term cost.

9.3.2.1 State

Let $\boldsymbol{H}(t)$ denote the historical load level dynamics of the edge nodes within the previous T^{step} time slots. It is a matrix with size $T^{\text{step}} \times N$ and contains the number of active queues of all edge nodes from time slot $t - T^{\text{step}}$ to $t - 1$. In particular, element (i, j) of matrix $\boldsymbol{H}(t)$ is denoted by $\{\boldsymbol{H}(t)\}_{i,j}$, which corresponds to the number of active queues at edge node j in time slot $t - T^{\text{step}} + i - 1$. That is, $\{\boldsymbol{H}(t)\}_{i,j} = B_j(t - T^{\text{step}} + i - 1)$. We assume that the edge nodes will broadcast the number of active queues at the end of each time slot. The number of active queues can be represented by a maximum of $\lfloor \log_2 M \rfloor + 1$ bits. For example, if $M = 1000$, then a maximum of 10 bits are needed.

At the beginning of time slot $t \in \mathcal{T}$, each mobile device $m \in \mathcal{N}$ observes the task size $\lambda_m(t)$, the number of time slots required to wait for processing and offloading (i.e., $w_m^{\text{comp}}(t)$ and $w_m^{\text{tran}}(t)$), the queue occupancy at all the edge nodes $\boldsymbol{q}_m^{\text{edge}}(t - 1) \triangleq (q_{m,n}^{\text{edge}}(t - 1), n \in \mathcal{N})$, and the historical load level $\boldsymbol{H}(t)$. The state can be represented as follows:

$$s_m(t) = \left(\lambda_m(t), w_m^{\text{comp}}(t), w_m^{\text{tran}}(t), \boldsymbol{q}_m^{\text{edge}}(t - 1), \boldsymbol{H}(t)\right). \tag{9.13}$$

Mobile device m can compute $\boldsymbol{q}_m^{\text{edge}}(t - 1)$ locally based on the tasks that have been offloaded to the edge nodes and the number of active queues at the edge nodes. Let $\mathcal{S}$ denote the finite and discrete space of the state. That is, $\mathcal{S} \triangleq \Lambda \times \{0, 1, \ldots, T\}^2 \times \mathcal{Q} \times \{0, 1, \ldots, M\}^{T^{\text{step}} \times N}$, where $\mathcal{Q}$ denotes the set of available queue occupancy at the edge nodes within T time slots.

9.3.2.2 Action

After mobile device m observes state $s_m(t)$ at the beginning of time slot t, if there is a new task arrival (i.e., $\lambda_m(t) > 0$), then the mobile device will choose an action for the task, denoted by $a_m(t)$. We consider an action space $\mathcal{A} = \{0\} \cup \mathcal{N}$. If the mobile device chooses to process the task locally, then $a_m(t) = 0$. If the mobile device offloads the task to edge node $n \in \mathcal{N}$, then $a_m(t) = n$. That is,

$$a_m(t) = \begin{cases} 0, & x_m(t) = 1, \\ n \text{ such that } y_{m,n}(t) = 1, \; x_m(t) = 0. \end{cases} \tag{9.14}$$

Note that we represent the task offloading decisions $x_m(t)$ and $\boldsymbol{y}_m(t)$ using $a_m(t)$ for the presentation simplicity of the algorithm.

9.3.2.3 Cost

Given the state $\boldsymbol{s}_m(t)$ and action $a_m(t)$, mobile device m will observe a cost after task $k_m(t)$ has either been processed or being dropped due to the task deadline. If task $k_m(t)$ has been processed, then we define the cost as the delay of task $k_m(t)$, i.e., $\text{Delay}_m(t)$. If the task $k_m(t)$ has been dropped, then we set the cost to be a constant penalty C_m, where C_m is larger than the maximum delay τ_m. Specifically, cost function $c_m(\boldsymbol{s}_m(t), a_m(t))$ is defined as follows:

$$c_m(\boldsymbol{s}_m(t), a_m(t)) = \begin{cases} \text{Delay}_m(t), & \text{if task } k_m(t) \text{ has been processed,} \\ C_m, & \text{if task } k_m(t) \text{ has been dropped.} \end{cases} \tag{9.15}$$

In the rest of this chapter, we will use the short form $c_m(t)$ to denote $c_m(\boldsymbol{s}_m(t), a_m(t))$.

Note that we focus on the unknown load level dynamics at the edge nodes. That is, a mobile device does not know the number of tasks offloaded by other mobile devices to an edge node a priori. Thus, when a mobile device makes an offloading decision for a task, it does not know the cost for choosing that decision. This leads to the necessity of using DRL to address the unknown and complex system dynamics.

9.3.2.4 Problem Formulation

For each mobile device $m \in \mathcal{M}$, the objective is to find an optimal policy $\pi_m^* : \mathcal{S} \to \mathcal{A}$ that minimizes the expected long-term cost. That is,

$$\pi_m^* = \arg\text{minimize}_{\pi_m} \;\; \mathbb{E}\left[\sum_{t \in \mathcal{T}} \gamma^{t-1} c_m(t) \,\middle|\, \pi_m \right]$$
$$\text{subject to} \;\; \text{constraints (9.1)–(9.4), (9.6), (9.7), (9.9)–(9.12),} \tag{9.16}$$

where the parameter $\gamma \in (0, 1]$ is a discount factor. This discount factor captures the discounted cost in the future. The expectation $\mathbb{E}[\cdot]$ is with respect to the time-varying parameters, e.g., the task arrivals and the task offloading decisions of other mobile devices.

9.3.3 Deep Q-Learning-Based Algorithm

In this section, we propose a DQL-based task offloading algorithm, under which the mobile devices can make their offloading decisions without knowing the system dynamics a priori. In this algorithm, each mobile device aims at learning a Q-value for each action given each state. The Q-value reveals the expected long-term cost of the mobile device given the state by selecting the associated action. The mapping from the state to the Q-value of each action is characterized by a neural network. With such a mapping, each mobile device can minimize its expected long-term cost by selecting the action with the minimum Q-value under its state.

In the following, we first present the neural network. Then, we propose the DQL-based algorithm.

9.3.3.1 Neural Network

For each mobile device, we use a neural network to characterize the mapping from each state to the Q-value of each action. The neural network contains six layers, as shown in Fig. 9.2. For the neural network of mobile device $m \in \mathcal{M}$, we use $\boldsymbol{\theta}_m$ to denote the parameter vector. This vector contains the biases of all neurons and the weights of all connections from the input layer to A&V layer (see Fig. 9.2).[2] In the following, we present each layer in detail.

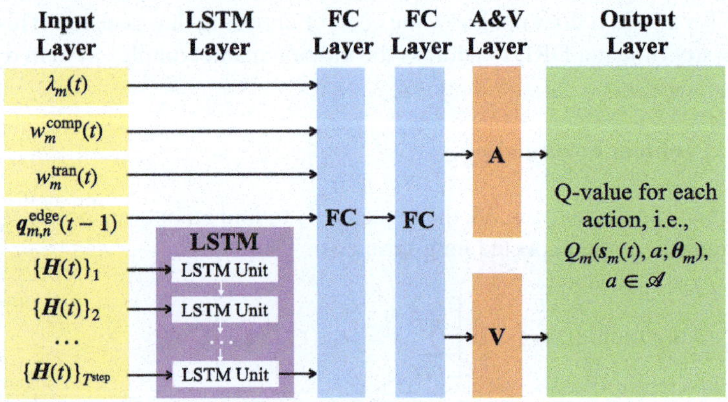

Fig. 9.2 The neural network of mobile device $m \in \mathcal{M}$

[2] The weights of the connections between the A&V layer and the output layer as well as the bias of the neurons in the output layer are given and fixed. Hence, we do not include them in the network parameter vector $\boldsymbol{\theta}_m$, as the vector $\boldsymbol{\theta}_m$ includes the parameters that are adjustable through learning in the DQL-based algorithm.

Input Layer In the neural network of any mobile device $m \in \mathcal{M}$, an input layer is responsible for taking the state (i.e., $s_m(t)$) as input to the neural network.

Long Short-Term Memory (LSTM) Layer After the input layer, the LSTM is in charge of predicting the load level dynamics at the edge nodes in the short-term future. In the LSTM layer, there is an LSTM network that has T^{step} LSTM units. The LSTM units are connected in sequence. Let $\{\boldsymbol{H}_m(t)\}_i$ denote the ith row of the historical load level dynamics matrix $\boldsymbol{H}_m(t)$. The ith LSTM unit takes $\{\boldsymbol{H}_m(t)\}_i$ as an input. These LSTM units can record the variations of the load level dynamics at the edge nodes from $\{\boldsymbol{H}_m(t)\}_1$ to $\{\boldsymbol{H}_m(t)\}_{T^{\text{step}}}$. The output of the last LSTM unit is connected to the next layer in the neural network. This output can reveal the information indicating the load level dynamics in the short-term future.

Fully Connected (FC) Layer There are two FC layers after the LSTM layer. These layers are in charge of mapping from the learned load level dynamics and the state information to the Q-value of each action. Each FC layer has a set of neurons with rectified linear unit (ReLU). In the first FC layer, each neuron has full connections to all neurons (except those related to $\boldsymbol{H}(t)$) in the input layer and the output of the LSTM network. In the second FC layer, each neuron is connected to all neurons in the first FC layer.

A&V Layer and Output Layer We include an A&V layer after the FC layers. This is inspired by dueling DQN technique [14]. The main idea is to separately estimate the *state-value* (i.e., the part of Q-value resulting from the state) and the *advantage-value* for each action (i.e., the part of Q-value resulting from the action). The Q-value of each action given a state is the combination of the associated state-value and the advantage-value of the action.

In particular, in the A&V layer, there are two networks, i.e., network A and network V. Network A contains $|\mathcal{A}|$ neurons, where $|\mathcal{A}| = 1 + N$ is the number of available actions. This network is in charge of estimating the advantage-value for each action $a \in \mathcal{A}$. Recall that $\boldsymbol{\theta}_m$ denotes the biases and weights from the input layer to the A&V layer. Given parameter vector $\boldsymbol{\theta}_m$, let $A_m(\boldsymbol{s}_m(t), a; \boldsymbol{\theta}_m)$ denote the advantage-value of action $a \in \mathcal{A}$ under state $\boldsymbol{s}_m(t) \in \mathcal{S}$. Network V contains one neuron. It is responsible to estimate the state-value. Given parameter vector $\boldsymbol{\theta}_m$, we denote $V_m(\boldsymbol{s}_m(t); \boldsymbol{\theta}_m)$ as the state-value of state $\boldsymbol{s}_m(t)$. Note that parameter vector $\boldsymbol{\theta}_m$ needs to be trained during the DQL-based algorithm.

The output layer determines the Q-value of each action $a \in \mathcal{A}$ given state $\boldsymbol{s}_m(t) \in \mathcal{S}$. Such a Q-value can be determined as follows [14]:

$$Q_m(\boldsymbol{s}_m(t), a; \boldsymbol{\theta}_m) = V_m(\boldsymbol{s}_m(t); \boldsymbol{\theta}_m) + \Bigg(A_m(\boldsymbol{s}_m(t), a; \boldsymbol{\theta}_m) - \frac{1}{|\mathcal{A}|} \sum_{a' \in \mathcal{A}} A_m(\boldsymbol{s}_m(t), a'; \boldsymbol{\theta}_m) \Bigg). \tag{9.17}$$

Specifically, the Q-value of an action is equal to the summation of the corresponding state-value and the additional advantage-value of the action, where the additional advantage-value is with reference to the average advantage-value among all actions.

9.3.3.2 Algorithm Design

We now present our proposed DQL-based task offloading algorithm. To reduce the computational loads at the mobile devices, we consider a setting where the edge nodes help mobile devices to perform training of the neural network. In particular, let $n_m \in \mathcal{N}$ denote the edge node that helps mobile device $m \in \mathcal{M}$ for training. This edge node can be the one that has the maximum transmission capacity with mobile device m. For edge node $n \in \mathcal{N}$, let $\mathcal{M}_n \subset \mathcal{M}$ denote the set of mobile devices for which the edge node performs training, i.e., $\mathcal{M}_n = \{m \mid n_m = n, m \in \mathcal{M}\}$.

Mobile device $m \in \mathcal{M}$ and edge node $n \in \mathcal{N}$ execute Algorithms 1 and 2, respectively. In particular, mobile device m collects experience $(s_m(t), a_m(t), c_m(t), s_m(t+1))$ for $t \in \mathcal{T}$ through the interaction with the MEC system. The associated edge node n_m maintains an experience replay D_m, which stores the experience of mobile device m. For presentation simplicity, we use the experience t of mobile device m to refer to $(s_m(t), a_m(t), c_m(t), s_m(t+1))$. Edge node n_m learns the mapping from each state to the Q-value of each action using the experience. Specifically, edge node n_m keeps two neural networks for mobile device m, including an evaluation network Net_m and a target network $Target\_Net_m$. Both neural networks follow the structure presented in Sect. 9.3.3.1. Nevertheless, they have different parameter vectors, i.e., $\boldsymbol{\theta}_m$ for Net_m and $\boldsymbol{\theta}_m^-$ for $Target\_Net_m$, and different functionalities. Edge node n_m aims at training the evaluation network Net_m to characterize the mapping from state to Q-values. During the training process, edge node n_m uses Net_m for action selection. It uses the target network $Target\_Net_m$ to approximate the expected long-term cost of each action given any state. We call the output of the target network as target Q-value. Edge node n_m will update the parameter vector of Net_m by minimizing the gap between the target Q-value and the Q-value under Net_m.

Algorithm 1 at Mobile Device $m \in \mathcal{M}$ The algorithm iterates for E episodes. At the beginning of each episode, mobile device $m \in \mathcal{M}$ initializes the state, i.e.,

$$s_m(1) = (\lambda_m(1), w_m^{\text{comp}}(1), w_m^{\text{tran}}(1), \boldsymbol{q}_m^{\text{edge}}(0), \boldsymbol{H}(1)). \tag{9.18}$$

We set $q_{m,n}^{\text{edge}}(0) = 0$ for all $n \in \mathcal{N}$ and set $H(1)$ as a zero matrix with size $T^{\text{step}} \times N$.

At the beginning of time slot $t \in \mathcal{T}$, if mobile device m has a new task arrival $k_m(t)$, then it will request the recent parameter vector of network Net_m, i.e., $\boldsymbol{\theta}_m$, through sending a parameter_request to edge node n_m. Note that in practical systems, the mobile device can choose not to request the parameter vector in every time slot with new task arrivals in order to reduce the signaling overhead. Although reducing such a frequency may degrade the convergence rate of the proposed

Algorithm 1 Deep Q-learning-based algorithm at mobile device $m \in \mathcal{M}$

for episode = $1, 2, \ldots, E$ **do**
  Initialize state $\boldsymbol{s}_m(1)$;
  for time slot $t \in \mathcal{T}$ **do**
    if mobile device m has a new task arrival $k_m(t)$ **then**
      Send a parameter_request to edge node n_m;
      Receive network parameter vector $\boldsymbol{\theta}_m$;
      Select an action $a_m(t)$ according to (9.19);
    end if
    Observe the next state $\boldsymbol{s}_m(t+1)$;
    Observe a set of costs $\{c_m(t'), t' \in \widetilde{\mathcal{T}}_{m,t}\}$;
    for each task $k_m(t')$ with $t' \in \widetilde{\mathcal{T}}_{m,t}$ **do**
      Send $(\boldsymbol{s}_m(t'), a_m(t'), c_m(t'), \boldsymbol{s}_m(t'+1))$ to edge node n_m;
    end for
  end for
end for

algorithm, we have evaluated empirically that the degradation of the convergence rate is minimal if the frequency is maintained within a certain range. For example, with the system setting in Sect. 9.3.4, requesting the parameter vector every 100 time slots leads to a similar convergence rate as requesting it every time slot.

With a probability of ϵ, mobile device m randomly selects an action in set $\mathcal{A}$. With a probability of $1-\epsilon$, given the recent state $\boldsymbol{s}_m(t)$, it selects an action that leads to the minimum Q-value according to $\boldsymbol{\theta}_m$. That is,

$$a_m(t) = \begin{cases} \text{a random action from } \mathcal{A}, & \text{with a probability of } \epsilon, \\ \arg\min_{a \in \mathcal{A}} Q_m(\boldsymbol{s}_m(t), a; \boldsymbol{\theta}_m), & \text{with a probability of } 1-\epsilon. \end{cases} \tag{9.19}$$

Then, at the beginning of time slot $t+1$, mobile device m can observe the next state $\boldsymbol{s}_m(t+1)$. Note that in our system setting, the processing of task $k_m(t)$ does not need to be accomplished within time slot t. Thus, at the beginning of time slot $t+1$, the cost $c_m(t)$ associated with task $k_m(t)$ may have not been observed. Due to the same reason, mobile device m may observe a set of costs associated with some tasks $k_m(t')$ arrived in time slot $t' \leq t$. Thus, we denote $\widetilde{\mathcal{T}}_{m,t} \subset \mathcal{T}$ as the set of time slots such that the tasks associated with those time slots have been either processed or dropped within time slot t. That is,

$$\widetilde{\mathcal{T}}_{m,t} = \{t' \mid t' = 1, 2, \ldots, t,\ \lambda_m(t') > 0,\ t' + \text{Delay}_m(t') - 1 = t\}, \tag{9.20}$$

where set $\widetilde{\mathcal{T}}_{m,t}$ can be an empty set for some $m \in \mathcal{M}$ and $t \in \mathcal{T}$. At the beginning of time slot $t+1$, mobile device m can observe a set of costs $\{c_m(t'), t' \in \widetilde{\mathcal{T}}_{m,t}\}$. For each task $k_m(t')$ with $t' \in \widetilde{\mathcal{T}}_{m,t}$, mobile device m sends the associated experience $(\boldsymbol{s}_m(t'), a_m(t'), c_m(t'), \boldsymbol{s}_m(t'+1))$ to edge node n_m. To reduce the signaling overhead, we consider a setting where mobile device m does not send matrices $\boldsymbol{H}(t')$ and $\boldsymbol{H}(t'+1)$ in states $\boldsymbol{s}_m(t')$ and $\boldsymbol{s}_m(t'+1)$. This is feasible because

Algorithm 2 Deep Q-learning-based algorithm at edge node $n \in \mathcal{N}$

1: Initialize neural network Net_m with random $\boldsymbol{\theta}_m$ for $m \in \mathcal{M}_n$;
2: Initialize neural network $Target\_Net_m$ with random $\boldsymbol{\theta}_m^-$ for $m \in \mathcal{M}_n$;
3: Initialize experience replay D_m for $m \in \mathcal{M}_n$ and Count $\leftarrow 0$;
4: **while** True **do**
5:     **if** receive a parameter_request from $m \in \mathcal{M}_n$ **then**
6:         Send the recent parameter vector $\boldsymbol{\theta}_m$ to mobile device m;
7:     **end if**
8:     **if** receive an experience $(\boldsymbol{s}_m(t), a_m(t), c_m(t), \boldsymbol{s}_m(t+1))$ from $m \in \mathcal{M}_n$ **then**
9:         Store $(\boldsymbol{s}_m(t), a_m(t), c_m(t), \boldsymbol{s}_m(t+1))$ in experience replay D_m;
10:        Sample a set of experiences (denoted by $\mathcal{I}$) from D_m;
11:        **for** each experience $i \in \mathcal{I}$ **do**
12:            Obtain experience $(\boldsymbol{s}_m(i), a_m(i), c_m(i), \boldsymbol{s}_m(i+1))$;
13:            Compute $\hat{Q}_{m,i}^{\text{Target}}$ according to (9.23);
14:        **end for**
15:        Set vector $\hat{\boldsymbol{Q}}_m^{\text{Target}} \leftarrow (\hat{Q}_{m,i}^{\text{Target}}, i \in \mathcal{I})$;
16:        Update $\boldsymbol{\theta}_m$ to minimize $L(\boldsymbol{\theta}_m, \hat{\boldsymbol{Q}}_m^{\text{Target}})$ in (9.21);
17:        Count $\leftarrow$ Count $+ 1$;
18:        **if** mod(Count, Replace_Threshold) $= 0$ **then**
19:            $\boldsymbol{\theta}_m^- \leftarrow \boldsymbol{\theta}_m$;
20:        **end if**
21:     **end if**
22: **end while**

we have assumed that the edge nodes broadcast their load level dynamics in each time slot.

Algorithm 2 at Edge Node $n \in \mathcal{N}$ Edge node $n \in \mathcal{N}$ first initializes neural networks Net_m and $Target\_Net_m$ and experience replay D_m for mobile device $m \in \mathcal{M}_n$. Then, it will wait for the messages from the mobile devices.

If edge node n receives a parameter_request from mobile device $m \in \mathcal{M}_n$, then it will forward the recent parameter vector $\boldsymbol{\theta}_m$ to mobile device m. If edge node n receives an experience from mobile device $m \in \mathcal{M}_n$, then it will store the experience in the experience replay D_m. After that, edge node n will update the parameter vector $\boldsymbol{\theta}_m$ of network Net_m according to steps 10–20 in Algorithm 2. The edge node first randomly samples I experiences from D_m. Let $\mathcal{I}$ denote the set of sampled experiences. For each experience $i \in \mathcal{I}$, edge node n will compute a target Q-value $\hat{Q}_{m,i}^{\text{Target}}$ (to be explained in the next paragraph) and update $\boldsymbol{\theta}_m$ by minimizing the following loss function:

$$L\left(\boldsymbol{\theta}_m, \hat{\boldsymbol{Q}}_m^{\text{Target}}\right) = \frac{1}{I} \sum_{i \in \mathcal{I}} \left(\hat{Q}_{m,i}^{\text{Target}} - Q_m(\boldsymbol{s}_m(i), a_m(i); \boldsymbol{\theta}_m) \right)^2, \tag{9.21}$$

where $\hat{\boldsymbol{Q}}_m^{\text{Target}} = (\hat{Q}_{m,i}^{\text{Target}}, i \in \mathcal{I})$. This loss function captures the difference between the target Q-value and the output of network Net_m for each experience $i \in$

$\mathcal{I}$. The edge node minimizes the loss function using backpropagation (see Section 6 in [33]).

Edge node n determines the target Q-value $\hat{Q}_{m,i}^{\text{Target}}$ for $i \in \mathcal{I}$ using double DQN technique [13]. This technique can improve the approximation of the expected long-term cost when compared with the conventional method (e.g., [12]). To determine the target Q-value, we denote a_i^{Next} as the action that leads to the minimum Q-value given the next state $\boldsymbol{s}_m(i+1)$ under network Net_m, i.e.,

$$a_i^{\text{Next}} = \arg\min_{a \in \mathcal{A}} Q_m(\boldsymbol{s}_m(i+1), a; \boldsymbol{\theta}_m). \tag{9.22}$$

The target Q-value $\hat{Q}_{m,i}^{\text{Target}}$ is computed as follows:

$$\hat{Q}_{m,i}^{\text{Target}} = c_m(i) + \gamma Q_m\left(\boldsymbol{s}_m(i+1), a_i^{\text{Next}}; \boldsymbol{\theta}_m^-\right). \tag{9.23}$$

Intuitively, the target Q-value is equal to the summation of the cost in experience i and the discount factor multiplied by the Q-value of action a_i^{Next} given the next state $\boldsymbol{s}_m(i+1)$ under network *Target_Net*$_m$. This value essentially approximates the expected long-term cost of action $a_m(i)$ given state $\boldsymbol{s}_m(i)$.

To keep the parameter vector $\boldsymbol{\theta}_m^-$ of *Target_Net*$_m$ up-to-date, edge node n updates $\boldsymbol{\theta}_m^-$ every several number of training rounds by copying the parameter vector $\boldsymbol{\theta}_m$ of network Net_m. Such updates make the target Q-value (which is derived based on *Target_Net*$_m$) a more accurate approximation of the expected long-term cost. We use Replace_Threshold to denote the corresponding number of training rounds, where mod$(\cdot)$ is the modulo operator in step 18 in Algorithm 2.

Discussion on Convergence Despite that we are able to prove the convergence of Q-learning algorithm, the convergence of a DQL-based algorithm is still an open problem. This is because the neural network is essentially an approximation of the mapping from state to Q-values. Due to such an approximation, the convergence may no longer be guaranteed. In this chapter, we empirically evaluate the convergence performance of the proposed algorithm in Sect. 9.3.4.

9.3.4 Performance Evaluation

We consider five edge nodes and 50 mobile devices. Table 9.1 shows the parameter settings of the MEC system and the hyperparameters of the proposed algorithm. As shown in Table 9.1, we consider a setting where each task has a deadline $\tau_m = 10$ time slots regardless of the task size. For example, for a video segment decoding task in live streaming, a video segment should always be decoded before a certain deadline to avoid rebuffering, where the deadline is independent of the number of image frames in the video segment. We set the penalty for dropped tasks C_m to 20

Table 9.1 Parameter settings of the MEC system

Parameter	Value	Parameter	Value
Δ	0.1 second	$f_m^{\text{device}}, m \in \mathcal{M}$	2.5 GHz [34]
$\lambda_m(t), m \in \mathcal{M}, t \in \mathcal{T}$	Discrete uniform distribution over set {2.0, 2.1, ..., 5.0} Mbits [35]	$r_{m,n}^{\text{tran}}, m \in \mathcal{M}, n \in \mathcal{N}$	14 Mbps [36]
$\rho_m, m \in \mathcal{M}$	0.297 Gigacycles per Mbits [35]	$f_n^{\text{edge}}, n \in \mathcal{N}$	41.8 GHz [34]
$\tau_m, m \in \mathcal{M}$	10 time slots (i.e., 1 second [37])	$C_m, m \in \mathcal{M}$	20
Task arrival probability	0.3	Discount factor	0.9
Learning rate	0.001	Batch size	16
ϵ	Decrement from 1 to 0.01		

for $m \in \mathcal{M}$, where this value is twice as large as the maximum delay of a processed task (i.e., 10 time slots). Such a penalty setting makes the cost of a dropped task be always larger than the cost of a processed task, under which the proposed DQL-based algorithm will avoid tasks being dropped by optimizing the task offloading decision. In addition, we consider a setting where the task arrival probability is a constant value [29]. Despite that the task arrival probability is fixed across time, the number of tasks offloaded to an edge node can be time varying and is unknown to any particular mobile device, due to the time-varying and unknown offloading decisions of other mobile devices. This leads to the unknown load level dynamics at the edge nodes and the necessity of using our proposed DQL-based approach.

In the following, we evaluate the algorithm convergence. Then, we compare the performance of our proposed algorithm with some existing algorithms.

9.3.4.1 Algorithm Convergence

Figure 9.3 shows the convergence of the average cost among mobile devices of our proposed algorithm under different hyperparameters. For comparison, we show the random policy, where mobile devices randomly select their actions.

Figure 9.3(a) shows the algorithm convergence under different values of batch size, i.e., the number of experience sampled in one training round (i.e., I). When the batch size is increased from 2 to 8, the average cost converges to a lower value. Further increasing the batch size to 32 does not make a significant difference. Thus, a small batch size (e.g., 8) is sufficient for achieving a satisfactory convergence.

Figure 9.3(b) shows the algorithm convergence under different values of learning rate, which is the step size for updating network Net_m. As shown in the figure, when the learning rate is equal to 0.001, the average cost converges relatively fast and converges to a smaller value when compared with the other values of learning rate.

On the other hand, in practical systems, the task arrival probability can be non-stationary. Under such a scenario, once the environment has changed, the proposed

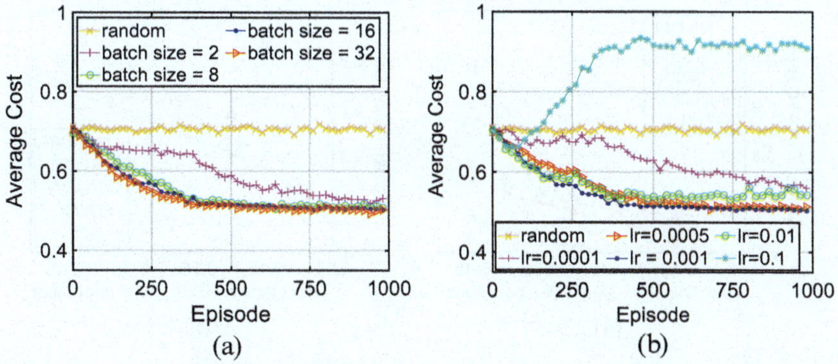

Fig. 9.3 Algorithm convergence under different: (**a**) batch size and (**b**) learning rate (denoted by "lr")

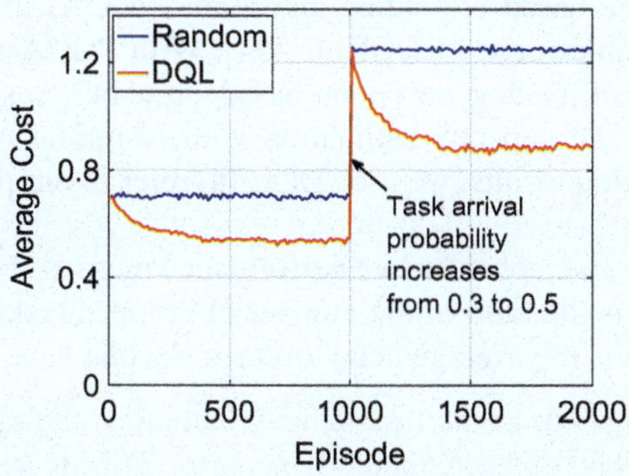

Fig. 9.4 An illustration of the algorithm performance with non-stationary task arrival probability

algorithm can adapt to it by resetting the probability of random exploration to be one in order to enable the random exploration again. Figure 9.4 shows an example of the performance of the proposed DQL-based algorithm with non-stationary task arrival probability. In this simulation, at around 1000 episodes, the task arrival probability increases from 0.3 to 0.5. Hence, the average cost of both the random policy and our proposed DQL-based algorithm are changed accordingly. As the episodes proceed, our proposed algorithm gradually adapts to the change of task arrival probability and converges again. We will leave it as future work to design an efficient algorithm for addressing frequent environmental changes. Candidate approaches include concept drift detection [38] and non-stationary reinforcement learning [39, 40].

9.3.4.2 Method Comparison

We compare our proposed algorithm with several benchmark methods. These include no offloading (denoted by No Offl.), random offloading (denoted by R.

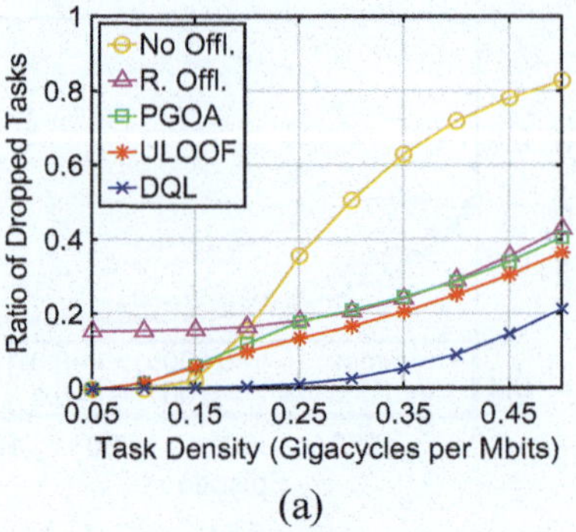

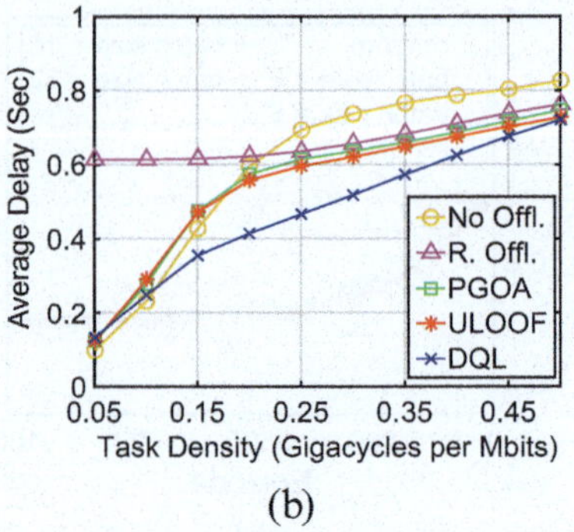

Fig. 9.5 Performance with different task densities: (**a**) ratio of dropped tasks and (**b**) average delay

Offl.), potential game-based offloading algorithm (PGOA) in [31], and user-level online offloading framework (ULOOF) in [34]. With PGOA and ULOOF, mobile devices make their offloading decisions based on a best response algorithm for potential game and the capacity estimation with historical observations, respectively. In the simulation results, we use "DQL" to refer to our proposed DQL-based algorithm.

In the simulations, we consider two performance metrics. The first is the ratio of dropped tasks. This is the ratio of the number of dropped tasks to the total number of tasks. The second is the average delay of the tasks that have been processed.

Task Density In Fig. 9.5, as the task density increases, the ratio of dropped tasks and the average delay of each method increase. This is because a larger task density implies a higher computational requirement of each task. As the task density increases from 0.05 to 0.25 Gigacycles per Mbits, the ratio of dropped tasks and average delay of our proposed algorithm increase less drastically than those of the benchmark methods. When the density is 0.25 Gigacycles per Mbits, our proposed algorithm maintains a ratio of dropped tasks of around 0.01 and an average delay of 0.47 second. As the task density further increases to 0.5 Gigacycles per Mbits, although all methods have a similar average delay, our proposed algorithm can reduce the ratio of dropped tasks by 41.4%–74.1% when compared with the benchmark methods.

Processing Capacity of Edge Node As shown in Fig. 9.6, under various values of the processing capacity of each edge node, our proposed algorithm can reduce the ratio of dropped tasks and the average delay when compared with the benchmark methods. The reduction of the ratio of dropped tasks is especially significant when the processing capacity of each edge node is small. When the processing capacity is 15 GHz, the proposed algorithm reduces the ratio of dropped tasks by at least 57.0% and reduces the average delay by at least 9.4% when compared with the benchmark methods. As the processing capacity further increases, both performance metrics converge, because further increasing the capacity does not reduce the delay of those

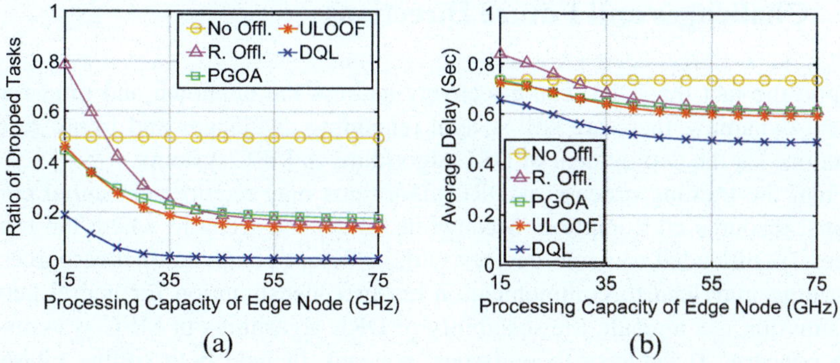

Fig. 9.6 Performance with different processing capacities of edge nodes: (**a**) ratio of dropped tasks and (**b**) average delay

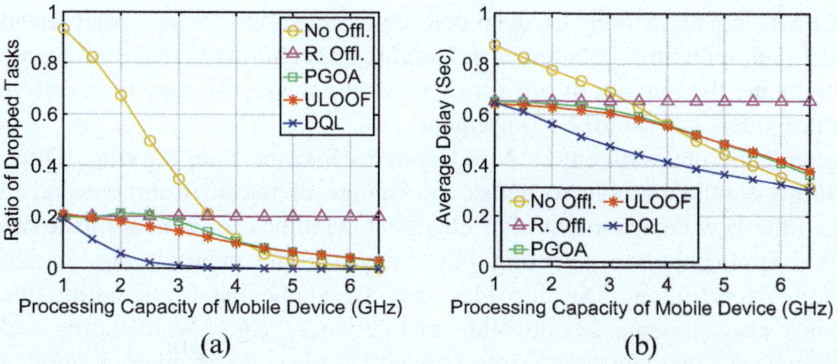

Fig. 9.7 Performance with different processing capacities of mobile devices: (**a**) ratio of dropped tasks and (**b**) average delay

tasks offloaded due to the limited transmission capacity. The converged ratio of dropped tasks and the average delay of our proposed algorithm is at least 84.3% and 17.2% less than those of the benchmark methods, respectively.

Processing Capacity of Mobile Devices In Fig. 9.7, as the processing capacity of each mobile device increases, our proposed algorithm has a more significant decrease in terms of the ratio of dropped tasks and average delay when compared with PGOA and ULOOF. When the processing capacity increases to 3.5 GHz, our proposed algorithm achieves a ratio of dropped tasks of 0.007, which is 93.9%–96.5% lower than those of the benchmark methods. Meanwhile, our algorithm achieves an average delay that is 31.4% and 29.4% lower than those of PGOA and ULOOF, respectively. As the processing capacity of each mobile device further increases, processing a task locally becomes optimal. Thus, our proposed algorithm tends to choose local processing and achieves a similar performance as no offloading.

9.4 Challenges and Future Directions

Despite the fact that DRL can effectively address the unknown and time-varying system dynamics, there are still several remaining challenges and future research directions for the deployment of DRL algorithms in MEC systems.

First, the training process of DRL algorithms may require substantial computational resource consumption. Meanwhile, under the scenario where the training process is offloaded to some devices with sufficient computational resources, the offloading may lead to communication resource consumption for neural network transmission. As a result, the scalability of DRL algorithms in MEC systems may be a concern. To address these issues, we may include both offline phase and online phase for DRL algorithms, where the offline phase is performed offline with powerful devices (e.g., cloud server). Transfer learning techniques [41] may be utilized for the online phase to make the neural network quickly adapt to the real-world environment. Moreover, deep compression techniques [42], such as network pruning (i.e., reducing the number of weights in neural networks) and quantization (i.e., reducing the number of bits for representing a weight), may be used to reduce the communication resource requirement.

Second, the environment in MEC systems may be time varying. Thus, DRL algorithms should be able to detect the change of the environment and quickly adapt to the new environment after changing. Methods for concept drift detection [38] are applicable for detecting the change of the environment. In addition, techniques for lifelong learning [43] may be applicable for handling the non-stationary environments. Meanwhile, existing works, e.g., [39, 40], proposed DRL algorithms for non-stationary reinforcement learning, which may be applicable to MEC systems.

Third, in MEC systems with a large number of mobile devices and edge nodes, there are potentials for the mobile devices and edge nodes to cooperate to learn the optimal policy through their interaction with the environments. In other words, the mobile devices and edge nodes may cooperatively train the neural networks in the DRL algorithms. Such a collaboration can alleviate the requirements for computational resources and improve the resource efficiency. Federated learning techniques [44] can enable the collaboration among mobile devices and edge nodes for neural network training and hence may be incorporated in DRL algorithms.

9.5 Conclusion

In this chapter, we provided an overview of the DRL algorithms for MEC systems. We introduced DRL fundamentals and then presented a case study on task offloading in MEC systems. In this case study, we focused on the unknown and time-varying load level dynamics at the edge nodes and proposed a DQL-based algorithm that enables the mobile devices to make task offloading decisions in a decentralized fashion. We conducted simulations and showed that the proposed algorithm can reduce the task delay and ratio of dropped tasks. Finally, we outlined the challenges and future research directions for DRL algorithms in MEC systems.

References

1. Y. Mao, C. You, J. Zhang, K. Huang, K.B. Letaief, A survey on mobile edge computing: the communication perspective. IEEE Commun. Surv. Tutorials **19**(4), 2322–2358, Fourth quarter (2017)
2. P. Porambage, J. Okwuibe, M. Liyanage, M. Ylianttila, T. Taleb, Survey on multi-access edge computing for Internet of things realization. IEEE Commun. Surv. Tutorials **20**(4), 2961–2991 (2018)
3. F. Bonomi, R. Milito, J. Zhu, S. Addepalli, Fog computing and its role in the Internet of things, in *Proc. ACM SIGCOMM Workshop on Mobile Cloud Computing (MCC), Helsinki*, August 2012
4. Z. Sanaei, S. Abolfazli, A. Gani, R. Buyya, Heterogeneity in mobile cloud computing: taxonomy and open challenges. IEEE Commun. Surv. Tutorials **16**(1), 369–392, First quarter (2014)
5. T. Qiu, J. Chi, X. Zhou, Z. Ning, M. Atiquzzaman, D.O. Wu, Edge computing in industrial Internet of things: architecture, advances and challenges. IEEE Commun. Surv. Tutorials **22**(4), 2462–2488, Fourth quarter (2020)
6. P. Ranaweera, A.D. Jurcut, M. Liyanage, Survey on multi-access edge computing security and privacy. IEEE Commun. Surv. Tutorials **23**(2), 1078–1124, Second quarter (2021)
7. T. Chen, Q. Ling, G.B. Giannakis, An online convex optimization approach to proactive network resource allocation. IEEE Trans. Signal Process. **65**(24), 6350–6364 (2017)
8. H. Shah-Mansouri, V.W.S. Wong, Hierarchical fog-cloud computing for IoT systems: a computation offloading game. IEEE Internet Things J. **5**(4), 3246–3257 (2018)
9. V. François-Lavet, P. Henderson, R. Islam, M.G. Bellemare, J. Pineau, An introduction to deep reinforcement learning. Found. Trends Mach. Learn. **11**(3–4), 219–354 (2018)
10. X. Wang, Y. Han, V.C.M. Leung, D. Niyato, X. Yan, X. Chen, Convergence of edge computing and deep learning: a comprehensive survey. IEEE Commun. Surv. Tutorials **22**(2), 869–904, Second quarter (2020)
11. R.S. Sutton, A.G. Barto, *Reinforcement Learning: An Introduction*, 2nd edn. (MIT Press, Cambridge, MA, 2018)
12. V. Mnih, K. Kavukcuoglu, D. Silver, A.A. Rusu, J. Veness, M.G. Bellemare, A. Graves, M. Riedmiller, A.K. Fidjeland, G. Ostrovski, Human-level control through deep reinforcement learning. Nature **518**(7540), 529–533 (2015)
13. H. van Hasselt, A. Guez, D. Silver, Deep reinforcement learning with double Q-learning, in *Proc. AAAI Conf. on Artificial Intelligence, Phoenix, AZ*, May 2016
14. Z. Wang, T. Schaul, M. Hessel, H. van Hasselt, M. Lanctot, N. de Freitas, Dueling network architectures for deep reinforcement learning, in *Proc. Int'l Conf. on Machine Learning (ICML), New York City, NY*, June 2016
15. M.G. Bellemare, W. Dabney, R. Munos, A distributional perspective on reinforcement learning, in *Proc. Int'l Conf. on Machine Learning (ICML), Sydney*, August 2017
16. T.P. Lillicrap, J.J. Hunt, A. Pritzel, N. Heess, T. Erez, Y. Tassa, D. Silver, D. Wierstra, Continuous control with deep reinforcement learning. Preprint, arXiv:1509.02971, July 2019
17. G. Barth-Maron, M.W. Hoffman, D. Budden, W. Dabney, D. Horgan, T.B. Dhruva, A. Muldal, N. Heess, T. Lillicrap, Distributed distributional deterministic policy gradients, in *Proc. Int'l Conf. on Learning Representations (ICLR), Vancouver*, April 2018
18. V. Mnih, A.P. Badia, M. Mirza, A. Graves, T. Lillicrap, T. Harley, D. Silver, K. Kavukcuoglu, Asynchronous methods for deep reinforcement learning, in *Proc. Int'l Conf. on Machine Learning (ICML), New York City, NY*, June 2016

19. Z. Wang, V. Bapst, N. Heess, V. Mnih, R. Munos, K. Kavukcuoglu, N. de Freitas, Sample efficient actor-critic with experience replay. Preprint, arXiv:1611.01224, July 2017
20. T. Haarnoja, A. Zhou, P. Abbeel, S. Levine, Soft actor-critic: off-policy maximum entropy deep reinforcement learning with a stochastic actor, in *Proc. Int'l Conf. on Machine Learning (ICML), Stockholm*, June 2018
21. S. Fujimoto, H. Hoof, D. Meger, Addressing function approximation error in actor-critic methods, in *Proc. Int'l Conf. on Machine Learning (ICML), Stockholm*, June 2018
22. J. Schulman, S. Levine, P. Abbeel, M. Jordan, P. Moritz, Trust region policy optimization, in *Proc. Int'l Conf. on Machine Learning (ICML), Lille*, June 2015
23. J. Schulman, F. Wolski, P. Dhariwal, A. Radford, O. Klimov, Proximal policy optimization algorithms. Preprint, arXiv:1707.06347, August 2017
24. C.B. Browne, E. Powley, D. Whitehouse, S.M. Lucas, P.I. Cowling, P. Rohlfshagen, S. Tavener, D. Perez, S. Samothrakis, S. Colton, A survey of Monte Carlo tree search methods. IEEE Trans. Comput. Intel. AI **4**(1), 1–43 (2012)
25. D. Silver et al., Mastering the game of Go with deep neural networks and tree search. Nature **529**(7587), 484–489 (2016)
26. A. Plaat, W. Kosters, M. Preuss, Model-based deep reinforcement learning for high-dimensional problems, a survey. Preprint, arXiv:2008.05598, December 2020
27. N.C. Luong, D.T. Hoang, S. Gong, D. Niyato, P. Wang, Y.-C. Liang, D.I. Kim, Applications of deep reinforcement learning in communications and networking: a survey. IEEE Commun. Surv. Tutorials **21**(4), 3133–3174, Fourth quarter (2019)
28. M. Tang, V.W.S. Wong, Deep reinforcement learning for task offloading in mobile edge computing systems. IEEE Trans. Mobile Comput. **21**(6), 1985–1997 (2020)
29. J. Liu, Y. Mao, J. Zhang, K.B. Letaief, Delay-optimal computation task scheduling for mobile-edge computing systems, in *Proc. IEEE Int'l Symp. on Information Theory (ISIT), Barcelona*, July 2016
30. X. Lyu, W. Ni, H. Tian, R.P. Liu, X. Wang, G.B. Giannakis, A. Paulraj, Distributed online optimization of fog computing for selfish devices with out-of-date information. IEEE Trans. Wirel. Commun. **17**(11), 7704–7717 (2018)
31. L. Yang, H. Zhang, X. Li, H. Ji, V. Leung, A distributed computation offloading strategy in small-cell networks integrated with mobile edge computing. IEEE/ACM Trans. Netw. **26**(6), 2762–2773 (2018)
32. A.K. Parekh, R.G. Gallager, A generalized processor sharing approach to flow control in integrated services networks: the single-node case. IEEE/ACM Trans. Netw. **1**(3), 344–357 (1993)
33. I. Goodfellow, Y. Bengio, A. Courville, *Deep Learning* (MIT Press, Cambridge, 2016)
34. J.L.D. Neto, S.-Y. Yu, D.F. Macedo, M.S. Nogueira, R. Langar, S. Secci, ULOOF: a user level online offloading framework for mobile edge computing. IEEE Trans. Mobile Comput. **17**(11), 2660–2674 (2018)
35. C. Wang, C. Liang, F.R. Yu, Q. Chen, L. Tang, Computation offloading and resource allocation in wireless cellular networks with mobile edge computing. IEEE Trans. Wirel. Commun. **16**(8), 4924–4938 (2017)
36. Speedtest Intelligence, Speedtest global index: Canada average mobile upload speed based on March 2021 data, https://www.speedtest.net/reports/canada/. Accessed 25 June 2021
37. X. Lyu, H. Tian, W. Ni, Y. Zhang, P. Zhang, R.P. Liu, Energy-efficient admission of delay-sensitive tasks for mobile edge computing. IEEE Trans. Commun. **66**(6), 2603–2616 (2018)
38. J. Lu, A. Liu, F. Dong, F. Gu, J. Gama, G. Zhang, Learning under concept drift: a review. IEEE Trans. Knowl. Data Eng. **31**(12), 2346–2363 (2019)
39. A. Xie, J. Harrison, C. Finn, Deep reinforcement learning amidst lifelong non-stationarity. Preprint, arXiv:2006.10701, June 2020
40. V. Lomonaco, K. Desai, E. Culurciello, D. Maltoni, Continual reinforcement learning in 3D non-stationary environments, in *Proc. IEEE/CVF Conference on Computer Vision and Pattern Recognition Workshops*, June 2020

41. F. Zhuang, Z. Qi, K. Duan, D. Xi, Y. Zhu, H. Zhu, H. Xiong, Q. He, A comprehensive survey on transfer learning. Proc. IEEE **109**(1), 43–76 (2021)
42. S. Han, H. Mao, W.J. Dally, Deep compression: compressing deep neural networks with pruning, trained quantization and Huffman coding, in *Proc. Int'l Conf. on Machine Learning (ICML), New York City, NY*, June 2016
43. Z. Chen, B. Liu, Lifelong machine learning. Synth. Lect. Artif. Intell. Mach. Learn. **12**(3), 1–207 (2018)
44. W.Y.B. Lim, N.C. Luong, D.T. Hoang, Y. Jiao, Y.-C. Liang, Q. Yang, D. Niyato, C. Miao, Federated learning in mobile edge networks: a comprehensive survey. IEEE Commun. Surv. Tutorials **22**(3), 2031–2063, Third quarter (2020)

Chapter 10
Mobile Computation Offloading with Hard Task Completion Times

Peyvand Teymoori, Arvin Hekmati, Terence D. Todd, Dongmei Zhao, and George Karakostas

10.1 Introduction

Mobile devices (MDs) are continuing to become more pervasive as personal computing platforms. This trend is coinciding with significant increases in mobile application features that benefit from tight interactions with fixed computation infrastructure. Due to their limited physical size however, mobile devices are inherently resource-constrained, especially from an energy and computational viewpoint. This has motivated a wide variety of recent research on mobile energy efficiency [1].

Mobile cloud computing has been introduced to help alleviate some of these shortcomings and to support the ever-increasing computation and storage demands for MDs [2, 3]. It has been estimated that tens of billions of cloud-based network edge devices will be deployed in the future to satisfy mobile demands. This will provide significant resources for performing computation-intensive and latency-critical mobile-centric tasks. Mobile computation offloading has been proposed as a way of decreasing MD energy use by dynamically offloading job execution to infrastructure-based cloud servers [4–8]. Access to the cloud servers is performed by the MD using wireless transmission.

P. Teymoori · T. D. Todd · D. Zhao (✉)
Department of Electrical and Computer Engineering, McMaster University, Hamilton, ON, Canada
e-mail: teymoorp@mcmaster.ca; todd@mcmaster.ca; dzhao@mcmaster.ca

A. Hekmati
Department of Computer Science, University of Southern California, Los Angeles, CA, USA
e-mail: hekmati@usc.edu

G. Karakostas
Department of Computing and Software, McMaster University, Hamilton, ON, Canada
e-mail: karakos@mcmaster.ca

L. Cai et al. (eds.), *Broadband Communications, Computing, and Control for Ubiquitous Intelligence*, Wireless Networks,
https://doi.org/10.1007/978-3-030-98064-1_10

Various architectures have been proposed for mobile computation offloading. Reference [9] originally proposed an architecture known as MAUI, which controls computation offloading for runtime .NET applications by formulating the offloading problem as a linear program. A similar architecture for Android applications has also been proposed in [4]. Computation offloading typically reduces execution energy consumption but usually incurs communication energy costs since the tasks to be executed (and their inputs/outputs) must be transferred over wireless network links. This may also incur additional latency. The tradeoff between energy saving and computing performance for single-device offloading has been studied extensively. For a single user offloading its entire application to the cloud, the tradeoff between energy saving and computing performance was studied in [1, 10, 11].

References [12–15] considered multi-user scenarios with a single application or user task, where the entire application is offloaded to the cloud. Unlike whole-application offloading, [4, 9, 16–18] have considered partitioning applications into multiple offloaded tasks. This work raises the issues of how to perform the partitioning, and which tasks should be offloaded given that there may be data dependencies between the tasks. There is also work that models MCO as a competitive game. In this type of system there may be resource contention among the devices, such as the case where they share communication channels and/or cloud servers [19–21].

In this chapter, we study MCO where job completion times are subjected to hard deadline constraints, i.e., the deadline constraints should never be violated. This objective will become increasingly important as mobile applications become more sophisticated and interact more closely with cloud job execution [22]. Hard deadlines are often difficult to achieve in mobile networks due to the randomness of the wireless channels used for the mobile/cloud data interactions. In harsh wireless conditions, for example, complete channel outage can even occur over extended time periods. In this chapter, we study the straightforward approach of permitting concurrent local and cloud offload execution when a job completion deadline must be respected. This is in contrast to the conventional computation offloading model where job execution is either local or remote [23, 24]. As is the case in conventional computation offloading, the objective is to reduce the MD energy needed for job execution. The chapter studies this problem for Markovian wireless channel models [25].

Online optimal algorithms are introduced that reduce the remote and local execution overlap so that energy wastage is minimized. Two job uploading schemes are introduced, continuous and discrete. In the case of continuous offloading, the MD will upload the entire job in one piece without interruption, and an OnOpt algorithm is used for deciding the offloading initiation time. In discrete computation offloading, the job is partitioned into a known number of parts, each part is uploaded separately, and a MultiOpt algorithm is used to decide the upload initiation time of each part during runtime. The optimality of the algorithms is proven by augmenting the underlying channel model so that it forms a time-dilated absorbing Markov process. Dynamic programming is then used to establish a test that determines whether a given job (for continuous uploading) or a part of the job (for discrete uploading) should be uploaded at the current time, or to wait for some future upload opportunity. The performance results presented use the Gilbert–Elliott channel model. Performance results show that the proposed algorithms can significantly improve MD energy consumption compared to the other approaches, and using MultiOpt further improves the energy consumption compared to OnOpt.

10.2 Continuous Offloading

10.2.1 System Description and Problem Formulation

We consider the execution of computational tasks (jobs) generated by a MD, either locally (by the device itself), or by offloading them on a remote cloud server through a wireless transmission channel. Each job could be a sub-task associated with multiple local/remote job execution components [2, 3]. We focus on a single task whose characteristics are known at its release time. Note that time is taken to be discrete, i.e., quantized into equal length *time slots* whose duration is normalized to 1. Time values are therefore referred to by their time slot indices. Note that the time slot duration is defined to accommodate the channel propagation model discussed in Sect. 10.2.2 and may contain multiple packet transmission times on the channel. Each job to be executed is characterized by the following:

t_R: *Release time* of the job, i.e., the time when the job is ready to start execution, either locally or via offloading.

t_D: *Hard deadline* of the job, i.e., the job execution results *must* be available at the MD by time t_D. $T_D = t_D - t_R + 1$ is the maximum number of time slots available for completing the job.

S_{up}: The number of bits transmitted through the uplink channel when uploading the job to the cloud.

S_{down}: The number of bits transmitted through the downlink channel when downloading job results from the cloud.

10.2.1.1 Local Execution

It is assumed that the energy cost and time needed to execute a job locally are known at the job release time, t_R, and these are defined by E_L and T_L, respectively. This assumption is often true and has been made in many computational offloading studies [1, 13, 21]. If the computation offloading algorithm elects to execute the job locally without any remote offloading, we must ensure that the job deadline is always satisfied. Therefore, local execution must start no later than $t_L = t_D - T_L + 1$, unless remote offload/execution results are available at the mobile device before t_L.

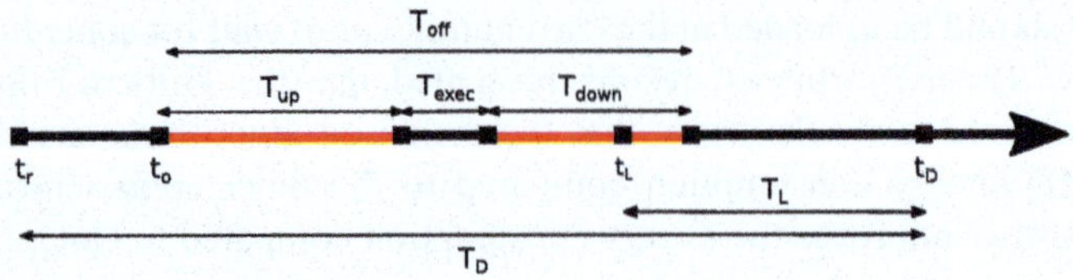

Fig. 10.1 Job computation timing for continuous offloading

10.2.1.2 Remote Execution

In the case of offloading a job, we will assume that, upon its release, the job is assigned an execution time T_{exec} by the cloud server, which is communicated to the MD (or is prescribed by, say, the contractual agreement between the user of the device and the cloud server operator). In addition, we assume that the user has been allocated capacity (such as recurring time slots) until the offload has completed. Therefore, if T_{up} and T_{down} are the time periods needed to upload the job to the cloud server and download its results to the device, respectively, the total offloading time T_{off} is given by $T_{off} = T_{up} + T_{exec} + T_{down}$. These components are shown in Fig. 10.1, where t_o is the remote offload initiation time. It is assumed that the channel uses bit rate adaptation to accommodate random variations in channel conditions. As a result, T_{up} is a random variable, dependent on the evolution of the uplink channel state as a given upload occurs. In what follows, it is assumed that the channel state can be modeled as a homogeneous discrete-time Markov process; the same holds for T_{down}.

In order to simplify our exposition, we will initially focus on the randomness induced by the Markovian uplink channel. In the following development, we therefore temporarily assume that all offloading deadlines, job sizes (in bits), and energy costs are related only to job uploading, i.e., $T_{off} \equiv T_{up}$ and $S \equiv S_{up}$, so that $T_{exec} = T_{down} = 0$.

Since the job's hard deadline constraint must always be satisfied, we propose its simultaneous cloud server offloading (if possible and beneficial) *and* its local execution. Given the stochastic nature of the transmission channel, deciding whether and when to offload (i.e., t_o in Fig. 10.1) depends on the estimation of offloading energy consumption *and* offloading time, in order to both minimize energy costs for the MD *and* satisfy the job deadline constraint.[1] Depending on these estimates, there are three possibilities for offloading at time slot t_o: (1) it certainly finishes before starting the local execution of the job, and, hence, local execution never starts, or (2) it finishes after starting the local execution of the job, and, possibly, *before* deadline t_D; then, the fraction of local execution energy cost incurred is equal to the fraction of T_L overlapping with the offloading (i.e., local execution is terminated if a

[1] Note that when offloading occurs, then $t_R \leq t_o \leq t_D$, and when $t_o > t_D$, then there has been no offloading, i.e., there is only local job execution.

remote offload response is received), or (3) it certainly finishes *after* deadline t_D, so it does not even start, and the total energy cost is equal to the local execution energy cost. Note that in the case of a deterministic channel, one can calculate exactly in which of these three cases the job falls. In this chapter, we analyze the problem of offloading with hard deadlines over a Markovian stochastic channel.

As in most of the related work references, we assume that the current state of the channel can be determined prior to making the decision to start an offload. This information can be learned in a variety of ways, such as via a short handshake with the base station at the start of the time slot.

Figure 10.1 represents the case of continuous offloading. The offload begins at t_o, and execution is completed $T_{up} + T_{exec} + T_{down}$ time slots later. To enforce the job deadline, local execution must begin at t_L if the mobile is still awaiting a remote response. At time $t_o + T_{up} + T_{exec} + T_{down}$, local execution is terminated provided that a remote offload response arrives before t_D.

Note that starting the local job execution at time slot t_L ensures the hard delay constraint of the task, if remote offloading is not received in time. Although this may result in both local and remote executions of the task, it will always satisfy the hard deadline, even if there is channel contention or extended channel outages. However, with the objective of minimizing the mean energy consumption of the MD, the proposed algorithm will reduce the possibility of both local and remote executions.

10.2.2 Markovian Channel and the Time-Dilated Absorbing Markov Model

In many studies, homogeneous Markov chains have been used to model random wireless channel conditions, and as is often assumed, the Markovian transition probabilities are taken to be known or have been learned dynamically [26–28]. Accordingly, we assume that the computation offloading occurs over a finite-state Markovian channel. In this case, the OnOpt (Online Optimal) algorithm proposed in Sect. 10.2.4 is an online computation offloading algorithm that attains the minimum expected execution energy for continuous offloading. As is commonly assumed, the channel data rate is defined by the Markovian channel state, and the receive signal-to-noise ratio (SNR) is such that errors due to random noise are negligible. When this is not the case, then the execution time constraint will still be satisfied by the OnOpt algorithm [28, 29].

In this section, we use the conventional channel state Markov chain (CSMC) to form a time-dilated absorbing Markov chain (TDAMC) that models the offloading over the channel. The resulting Markov process is used by OnOpt in order to compute its energy and offloading time estimates, and by our analysis, in order to show its optimality. As mentioned above, we focus on T_{up}, ignoring T_{exec} and T_{down} (cf. Fig. 10.1); hence, T_{off} and S below refer to T_{up} and S_{up}, respectively.

In the CSMC, and starting from the current time slot t_s, the channel conditions will evolve from one time slot to the next according to a homogeneous finite-state Markov chain. We denote the set of possible channel states by $\mathcal{M}$, where $M = |\mathcal{M}|$ is the number of states in the CSMC. As discussed previously, the radio transmit power is fixed, and bit rate adaptation is used to adjust to varying channel conditions. Therefore, each state in the CSMC has an associated bit rate that gives the number of bits per time slot that can be uploaded when offloading occurs in that state. In a general Markov chain model, the CSMC transition matrix is defined as $\mathbb{P} = [P_{i,j}]$, where P_{ij} is the probability of transitioning to channel state j in the next time slot, given that the channel is currently in state i. Unfortunately, CSMC is memoryless as far as the state of offloading and channel conditions are concerned; in order to incorporate them into our model, we form a new Markov chain, referred to as a *time-dilated absorbing Markov chain (TDAMC)*. We are again interested in the evolution of the system starting at the current time slot t_s, and running until the computation has completed, either locally or via offloading. The state of the channel in each TDAMC state at time $t \geq t_s$ is represented by X_t, where $X_t \in \mathcal{M}$. However, unlike the CSMC, the TDAMC incorporates t and other information into its structure.

The TDAMC models the job offloading progress if the latter is initiated at the current time slot t_s. It is a rooted tree, constructed as follows: The root state is the channel state X_{t_s} at current time slot t_s; since this is the current time slot, X_{t_s} is known. At each subsequent time slot, the Markov chain tree branches forward, according to the transitions possible from the current state (X_{t_s}, initially) to other CSMC states. At each step along a given tree branch, the number of job bits transmitted is determined by the bit rate associated with the channel state in question. This construction continues along each branching tree path until the number of bits offloaded reaches the job upload size, $S = S_{up}$. At that point, the state reached in the TDAMC is defined as a Markov chain *absorbing* state, i.e., it has a self-transition with probability 1. From this construction, it can be seen that the TDAMC includes all possible paths that lead to a successful job offload and that all of the states are either transient or absorbing. Eventually, all paths terminate in an absorbing state, and the energy cost of that path is proportional to its length, i.e., the number of time slots needed.

An example of a TDAMC is shown in Fig. 10.2, for $t_s = 1$. It is constructed from a two-state Gilbert–Elliott channel [26, 30], which is modeled by a CSMC with $\mathcal{M} = \{G, B\}$ (i.e., with "Good" and "Bad" states, respectively), and transition probability matrix

$$\begin{bmatrix} P_{GG} & P_{GB} \\ P_{BG} & P_{BB} \end{bmatrix},$$

i.e., $P_{1,1} = P_{GG}$, $P_{1,2} = P_{GB}$, $P_{2,1} = P_{BG}$ and $P_{2,2} = P_{BB}$. In each time slot, the TDAMC transitions to a new state in accordance with these transition probabilities. For clarity, each channel state in the figure is subscripted with its level time and the index of the subtree it belongs to. For example, $G_{3,2}$ indicates that the channel state at level $t = 3$ and subtree 2 is Good. The TDAMC shows that at $t = 3$, the channel

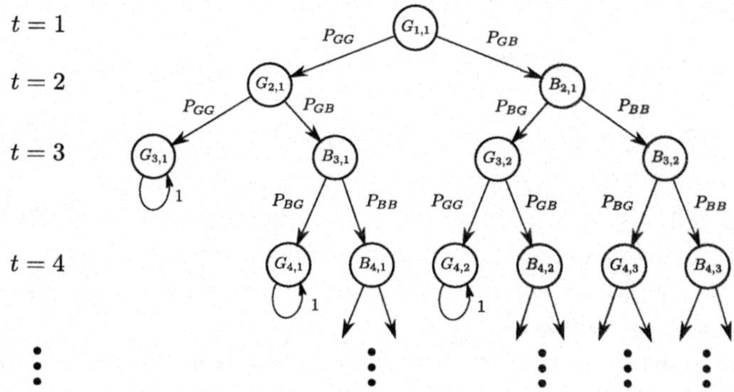

Fig. 10.2 Time-dilated absorbing Markov chain example

can remain in the G state, i.e., $G_{4,2}$ or transition to the B state, i.e., $B_{4,2}$ with the given CSMC transition probabilities. Each state of the TDAMC defines the number of bits that can be offloaded during a time slot while in that state. In the example of Fig. 10.2, when the channel state is G, the number of payload bits is defined by the number of bits that can be carried on the channel during a good (i.e., high bit rate) channel state. In the general case, when the channel is in state X_t at time t, the number of child states at $t + 1$ is given by the number of non-zero values in the same row of the original CSMC transition matrix. In Fig. 10.2, each state continues to branch downward until the number of offloaded bits for a given branch reaches the total number needed for the offload. At that point, the branch ends in a Markov chain absorbing state discussed previously. In Fig. 10.2, states $G_{3,1}$, $G_{4,1}$, and $G_{4,2}$ are the absorbing states.

The non-absorbing states in the TDAMC are clearly all transient states. We define $\mathcal{A}$ to be the set of absorbing states and $\mathcal{T}$ to be the set of transient states in the TDAMC. For an absorbing Markov chain, by labeling the transient states first, the resulting transition matrix can be written in the following form [31]:

$$\mathbb{P}_{\text{TDAMC}} = \begin{bmatrix} Q & R \\ \mathbf{0} & I_{|\mathcal{A}|} \end{bmatrix}. \tag{10.1}$$

In $\mathbb{P}_{\text{TDAMC}}$, the $|\mathcal{T}| \times |\mathcal{T}|$ sub-matrix Q contains the probabilities of transitioning between transient states before the job upload is completed. The $|\mathcal{T}| \times |\mathcal{A}|$ sub-matrix R contains the probabilities of transitioning from a transient state to an absorbing state, indicating that the job upload is finished. $\mathbf{0}$ is an $|\mathcal{A}| \times |\mathcal{T}|$ zero matrix and $I_{|\mathcal{A}|}$ is an $|\mathcal{A}| \times |\mathcal{A}|$ identity (i.e., absorbing) matrix.

Q contains the entries of the original CSMC transition matrix that give the transition probabilities of each state k when it transits to a state in $\{s_k, s_k + 1, \ldots, f_k\}$, and, for our TDAMC, it has the following form:

$$Q = \begin{bmatrix} 0 & P_{1,s_1} & \cdots & P_{1,f_1} & 0 & \cdots & 0 & \cdots & 0 \\ 0 & 0 & \cdots & 0 & P_{2,s_2} & \cdots & P_{2,f_2} & \cdots & 0 \\ \vdots & \vdots & & \vdots & \vdots & & \vdots & & \vdots \\ 0 & 0 & \cdots & 0 & 0 & \cdots & 0 & \cdots & 0 \end{bmatrix}.$$

It can be seen that Q is upper triangular, as expected, since all states are transient and can be visited at most once. The (possibly) non-zero transition probabilities shown in row one, for example, give the probability of transitioning to all possible $t = 2$ channel states and so on.

With the above construction and using results from the theory of absorbing Markov chains, various statistics can be computed by first forming the fundamental matrix

$$N = (I - Q)^{-1}. \tag{10.2}$$

For example, entry (i, j) of N gives the expected number of times that the TDAMC is in transient state j if the system is started in transient state i.

Due to the structure of our TDAMC, the computation needed in Eq. (10.2) can be greatly simplified. Note that N^{-1} is still an upper triangular matrix with all the diagonal entries equal to one and can be decomposed as follows:

$$N^{-1} = N_{|\mathcal{T}|} N_{|\mathcal{T}|-1} N_{|\mathcal{T}|-2} \cdots N_1,$$

where

$$N_k = \begin{bmatrix} 1 & 0 & \cdots & 0 & n_{1,k} & \cdots & 0 \\ 0 & 1 & \cdots & 0 & n_{2,k} & \cdots & 0 \\ \vdots & \vdots & \ddots & \vdots & \vdots & \ddots & \vdots \\ 0 & 0 & \cdots & 1 & n_{k-1,k} & \cdots & 0 \\ 0 & 0 & \cdots & 0 & 1 & \cdots & 0 \\ \vdots & \vdots & \ddots & \vdots & \vdots & \ddots & \vdots \\ 0 & 0 & \cdots & 0 & 0 & \cdots & 1 \end{bmatrix}.$$

N_k is an atomic triangular matrix whose inverse is given by

$$N_k^{-1} = \begin{bmatrix} 1 & 0 & \cdots & 0 & -n_{1,k} & \cdots & 0 \\ 0 & 1 & \cdots & 0 & -n_{2,k} & \cdots & 0 \\ \vdots & \vdots & \ddots & \vdots & \vdots & \ddots & \vdots \\ 0 & 0 & \cdots & 1 & -n_{k-1,k} & \cdots & 0 \\ 0 & 0 & \cdots & 0 & 1 & \cdots & 0 \\ \vdots & \vdots & \ddots & \vdots & \vdots & \ddots & \vdots \\ 0 & 0 & \cdots & 0 & 0 & \cdots & 1 \end{bmatrix}$$

Then

$$N = (N_{|\mathcal{T}|}N_{|\mathcal{T}|-1}N_{|\mathcal{T}|-2}\cdots N_1)^{-1} = N_1^{-1}N_2^{-1}N_3^{-1}\cdots N_{|\mathcal{T}|}^{-1}.$$

Note that each column of the Q matrix has only one non-zero element. Therefore, N^{-1} will have only two non-zero elements in each column. Similarly, in N_k only one of the $n_{1,k}, n_{2,k}, \ldots, n_{k-1,k}$ is non-zero. Therefore, the multiplication can be done efficiently.

The absorption probabilities for all absorbing states can be obtained by

$$W = NR, \tag{10.3}$$

where W is a $|\mathcal{T}| \times |\mathcal{A}|$ matrix and $W[i, j]$ gives the probability that a particular absorbing state j will be reached if the system starts in transient state i. Using this procedure, we can thus compute the various probabilities of absorption for each absorbing state, given knowledge of the starting state. Therefore, we can obtain the probability of finishing the offload for every possible offloading time T_{off} by summing all of the absorbing state probabilities that have the same TDAMC path length. We define $P_t(T, x)$ to be the probability of offloading in exactly T time slots, when offloading starts at time t with the channel in state $X_t = x$. Then

$$P_t(T_{off}, x) = \textstyle\sum_{j \in \mathcal{S}} W[x, j] \tag{10.4}$$

where $\mathcal{S}$ are all of the entries of the matrix where the offloading time is equal to T_{off}. Note that $P_t(T, x) = 0$ when it is impossible to offload in a period of exactly T time slots when offloading at t with the channel in state $X_t = x$, i.e., T is shorter (longer) than the shortest (longest) time needed to offload, under the best (worst) channel conditions. $P_t(T, x)$ is critical for computing the expected cost of offloading used by the algorithm OnOpt.

The ability to compute the $P_t(T, x)$ values allows for the computation of the energy costs for both offloading and local execution. If offloading the job (of bit size S) starts at time slot t, its expected transmission energy is calculated as follows, depending on whether:

- Offloading is certainly completed ($1 \leq t < t_D - \frac{S}{B_{min}} + 1$), in which case the energy spent is proportional to T_{off}.
- Offloading may or may not be completed within the deadline ($t_D - \frac{S}{B_{min}} + 1 \leq t \leq t_D$), in which case the energy cost is T_{off} or $t_D - t + 1$, respectively (clearly the deadline t_D is the last time slot where offloading can be done).

Noting that $P_t(T_{off}, x) = 0$ when $T_{off} < \frac{S}{B_{max}}$ or $T_{off} > \frac{S}{B_{min}}$, the expected offloading energy cost when offloading starts at time slot t with the channel in state x is given by (10.5).

$$E_{off}(t,x) = \begin{cases} E_{tr} \sum_{T_{off}=\frac{S}{B_{max}}}^{\frac{S}{B_{min}}} P_t(T_{off},x)T_{off}, & 1 \le t < t_D - \frac{S}{B_{min}} + 1 \\ E_{tr}\left(\sum_{T_{off}=\frac{S}{B_{max}}}^{t_D-t} P_t(T_{off},x)T_{off} + \right. & \\ \left. \sum_{T_{off}=t_D-t+1}^{\frac{S}{B_{min}}} P_t(T_{off},x)(t_D-t+1)\right), & t_D - \frac{S}{B_{min}} + 1 \le t \le t_D \\ 0 & t > t_D \end{cases} \tag{10.5}$$

$$E_L(t,x) = \begin{cases} \sum_{T_{off}=t_L-t+1}^{\frac{S}{B_{min}}} P_t(T_{off},x)\left(\frac{\min\{t_D+1,t+T_{off}\}-t_L}{T_L} E_L\right), & 1 \le t < t_L \\ \sum_{T_{off}=\frac{S}{B_{max}}}^{\frac{S}{B_{min}}} P_t(T_{off},x)\left(\frac{\min\{t_D+1,t+T_{off}\}-t_L}{T_L} E_L\right), & t_L \le t \le t_D \\ E_L & t > t_D. \end{cases} \tag{10.6}$$

Recall that local execution is postponed until the very last moment, i.e., time slot $t_L = t_D - T_L + 1$, where T_L is the number of time slots needed by the task to execute locally. A central idea is that, although local execution is always initiated (if offloading has not completed earlier) at time t_L, in order to guarantee completion within the deadline, offloading will be decided in such a way so that it will (hopefully) terminate *before* t_D, thus saving us the energy cost of the remaining local execution. The overlap time (when such exists) between offloading at time t and local execution is $\min\{t_D+1, t+T_{off}\} - t_L$. By recalling that E_L is the energy cost of complete local execution of the task, the local execution energy cost will be 0 if there is no overlap, or a fraction $\frac{\min\{t_D+1,t+T_{off}\}-t_L}{T_L}$ of E_L if there is. Hence, we obtain that the expected local execution cost when offloading starts at time t with the channel in state x is given by (10.6). In the first case, there will be overlap only for $T_{off} \ge t_L - t + 1$, while in the second, there is always overlap, since $t - t_L + T_{off} > 0$.

Note that the above development was presented by taking into account only the random job uploading process. These results are easily extended to include both the (deterministic) cloud execution, i.e., T_{exec}, and a Markovian random downlink channel, i.e., $T_{off} = T_{up} + T_{exec} + T_{down}$ and $S = S_{up} + S_{down}$. This is done as follows. The TDAMC of Fig. 10.2, which models the uploading of S_{up} bits, is extended by branching out from each (previously) absorbing state for T_{exec} transition steps. This is followed by branching out according to a process similar to the TDAMC of Fig. 10.2, which then models the downloading of S_{down} bits. The resulting Markov process therefore tracks the channel throughout all three offloading periods, i.e., upload, remote execution, and download, shown in Fig. 10.1.

The number of states in the TDAMC increases with both the number of states in the CSMC and the job completion deadline. As the number of states in the TDAMC increases, the complexity of calculating the average offloading time and energy based on the above analysis also increases. A general analysis about the complexity is difficult because it depends on the specific structure of the CSMC. In [32], different approximation methods have been provided that help reduce the computation complexity by taking advantage of the special structure of the Gilbert–Elliott channel model without significantly compromising the accuracy.

10.2.3 *Offline Bound*

In this subsection, an offline lower bound on MD energy is derived. This bound is used in Sect. 10.4 for performance comparisons with various online computation offloading algorithms. Since the bound is offline, we assume that the wireless channel states are known for all future time slots. When a job is released, the bound then chooses the job offload time so that its deadline is met and the energy needed is minimized.

Let t_o be the time to start offloading, given that we know the bit rate B_t (in bits per time slot) at all times $1 \leq t \leq t_D$ (recall that t_R is taken to be 1). Let $t_f(t_o)$ be defined as the offload finishing time when offloading starts at t_o. Then t_o can be found by solving the following IP.

$$\min_{t_o} \quad \frac{\max(t_o, t_L) - t_L}{T_L} E_L + \sum_{t=t_o}^{t_f(t_o)} e_t \tag{10.7}$$

$$s.t. \quad \frac{\max(t_o, t_L) - t_L}{T_L} E_L + \sum_{t=t_o}^{t_f(t_o)} e_t \leq E_L \tag{10.8}$$

$$1 \leq t_o \leq t_D. \tag{10.9}$$

Objective (10.7) consists of two terms. The first is the local execution energy cost incurred before offloading starts. If $t_o < t_L$, this term is zero, which means that there has been no local execution to that point; otherwise, $\frac{t_o - t_L}{T_L} E_L$ is the energy that has been expended by local execution energy before t_o. The second term in (10.7) is the total energy consumption after offloading starts where e_t is the energy expended in time slot t. When $t_o < t < t_L$, each e_t includes only the offloading energy; and when $t \geq t_L$, both offloading and local execution are performed at time slot t. Therefore, e_t is given as

$$e_t = \begin{cases} E_{tr}, & t < t_L \\ E_{tr} + \frac{E_L}{T_L}, & t \geq t_L, \end{cases} \tag{10.10}$$

where E_{tr} is the energy cost per time slot for transmitting on the channel. Constraint (10.8) ensures that the energy used in offloading does not exceed that of executing the job locally. Note that if the IP is infeasible, then there is no feasible offloading start time t_o, i.e., it is best to execute locally without offloading.

10.2.4 OnOpt (Online Optimal) Algorithm

In this section, we define an online offloading algorithm, OnOpt, which is then shown to be energy optimal. A high-level description of the algorithm is as follows: At each time slot t_s (starting from T_R), the algorithm considers the TDAMC model for starting offloading at current time t_s. It computes (based on the TDAMC) the optimal offloading starting time $t^* \geq t_s$, by formulating the problem as a Markovian optimal stopping problem. If $t^* = t_s$, then offloading is started immediately at time t_s. Otherwise, the algorithm waits till time slot $t_s + 1$, to repeat the above process.

Suppose that the current time slot is t_s, and consider the corresponding TDAMC rooted at current state $X_{t_s} = y$. In order to compute the optimal time slot for starting offloading (if offloading turns out to be more beneficial, in expectation, than executing the task solely locally), we need to compute the *offloading starting time* t^* that satisfies the following optimization problem:

$$\begin{aligned} v_{t_s}(y) &= \min_{t: t_s \leq t \leq t_D+1} E[g_t(X_t)|X_{t_s} = y] \\ &= \min_{t: t_s \leq t \leq t_D+1} \textstyle\sum_{z \in M} Pr[X_t = z|X_{t_s} = y] g_t(z), \end{aligned} \tag{10.11}$$

where X_{t_s} is the current channel state, and $g_t(x)$ is the expected total energy cost if offloading starts at time slot t with channel state $X_t = x$. The choice of $t = t_D + 1$ in (10.11) corresponds to no offloading, in which case (10.5) and (10.6) imply a total cost of E_L. Then, for $t_s \leq t \leq t_D$,

$$g_t(x) = E_{off}(t, x) + E_L(t, x), \tag{10.12}$$

where $E_{off}(t, x)$, $E_L(t, x)$ are the expected offloading and local execution costs, respectively, as defined in (10.5) and (10.6), when offloading starts at time t with the channel in state $X_t = x$.

The optimization problem (10.11) is inherently an offline problem, while the algorithm we would like to use is inherently an online one, in the sense that at every time slot it has to decide whether to offload or not, *given the history of channel states it has encountered so far.* Such an algorithm is defined by the following recursion, which can be solved using dynamic programming (DP), i.e.,

$$V_t(x) = \begin{cases} E_L, & t \geq t_D \\ \min\{g_t(x), E[V_{t+1}|X_t = x]\}, & t = t_s, \ldots, t_D - 1. \end{cases} \tag{10.13}$$

Note that $V_t(x)$ is the minimum between the expected total cost of offloading at the current time slot t, and the expected cost of postponing that decision to time slot $t + 1$, given that the channel state at time t is x, and $E[V_{t+1}|X_t = x]$ is the expectation of $V_{t+1}(X_{t+1})$ over all possible X_{t+1}, under the condition that $X_t = x$, i.e.,

$$E[V_{t+1}|X_t = x] = \sum_{y \in \mathcal{M}} Pr[X_{t+1} = y|X_t = x]V_{t+1}(y).$$

Note that (10.13) implies an *online* algorithm, namely, at every time t_s we can compute $V_{t_s}(X_{t_s})$, and if its minimum is achieved by $g_{t_s}(X_{t_s})$, then start offloading now (at t_s); otherwise, wait till the next time slot $t_s + 1$ and repeat. This online algorithm OnOpt is given in Algorithm 1.

Algorithm 1 OnOpt (online optimal) algorithm

Require: Local execution starting time t_L, local execution energy E_L, job deadline t_D, and job size S.

$t_o := \infty$ ▷ Offloading initially disabled (t_o is offload start time)
for all $t_s \in \{1, \ldots, t_D\}$ **do**
  if $t_s < t_o$ **then**
    $\overline{c_{t_s}} := g_{t_s}(x)$ ▷ Expected energy cost of offloading at t_s.
    $\overline{c_{t_s+1}} := E[V_{t_s+1}|X_{t_s} = x]$ ▷ Expected energy cost of waiting until $t_s + 1$.
    if $\overline{c_{t_s}} \leq \overline{c_{t_s+1}}$ **then**
      $t_o := t_s$ ▷ Start offloading.
    end if
  else if offloading terminates at t_s **then**
    Abort local execution (if active). ▷ Remote offload response has been received.
    return
  end if
  if $t_s = t_D - T_L + 1$ **then**
    Start local execution. ▷ Ensure that the job deadline is satisfied.
  end if
  if $t_s = t_D$ **then**
    Abort remote offload (if active). ▷ Local execution has completed.
    return
  end if
end for

We prove the optimality of OnOpt by proving that its online decisions coincide with the solution of minimization problem (10.11).

Lemma 10.1 *Let t^* be the time OnOpt decides to offload (or $t_D + 1$ if no offloading happens), i.e.,*

$$t^* = \arg\min_{t_R \leq t \leq t_D+1}\{V_t(x) = g_t(x)\}.$$

Then $v_{t^*}(x) = V_{t^*}(x)$, where $X_{t^*} = x$.

Proof In order to prove the lemma, we can first adapt a classic result from Stopping Theory (e.g., Theorem 1.7 in [33]) in order to prove that

$$v_{t_s}(y) = V_{t_s}(y), \ \forall y \tag{10.14}$$

for every time slot $t_R \leq t_s \leq t_D + 1$. Equation (10.14) implies that at every time t_s, the decision to offload or not at t_s based on V_{t_s} coincides with the optimal action at t_s dictated by the optimal solution of the offline problem (10.11), up to (and including) time t^*, when OnOpt offloads (or $t^* = t_D + 1$ if no offloading happens) for every problem (10.11) defined for every $t_R \leq t_s \leq t_D$.

Lemma 10.1 implies that OnOpt is optimal. Note that this result is true for any Markovian channel. The algorithm is given the local execution starting time t_L, local execution energy E_L, job deadline t_D, and job size S. It then arranges for the job to be executed either locally or by remote offloading (or both, if needed). Initially, the remote offload is disabled by setting t_o to a value greater than t_D in Line 1. At each time slot t_s with the channel at state $X_{t_s} = x$, we test if $t_s < t_o$, i.e., no offload has been initiated for the job. Then both $g_{t_s}(x)$ and $E[V_{t_s+1}|X_{t_s} = x]$ are computed (using (10.12) and using DP to solve (10.13), respectively). If $g_{t_s}(x) \leq E[V_{t_s+1}|X_{t_s} = x]$, then the offload begins at time t_s, i.e., $t_o = t_s$, since in this case $t^* = t_s$ from Lemma 10.1. If offloading finishes before a local execution finishes, then local execution is terminated (Line 11). At Line 13, we check to see if local execution should start so that the job's deadline can be guaranteed. Similarly, Line 16 tests if the local job has completed. In that case, any remote offload in progress will be terminated.

10.3 Multi-part Offloading

The OnOpt algorithm makes a single offloading decision. In this section, the task to be remotely executed is segmented into K parts, with K associated upload initiation time decisions. We assume that both K and the bit sizes of the K parts are predetermined. Splitting the uploading data into multiple parts helps further reduce the energy consumption of the MD when wireless channel conditions change during the computation offload, and this energy benefit increases with K. Since the task is uploaded in separate parts, separate offload initiation time decisions are needed for each, so that MD energy consumption is minimized. These decisions have to be made using CLE, so that the system always satisfies the given hard task execution time constraint.

The same Markovian wireless channel model is used as before. Under this assumption, a new computation offloading algorithm (MultiOpt) is introduced for multi-part offloading. MultiOpt is shown to be energy optimal, in the sense that no other CLE online computation offloading algorithm can achieve a lower mean MD energy consumption. This is shown by creating a new Markov process, which incorporates time in the given Markovian channel model, and then by using optimal Markovian Stopping Theory.

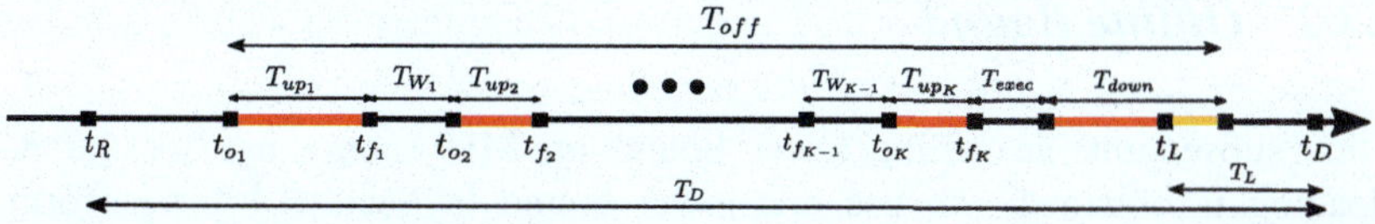

Fig. 10.3 Job offloading timing parameters

10.3.1 Problem Formulation

Compared to the continuous offloading case, there are some differences related to the K-Part offloading scenario that are discussed in what follows. We assume that each job can be split into K parts for offloading, each with a (known) number of bits to be transmitted through the uplink channel. S_{up_i} is the bit size of the ith piece, where $S_{up} = S_{up1} + S_{up2} + \cdots + S_{upK}$. Splitting the upload in this way can be advantageous when channel conditions change during the offload. For example, it may be better, energy-wise, to delay further uploading when channel conditions worsen, hoping that it will improve in time to complete the computation offload.

The K-part offloading timing sequence is shown in Fig. 10.3. The first job piece begins uploading at t_{o_1} and ends at t_{f_1}, and then there are T_{W_1} time slots before the second piece begins uploading at t_{o_2}, and so on, until the Kth piece is uploaded. Note that, by definition, if $t_{o_1} > t_D$, then there is only local execution. It is assumed that the uplink channel uses bit rate adaptation to accommodate random variations in channel conditions. As a result, time intervals $T_{up_i} = t_{f_i} - t_{o_i} + 1,\ i = 1, \ldots, K$, are random variables, dependent on the evolution of the uplink channel state as a given upload occurs. After its uploading is complete, the job is executed on the server in T_{exec} time slots, and its execution results are downloaded to the MD in T_{down} time slots. We assume that T_{exec} is assigned when the job is released, by the cloud server, which is communicated to the mobile device (or is prescribed by, say, the contractual agreement between the user of the device and the cloud server operator). We assume that T_{down} is also communicated to the MD at this time; more generally, we can treat the downloaded results as the $K + 1$'th job piece transmitted over the channel, but we will avoid doing this, in order to simplify our presentation. We assume that power control is used on the downlink, so that T_{down} is known before the upload. Therefore, the total offloading time T_{off} is given by

$$T_{off} = \sum_{i=1}^{K} T_{up_i} + \sum_{i=1}^{K-1} T_{W_i} + T_{exec} + T_{down}, \tag{10.15}$$

where T_{W_i} is the number of time slots that elapse after uploading the ith piece and before uploading the $(i + 1)$th piece. If the downloaded results are received before deadline t_D, any local execution of the job is automatically terminated.

In what follows, we define

$$T_{rest} = T_{exec} + T_{down}, \tag{10.16}$$

and we set $t_R = 1$ in order to reduce the symbol clutter of the presentation.

10.3.2 Offline Bound

In this subsection, an offline lower bound on MD energy use for the K-part offloading is obtained. We use this lower bound in Sect. 10.4 for performance comparisons. Since the bound is offline, we have complete knowledge of future wireless channel states in advance, i.e., we know the bit rate (in bits per time slot) at all times $1 \leq t \leq t_D$. When a job is released, the bound chooses the job offload times so that its deadline is satisfied, and the energy needed to offload the job is minimized. Let $t_{f_i}(t_{o_i})$, $1 \leq i \leq K$, be the upload finishing time as a function of the uploading starting time t_{o_i} for the ith part, and define $t_{f_0}(t_{o_0}) = 0$. E_{tr} and E_{rc} are the energy costs per time slot for channel transmitting and receiving, respectively. The optimal values for t_{o_i} are found by solving integer programming (IP) program (10.17)–(10.19). The first term in (10.17) is the local execution energy cost incurred, the second term accounts for the energy cost of uploading the K job parts, and the last term is the cost for downloading the results from the cloud. Constraint (10.18) ensures that the energy used in offloading does not exceed that of executing the job locally. Note that if the IP is infeasible, then there are no feasible uploading start times t_{o_i}, i.e., it is best to execute the job solely locally without offloading.

$$\min_{t_{o_1},\ldots,t_{o_K}} \frac{\max(t_{f_K}(t_{o_K}) + T_{rest}, t_L) - t_L}{T_L} E_L + E_{tr} \sum_{i=1}^{K} (t_{f_i}(t_{o_i}) - t_{o_i} + 1) + E_{rc} T_{down} \tag{10.17}$$

$$s.t. \quad \frac{\max(t_{f_K}(t_{o_K}) + T_{rest}, t_L) - t_L}{T_L} E_L + E_{tr} \sum_{i=1}^{K} (t_{f_i}(t_{o_i}) - t_{o_i} + 1) + E_{rc} T_{down} \leq E_L \tag{10.18}$$

$$t_{f_{i-1}}(t_{o_{i-1}}) + 1 \leq t_{o_i} \leq t_D, \quad i = 1, \ldots, K. \tag{10.19}$$

10.3.3 The Time-Dilated Absorbing Markov Model

In this subsection, we will extend the TDAMC in Sect. 10.2.2 produced by following the evolution of the channel, starting from an initial state at time $t = 1$, and branching out from each state according to the transition probabilities of the CSMC. We will denote by X_t a state in this Markov chain, reached after running the channel for t time slots. We will consider subtrees of this TDAMC (such as $TDAMC_1$ below), endowed with energy costs and absorbing states.

The part of the TDAMC that models the offloading progress if the uploading of S_{up_1} is initiated at a time slot t_s will be denoted as $TDAMC_1$. An example of $TDAMC_1$ is shown in Fig. 10.4. To simplify the exposition, the diagram shows the

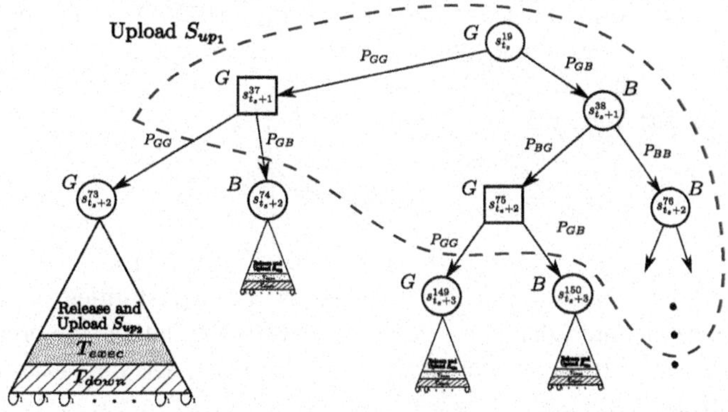

Fig. 10.4 $TDAMC_1$ when offloading S_{up_1} starts at time t_s

two-state Gilbert–Elliott channel case, but the procedure is valid for any Markovian channel. For clarity, each state s_t^a in the figure is subscripted by its time slot t and superscripted by a unique identifier a that distinguishes it from the other channel states reachable after t time slots. Hence, the $TDAMC_1$ of Fig. 10.4 models the offloading process initiated at time slot t_s, when the channel state that has been reached at that time is $s_{t_s}^{19}$. The bit rate at each state is also indicated.

In general, $TDAMC_1$ is a rooted subtree of the TDAMC, constructed as follows: The root state is the (known) channel state X_{t_s} at current time slot t_s. At each subsequent time slot, the Markov chain tree branches forward, according to the transitions possible from the current state (X_{t_s}, initially) to other TDAMC states. At each state, the number of job bits transmitted is determined by the bit rate associated with that state. The branching continues to create all possible paths of states needed to upload S_{up_1} bits, up to some state $X_{t_{f_1}}$ corresponding to upload finishing time t_{f_1} for each path from the root (such as $s_{t_s+1}^{37}$, $s_{t_s+2}^{75}$, represented by squares in Fig. 10.4). At time $t_{f_1} + 1$, the second part S_{up_2} is released. Continuing the branching of the TDAMC, and after a possible waiting period, the uploading of S_{up_2} commences, followed by the rest of the K pieces, and the job execution in the cloud in time T_{exec}, and the downloading of the results in time T_{down}, ending in an absorbing state (this part of the offloading for $K = 2$ is depicted in Fig. 10.4 as subtrees hanging from states $s_{t_s+2}^{73}$, $s_{t_s+2}^{74}$, $s_{t_s+3}^{149}$, $s_{t_s+3}^{150}$). The optimal waiting time for each path, i.e., the waiting times that optimize the total (over all paths) expected energy cost for uploading $S_{up_2}, \ldots, S_{up_K}$, is solved in Sect. 10.3.4. Then the energy cost of each subtree is the optimal expected cost (over all paths) of completing offloading, when uploading S_{up_1} finishes in time slot t_{f_1} and state $X_{t_{f_1}}$. In fact, $TDAMC_1$ does not need to extend all the way into these subtrees but can treat states $X_{t_{f_1}+1}$ as absorbing states, each with cost equal to the energy cost of its own subtree. This process can obviously be repeated, in order to build the corresponding $TDAMC_i$ for any piece i.

The probability of uploading S_{up_i} bits in T_{up_i} time slots, starting at time slot t_{o_i}, and a state $X_{t_{o_i}}$, for $i = 1, \ldots, K$, can be calculated by building a separate

$TDAMC_i$, with $\mathcal{A}_i$ as the set of absorbing states, and $|\mathcal{T}_i|$ as the set of transient states, for $i = 1, \ldots, K$. It encodes the evolution of the channel starting at time slot t_{o_i} and state $X_{t_{o_i}}$, and until S_{up_i} bits are uploaded, at which point an absorbing state is reached. Its transition matrix can be written [31] as

$$\mathbb{P}_i = \begin{bmatrix} Q_i & R_i \\ \mathbf{0} & I_{|\mathcal{A}_i|} \end{bmatrix}, \quad (10.20)$$

where the $|\mathcal{T}_i| \times |\mathcal{T}_i|$ sub-matrix Q_i contains the probabilities of transitioning between transient states, the $|\mathcal{T}_i| \times |\mathcal{A}_i|$ sub-matrix R_i contains the probabilities of transitioning from a transient state to an absorbing state, and $I_{|\mathcal{A}_i|}$ is an $|\mathcal{A}_i| \times |\mathcal{A}_i|$ identity matrix.

The theory of absorbing Markov chains implies that various statistics can be computed by forming the fundamental matrix $N_i = (I - Q_i)^{-1}$, where $N_i[l, m]$ gives the expected number of times that $TDAMC_i$ is in transient state m if the system is started in transient state l. Given the structure of $TDAMC_i$, N_i can be easily decomposed and calculated as in Sect. 10.2.2, since the particular structure of matrices Q_i, N_i, N_i^{-1} is the same simple one as in Sect. 10.2.2. The absorption probabilities matrix W_i for all absorbing states are given by

$$W_i = N_i R_i, \quad (10.21)$$

where W_i is a $|\mathcal{T}_i| \times |\mathcal{A}_i|$ matrix, and $W_i[l, m]$ gives the probability that absorbing state m will be reached when starting in transient state l. Therefore, the probability of uploading the ith part with size S_{up_i} in T_{up_i} time slots, starting at time slot t_{o_i} and state $X_{t_{o_i}}$, is

$$P_{t_{o_i}}(S_{up_i}, T_{up_i}, X_{t_{o_i}}) = \sum_{j \in \mathcal{S}_i} W_i[X_{t_{o_i}}, j], \quad (10.22)$$

where $\mathcal{S}_i$ is the set of absorbing states in $TDAMC_i$ reached by a path of length $T_{up_i} + 1$ from the root $X_{t_{o_i}}$.

For $i = 1, 2, \ldots, K - 1$, if the uploading of S_{up_i} starts at time slot t_{o_i}, the expected offloading energy cost when offloading starts at time slot t_{o_i} in state $X_{t_{o_i}}$ and finishes exactly in T_{up_i} time slots or at t_D (whichever comes first) is given by Eq. (10.23), where E_{tr} is the transmission energy of the MD during one time slot, while the expected energy cost of uploading S_{upK} is given by Eq. (10.24), where E_{rc} is the energy consumption of the MD during one time slot when receiving from the server. Then, the expected offloading energy cost E_{off_i} for $i = 1, 2, \ldots, K$ is computed by Eq. (10.25), where B_{max} and B_{min}, respectively, are the bit rates at the best and worst channel states.

Similarly, the local execution energy cost corresponding to the uploading of the ith job piece, for $i = 1, 2, \ldots, K-1$, is given by (10.26),[2] while the local execution

[2] We set $t_{f0} = t_R - 1$.

energy cost due to the Kth job piece and the rest of offloading time T_{rest} is given by (10.27). Then, the expected local execution energy cost E_{L_i} for $i = 1, 2, \ldots, K$ is computed by Eq. (10.28).

$$\hat{E}_{offi}(S_{upi}, T_{upi}, t_{oi}) = E_{tr}[\min\{t_D, t_{o_i} + T_{up_i} - 1\} - \min\{t_D, t_{o_i} - 1\}] \tag{10.23}$$

$$\hat{E}_{offK}(S_{upK}, T_{upK}, t_{oK}) = E_{tr}[\min\{t_D, t_{oK} + T_{upK} - 1\} - \min\{t_D, t_{oK} - 1\}] \tag{10.24}$$
$$+E_{rc}[\min\{t_D, t_{oK} + T_{rest} - 1\} - \min\{t_D, t_{oK} + T_{exec} - 1\}]$$

$$E_{offi}(S_{upi}, X_{t_{o_i}}) = \sum_{T_{up_i}=\frac{S_{up_i}}{B_{max}}}^{\frac{S_{up_i}}{B_{min}}} P_{t_{o_i}}(S_{upi}, T_{upi}, X_{t_{o_i}})\hat{E}_{offi}(S_{upi}, T_{upi}, t_{oi}) \tag{10.25}$$

$$\hat{E}_{L_i}(T_{upi}, t_{f_{i-1}}, t_{oi}) = \begin{cases} 0, & t_{o_i} \leq t_L - T_{up_i} \text{ or } t_{f_{i-1}} \geq t_D \\ \frac{\min\{t_D, t_{o_i} + T_{up_i} - 1\} - \max\{t_L, t_{f_{i-1}} + 1\} + 1}{T_L} E_L, & \text{otherwise} \end{cases} \tag{10.26}$$

$$\hat{E}_{LK}(T_{upK}, t_{fK-1}, t_{oK}) = \begin{cases} 0, & t_{o_i} \leq t_L - (T_{up_i} + T_{rest}) \text{ or } t_{f_{i-1}} \geq t_D \\ \frac{\min\{t_D, t_{o_K} + T_{upK} + T_{rest} - 1\} - \max\{t_L, t_{f_{K-1}} + 1\} + 1}{T_L} E_L, & \text{otherwise} \end{cases} \tag{10.27}$$

$$E_{L_i}(S_{upi}, t_{f_{i-1}}, X_{t_{o_i}}) = \sum_{T_{up_i}=\frac{S_{up_i}}{B_{max}}}^{\frac{S_{up_i}}{B_{min}}} P_{t_{o_i}}(S_{upi}, T_{upi}, X_{t_{o_i}})\hat{E}_{L_i}(T_{upi}, t_{f_{i-1}}, t_{oi}). \tag{10.28}$$

10.3.4 Optimal Algorithm for K-Part Offloading

In this subsection, we use the TDAMCs constructed in Sect. 10.3.3 and the theory of optimal stopping for Markov decision processes, in order to define optimal offloading algorithms, and prove their optimality. A high-level description of algorithm MultiOpt (cf. Algorithm 2) is as follows: Starting from time slot $t_R = 1$ (the release time of the job), at each time slot t_s, the algorithm considers $TDAMC_1$ in order to determine the expected cost of the whole offloading process if uploading S_{up1} commences at the current time t_s. If that cost is less than the expected offloading cost when the algorithm waits one more time slot, then $t^*_{o1} = t_s$ (offloading S_{up1} commences); otherwise, the algorithm postpones its decision for time slot $t_s + 1$. Once the uploading of S_{up1} finishes, for the rest of the parts, the algorithm repeats the same decision process at every time slot (using $TDAMC_i$ to compute expected costs), to determine the time t^*_{oi} of starting uploading S_{upi} for $i = 2, 3, \ldots, K$.

At any time slot t_s (and state X_{t_s}), and given that pieces $1, 2, \ldots, i - 1$ have already been uploaded, MultiOpt decides the uploading starting time $t^*_{oi} \geq t_s$ for S_{upi}. Its decisions t^*_{oi} for $i = 1, 2, \ldots, K$ are optimal iff they are the solutions of the minimization problems (one for each i) (10.29)–(10.30),

$$v_i(t_{f_{i-1}}, X_{t_s}) = \begin{cases} 0, & t_s > t_{f_{i-1}} \geq t_D \\ \min_{t_s \leq t_{o_i} \leq t_D+1} \sum_{X_{t_{o_i}} \in S_i} Pr[X_{t_{o_i}} | X_{t_s}] & \\ \Big(E_{off_i}(S_{up_i}, X_{t_{o_i}}) + E_{L_i}(S_{up_i}, t_{f_{i-1}}, X_{t_{o_i}}) & \\ + \sum_{X_{t_{f_i}+1} \in \hat{S}_i} W_i[X_{t_{o_i}}, X_{t_{f_i}+1}] v_{i+1}(t_{f_i}, X_{t_{f_i}+1})\Big), & t_{f_{i-1}} < t_s \leq t_D \end{cases} \tag{10.29}$$

$$v_K(t_{f_{K-1}}, X_{t_s}) = \begin{cases} 0, & t_s > t_{f_{K-1}} \geq t_D \\ \min_{t_s \leq t_{o_K} \leq t_D+1} \sum_{X_{t_{o_K}} \in S_K} Pr[X_{t_{o_K}} | X_{t_s}] & \\ \Big(E_{off_K}(S_{up_K}, X_{t_{o_K}}) + E_{L_K}(S_{up_K}, t_{f_{K-1}}, X_{t_{o_K}})\Big), & t_{f_{K-1}} < t_s \leq t_D, \end{cases} \tag{10.30}$$

where S_i is the set of states reachable after running the channel until time slot t_{o_i}, $\hat{S}_i$ is the set of absorbing states of $TDAMC_i$ rooted at $X_{t_{o_i}}$, and $v_{i+1}(t_{f_i}, X_{t_{f_i}+1})$ is the optimal expected energy cost for the rest of the offloading when S_{up_i} finishes uploading at time t_{f_i}, i.e., the cost of the absorbing state $X_{t_{f_i}+1}$ of $TDAMC_i$. In (10.29)–(10.30), $g_i(S_{up_i}, t_{f_{i-1}}, X_{t_{o_i}})$ is the expected energy cost of uploading S_{up_i}, if uploading of $S_{up_{i-1}}$ finishes at $t_{f_{i-1}}$ and uploading S_{up_i} starts at time slot t_{o_i} and state $X_{t_{o_i}}$, and is given by

$$g_i(S_{up_i}, t_{f_{i-1}}, X_{t_{o_i}}) = E_{off_i}(S_{up_i}, X_{t_{o_i}}) + E_{L_i}(S_{up_i}, t_{f_{i-1}}, X_{t_{o_i}}) + \sum_{X_{t_{f_i}+1} \in S_i} W_i[X_{t_{o_i}}, X_{t_{f_i}+1}] V_{i+1}(t_{f_i}, X_{t_{f_i}+1}), \quad i = 1, \ldots, K \tag{10.31}$$

and

$$g_K(S_{up_K}, t_{f_{K-1}}, X_{t_{o_K}}) = E_{off_K}(S_{up_K}, X_{t_{o_K}}) + E_{L_K}(S_{up_K}, t_{f_{K-1}}, X_{t_{o_K}}). \tag{10.32}$$

Note that we allow the algorithm to decide not to offload or stop offloading if this is to its benefit, by allowing uploading decisions to take the value $t_D + 1$.[3]

For every time slot t_{o_i} and state $X_{t_{o_i}}$, we define the expected cost $V_i(t_{f_{i-1}}, X_{t_{o_i}})$ recursively in (10.33), for $i = 1, \ldots, K$.

$$V_i(t_{f_{i-1}}, X_{t_{o_i}}) = \begin{cases} 0, & t_{o_i} > t_{f_{i-1}} \geq t_D \\ \frac{t_D - \max\{t_{f_{i-1}}+1, t_L\}+1}{T_L} E_L, & t_{o_i} \geq t_D > t_{f_{i-1}} \\ \min\{g_i(S_{up_i}, t_{f_{i-1}}, X_{t_{o_i}}), E[V_i(t_{f_{i-1}}, X_{t_{o_i}+1}) | X_{t_{o_i}}]\}, & t_D > t_{o_i}. \end{cases} \tag{10.33}$$

[3] Equations (10.23), (10.24), (10.26), and (10.27) have been set up to reflect this.

$V_i(t_{f_{i-1}}, X_{t_{o_i}})$ can be computed using dynamic programming (DP), and it is the minimum between the expected total cost of starting uploading S_{up_i} at time slot t_{o_i} and state $X_{t_{o_i}}$, and the expected cost of postponing that decision to time slot $t_{o_i} + 1$

$$E[V_i(t_{f_{i-1}}, X_{t_{o_i}+1})|X_{t_{o_i}}] = \sum_{X_{t_{o_i}+1} \in \mathcal{T}_i} Pr[X_{t_{o_i}+1}|X_{t_{o_i}}] V_i(t_{f_{i-1}}, X_{t_{o_i}+1}),$$

where $\mathcal{T}_i$ is the set of states reachable after running the channel for $t_{o_i} + 1$ time slots. Note that (10.33) implies an online algorithm that at any time t_{o_i} and state $X_{t_{o_i}}$ decides whether it should start uploading S_{up_i} (if the min is attained by g_i), or should otherwise wait.

We prove optimality by (reverse) induction. It is well known (e.g., Theorem 1.7 in [33]) that policy V_K is *optimal*, i.e., solves the original problem (10.30), since

$$v_K(t_{f_{K-1}}, X_{t_K}) = V_K(t_{f_{K-1}}, X_{t_K}), \ \forall t_K > t_{f_{K-1}}, X_{t_K}. \tag{10.34}$$

Hence, the following holds:

Lemma 10.2 ([33]) *The optimal time for starting uploading S_{up_K} is*

$$t^*_{o_K} = \arg\min_{t_{f_{K-1}} < t_{o_K} \le t_D} \{V_K(t_{f_{K-1}}, X_{t_{o_K}}) = g_K(S_{up_K}, t_{f_{K-1}}, X_{t_{o_K}})\}. \tag{10.35}$$

Assuming that decisions $t^*_{o_K}, t^*_{o_{K-1}}, \ldots, t^*_{o_{i+1}}$ are optimal, i.e.,

$$v_k(t_{f_{k-1}}, X_{t_k}) = V_k(t_{f_{k-1}}, X_{t_k}), \ \forall t_k > t_{f_{k-1}}, X_{t_k} \tag{10.36}$$

holds for $k = K, K-1, \ldots, i+1$, we prove that the ith decision $t^*_{o_i}$ of MultiOpt is also optimal. Note that (10.29) becomes (10.37).

$$v_i(t_{f_{i-1}}, X_{t_s}) = \begin{cases} 0, \quad t_s > t_{f_{i-1}} \ge t_D \\ \min\limits_{t_s \le t_{o_i} \le t_D+1} \sum\limits_{X_{t_{o_i}} \in \mathcal{S}_i} Pr[X_{t_{o_i}}|X_{t_s}]\Big(E_{off_i}(S_{up_i}, X_{t_{o_i}}) + E_{L_i}(S_{up_i}, t_{f_{i-1}}, X_{t_{o_i}}) \\ + \sum\limits_{X_{t_{f_i}+1} \in \hat{\mathcal{S}}_i} W_i[X_{t_{o_i}}, X_{t_{f_i}+1}] V_{i+1}(t_{f_i}, X_{t_{f_i}+1})\Big), \quad t_{f_{i-1}} < t_s \le t_D. \end{cases} \tag{10.37}$$

Theorem 1.7 in [33] can be applied again to show:

Lemma 10.3 ([33]) *The optimal time for starting uploading S_{up_i} is*

$$t^*_{o_i} = \arg\min_{1 \le t_{o_i} \le t_D} \{V_i(t_{f_{i-1}}, X_{t_{o_i}}) = g_i(S_{up_i}, t_{f_{i-1}}, X_{t_{o_i}})\}. \tag{10.38}$$

Lemmata 10.2 and 10.3 imply that the online algorithm MultiOpt (Algorithm 2) is optimal for general Markovian channels.

Algorithm 2 MultiOpt (multi-decision online optimal)

Require: Local execution starting time t_L, local execution energy E_L, job deadline t_D, and job sizes $S_{up_1}, S_{up_2}, \ldots, S_{up_K}$.

$i = 1$
for all $t = 1, \ldots, t_D$ **do**
    if job finished uploading **then**
        Break
    end if
    if currently uploading **then**
        Continue
    end if
    if min in (10.33) is g_i **then**
        start uploading part i ▷ part $i-1$ has been uploaded
        $i = i + 1$ ▷ but i has not started uploading
    end if
end for

The proposed online offloading decision algorithms may be run at either the MD or the cloud server. In the former case, running the algorithm takes the energy and time of the MD, which may affect the offloading decisions. In the latter case, the MD passes the job-related parameters to the server, which runs the algorithm and passes the decision to the device. In this case, running the algorithms does not consume the energy of the MD, while additional time is required to communicate with the server and run the algorithms at the server. The communication time is normally small in both directions due to the small amount of data that should be transmitted, and the computation time to run the algorithm is also short due to the high processing speed at the server. Therefore, this additional time can be ignored when making the offloading decisions. We consider the latter case in our simulation.

10.4 Numerical Results

In this section, computer simulation is used to study the performance of the proposed MultiOpt algorithms for $K = 1, 2, 3$, and 4. The OnOpt algorithm is a special version of MultiOpt when $K = 1$. For comparison, we also plot the *offline bound* given in Sects. 10.2.3 and 10.3.2 and performance of *Local Execution* and two other algorithms, referred to as *immediate offloading* and *multi-threshold*. These algorithms all employ CLE to ensure that job execution time constraints are satisfied. The local execution algorithm executes the entire job locally without doing any offloading. For the immediate offloading algorithm, a job is offloaded immediately at the release time if $S_{up}/B_{max} + T_{rest} \leq T_D$; otherwise, the job is executed locally without offloading since offloading cannot be completed before the job deadline even with contiguous best wireless channel states. For the multi-threshold algorithm, uploading for the first piece starts at the first time slot when the channel condition is above a given threshold, if the remaining time before the job completion deadline is at least $S_{up}/B_{max} + T_{rest}$; otherwise no offloading is performed for the entire job. For the Gilbert–Elliott channel, any threshold between the good and bad states can be used, i.e., uploading starts at the first time slot with the good channel state. After the $(i-1)$th piece is uploaded, uploading for the ith piece starts as soon as the channel state becomes above the threshold, if the remaining time before the job completion deadline is no less than $\sum_{k=i}^{K} S_{upk}/B_{\max} + T_{rest}$; otherwise, uploading is stopped. When $K = 1$, the multi-threshold algorithm is referred to as *channel threshold*. In both the immediate offloading and the multi-threshold (channel threshold) algorithms, local execution starts at time slot t_L if offloading (includes uploading to, remote execution at, and downloading from the server) is not completed at time slot $t_L - 1$, i.e., they ensure that the job deadline is satisfied.

In this simulation, the job size S_{up}, i.e., the total amount of data to be uploaded, is split into K equal parts, i.e., $S_{up1} = S_{up2} = \cdots = S_{upK} = S_{up}/K$. The default parameters used in the simulations are given as follows. Each time slot is 1 ms. The transmit and receive powers are 1 W and 0.5 W, respectively, which means that the transmission and receive energies during each time slot are $E_{tr} = 1$ mJ and $E_{rc} = 0.5$ mJ, respectively. In our offloading results, we have used the well-known Gilbert–Elliott channel model. This model is often used to characterize burst noise effects in wireless links, where the channel can abruptly transition between good and bad conditions. We assume that $P_{BB} = 1 - P_{GG}$ for the channel state transition probabilities. In this case, $P_{GB} = P_{BB}$, $P_{BG} = P_{GG}$, the equilibrium channel state probabilities are given by $P_g = P_{GG}$ and $P_b = P_{BB}$, and P_{GG} can be used as a measure of the average channel quality. The data transmission rates are $B_b = 1$ Mbps and $B_g = 10$ Mbps, or $B_b = 1$ kb per time slot and $B_g = 10$ kb per time slot. Note that in the two-state channel model, $B_{max} = B_g$ and $B_{min} = B_b$. In order to produce the results below, each value of average energy consumption is obtained by the averaging of 1500 random i.i.d. runs of the wireless channel, which follows the Gilbert–Elliott model described above.

10.4.1 Simulation Set 1

In this section, we consider a job with $D = 10$ M CPU cycles and $T_D = 60$ time slots. The local execution energy per CPU cycle is $v_l = 2 \times 10^{-6}$ mJ, and the local computation power is $f_l = 1$ M CPU cycles per time slot. Therefore, the local execution time is $T_L = D/f_l = 10$ time slots, and the local energy consumption $E_L = v_l D = 20$ mJ. We consider that the remote execution time is $T_{exec} = 1$ time slot, i.e., the remote processing speed is 10 times of the local processing. The download time T_{down} is assumed to be 1 time slot.

Figure 10.5(a) shows the average energy consumption of the MD as the data size S_{up} increases. The energy used by local execution is constant for all S_{up}. When S_{up} is smaller, the delay constraint is less stringent, and it is more likely for offloading (without local execution) to meet the delay constraint due to a shorter

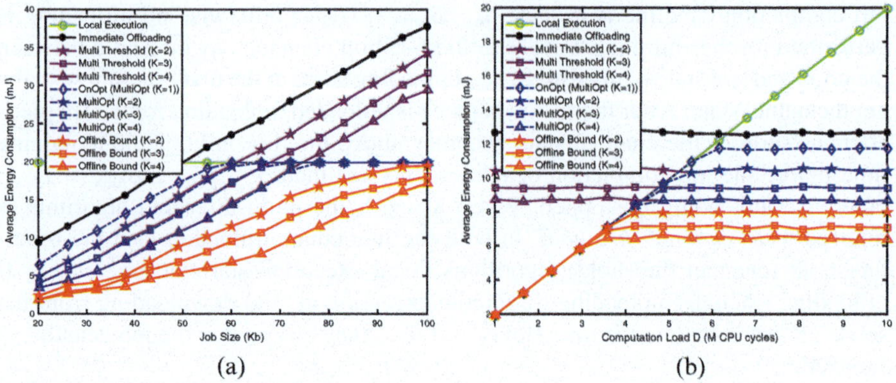

Fig. 10.5 Average energy consumption versus data size S_{up} and D. (**a**) $P_{GG} = 0.2$. (**b**) $P_{GG} = 0.5$

channel uploading time. In this case, the multi-threshold and MultiOpt algorithms have approximately the same average energy consumption, since it is more likely for the MultiOpt algorithm to decide starting the uploading as soon as the channel state is good, making it to start the upload at the same time slot with the multi-threshold algorithm. Compared to OnOpt, the energy consumption of MultiOpt with $K > 1$ is lower, indicating that splitting the job uploading into multiple pieces brings more flexibility that helps the MD to avoid uploading during bad channel states; on the other hand, since OnOpt uploads the job continuously, its uploading is more likely to encounter bad channel states and therefore takes a longer time and consumes more energy. The immediate offloading algorithm consumes higher energy as compared to the other algorithms because there is a certain probability to encounter bad channel state at the job release time and the following time slots, and the probability becomes higher when P_{GG} is lower. As S_{up} increases, a longer time is needed for wireless transmissions, and the offline bound and the MultiOpt algorithms may decide not to offload, resulting in the same energy consumption as local execution, while the immediate offloading and multi-threshold algorithms waste energy consumption by offloading unnecessarily and result in much higher energy consumption.

Comparing the MultiOpt algorithm with $K = 1$ (i.e., the OnOpt algorithm), 2, 3, and 4, we can see that splitting the job into multiple pieces helps reduce the energy consumption of the mobile device, since doing this can avoid uploading during some bad channel states, provided the job completion deadline can be satisfied.

Figure 10.5(b) shows the average energy consumption of the MD versus the amount of computation load D. Deadline T_D is set to 40 time slots. When D is small, the MultiOpt algorithm does not offload because the local execution energy is low and less than the energy needed to upload the data. As D increases, the energy required for local execution increases, and it becomes more likely that offloading consumes less MD energy than local execution. The energy consumption for

MultiOpt becomes constant when D is sufficiently large. This is because the delay constraint is relatively loose in the simulated system, which allows offloading to be completed before t_L. Therefore, when D is relatively large, the energy consumption is the same as the energy consumption for wireless transmissions, which does not depend on D. The figure also shows that MultiOpt can save mobile device energy by splitting the job into multiple pieces and uploading separately.

10.4.2 Simulation Set 2

In this set of results, we use the application parameters for x264 (H.264) encoding from [34] and consider a job with $S_{up} = 60$ Kb, $D = 18$ M CPU cycles, and $T_D = 60$ time slots. The local execution energy per CPU cycle is $v_l = 1.5 \times 10^{-6}$ mJ, and the local computation power is $f_l = 600$ M CPU cycles per second or $f_l = 0.6$ M CPU cycles per time slot. Therefore, the local execution time is $T_L = D/f_l = 30$ time slots, and the local energy consumption $E_L = v_l D = 27$ mJ. The remote execution time T_{exec} is 3 time slots, and the download time T_{down} is 1 time slot.

Figure 10.6(a) shows the average energy consumption of different algorithms as P_{GG} varies. The offline bound is close to the energy consumption of local execution only when P_{GG} is close to 0, in which case the channel is almost always in the bad state and local execution is almost always the optimum choice. The immediate offloading and multi-threshold algorithms result in much higher energy consumption when P_{GG} is small. Since the channel is in the bad state in most time slots, uploading data requires a long time. Therefore, there is a high probability that offloading cannot meet the delay constraint and/or consumes high energy; furthermore, due to the long uploading time, there may be a long overlap time between offloading and local execution. As a result, energy is unnecessarily wasted in the immediate offloading and multi-threshold algorithms by performing

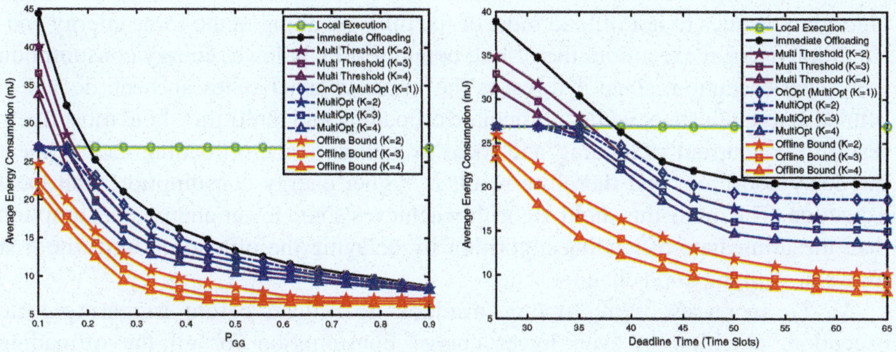

(a) (b)

Fig. 10.6 Average energy consumption versus P_{GG} and T_D. **(a)** $P_{GG} = 0.3$. **(b)** $P_{GG} = 0.3$

offloading. As P_{GG} increases, the possibility that offloading can meet the deadline increases, so that less local execution is performed, and the total energy consumption decreases for all the offloading algorithms. The multi-threshold algorithm consumes slightly lower energy than immediate offloading. By delaying the uploading (of each piece of the job) until the channel state becomes good, it reduces the transmission time and saves energy consumption. The difference is more obvious when P_{GG} is smaller, since the probability of encountering bad channel states is higher.

When P_{GG} is low, the MultiOpt (including OnOpt) algorithm chooses to not offload and therefore results in the same energy consumption as local execution; and when P_{GG} is larger, the algorithm more likely chooses to offload, since channel conditions become better and a shorter time and less energy are needed to offload. Figure 10.6(a) also shows that the energy consumption of MultiOpt is lower when K is larger, if the mobile device decides to offload, since splitting the job into more pieces brings more flexibility that helps the MD avoid transmissions in bad channel conditions.

By comparing the MultiOpt and multi-threshold algorithms, we can see that for given K, when P_{GG} is relatively large, the two algorithms consume almost the same energy. This is because the channel condition in general is good, so that the time required for uploading the data is relatively short, and the time required to complete offloading is much less than T_D. For the MultiOpt algorithm, if the decision is to offload the next piece of the job, it is most likely the first time slot with a good channel state, which is the same as the multi-threshold algorithm. The gap between the MultiOpt algorithm and the offline bound is due to the fact that the MultiOpt algorithm can only use statistical channel information, while the offline bound has knowledge of the channel states of all future time slots.

For all the offloading solutions, the MD energy consumption decreases as P_{GG} increases, since the probability of having the good channel state increases, which reduces the time needed to upload the data and makes it more likely for offloading to meet the delay constraint and consume less energy.

Figure 10.6(b) shows the average energy consumption of the algorithms as the job deadline T_D changes. When T_D is small, the MultiOpt (including OnOpt) algorithm decides to not offload most of the time, resulting in the same energy consumption as local execution; the offline bound results in lower energy consumption than local execution, since it foresees the future channel states and can decide to offload at a future state; while, immediate offloading and multi-threshold most likely result in concurrent offloading and local execution, since offloading cannot meet the delay constraint and therefore result in higher energy consumption than local execution. The multi-threshold algorithm achieves some lower energy consumption than the immediate offloading algorithm by delaying the offloading until the first time slot with the good channel state.

As T_D increases, more time is available to offload before triggering local execution, resulting in even lower energy consumption for all the offloading algorithms. When T_D is sufficiently large, all the offloading algorithms can achieve lower average energy consumption than using local execution. For given K, the

MultiOpt and the multi-threshold algorithms result in the same average energy consumption.

10.5 Summary

We have considered the problem of mobile computation offloading over stochastic wireless transmission channels, where task execution times are subjected to hard deadline constraints. We have shown that using concurrent local execution, hard task completion times can always be guaranteed, and the OnOpt and MultiOpt algorithms achieve the optimum average MD energy consumption when the uploading is performed continuously and using multiple parts, respectively. With higher computation complexity in deciding the uploading time, the MultiOpt algorithm achieves lower energy consumption compared to OnOpt.

Acknowledgments © 2020 IEEE. Part of this chapter is reprinted, with permission, from Optimal Mobile Computation Offloading with Hard Deadline Constraints, by Arvin Hekmati, Peyvand Teymoori, Terence D. Todd, Dongmei Zhao, and George Karakostas, published in IEEE Transactions on Mobile Computing, Sept. 2020, pp. 2160–2173.

Part of this chapter is reprinted from Optimal Multi-part Mobile Computation Offloading with Hard Deadline Constraints, by Arvin Hekmati, Peyvand Teymoori, Terence D. Todd, Dongmei Zhao, and George Karakostas, published in Computer Communications, 160(2020), pp. 614–622, Copyright 2020, with permission from Elsevier.

References

1. K. Kumar, Y.-H. Lu, Cloud computing for mobile users: can offloading computation save energy? IEEE Comput. **43**(4), 51–56 (2010)
2. C. You, K. Huang, H. Chae, Energy efficient mobile cloud computing powered by wireless energy transfer. IEEE J. Selec. Areas Commun. **34**, 1757–1771 (2016)
3. M. Chiang, T. Zhang, Fog and IoT: an overview of research opportunities. IEEE Int. Things J. **3**, 854–864 (2016)
4. B.-G. Chun, S. Ihm, P. Maniatis, M. Naik, A. Patti, CloneCloud: elastic execution between mobile device and cloud, in *Proceedings of the Sixth Conference on Computer Systems*, EuroSys'11 (New York, ACM, 2011), pp. 301–314
5. B.-G. Chun, P. Maniatis, Augmented smartphone applications through clone cloud execution, in *Proceedings of the 12th Conference Hot Topics Operating Systems* (2009), p. 8
6. M. Satyanarayanan, P. Bahl, R. Cáceres, The case for VM-based cloudlets in mobile computing. IEEE Pervasive Comput. **8**, 14–23 (2009)
7. G. Huerta-Canepa, D. Lee, A virtual cloud computing provider for mobile devices, in *Proceedings of the 1st ACM Workshop on Mobile Cloud Computing Services: Social Networks and Beyond* (2010), p. 6
8. H. Ba, W. Heinzelman, C.-A. Janssen, J. Shi, Mobile computing-A green computing resource, in *Proceedings of IEEE Wireless Communications and Networking Conference (WCNC)* (2013), pp. 4451–4456

9. E. Cuervo, A. Balasubramanian, D.-K. Cho, A. Wolman, S. Saroiu, R. Chandra, P. Bahl, MAUI: Making smartphones last longer with code offload, in *Proceedings of ACM International Conference on Mobile Systems, Applications, and Services (MobiSys)* (2010), pp. 49–62
10. Y. Wen, W. Zhang, H. Luo, Energy-optimal mobile application execution: Taming resource-poor mobile devices with cloud clones, in *Proceedings of IEEE International Conference on Computer Communications (INFOCOM)* (2012), pp. 2716–2720
11. O. Muñoz, A. Pascual-Iserte, J. Vidal, Optimization of radio and computational resources for energy efficiency in latency-constrained application offloading. IEEE Trans. Vehic. Technol. **64**, 4738–4755 (2015)
12. R. Kaewpuang, D. Niyato, P. Wang, E. Hossain, A framework for cooperative resource management in mobile cloud computing. IEEE J. Selec. Areas Commun. **31**, 2685–2700 (2013)
13. X. Chen, Decentralized computation offloading game for mobile cloud computing. IEEE Trans. Parallel Distributed Syst. **26**, 974–983 (2015)
14. S. Sardellitti, G. Scutari, S. Barbarossa, Joint optimization of radio and computational resources for multicell mobile-edge computing. IEEE Trans. Signal Inf. Process. Over Netw. **1**, 89–103 (2015)
15. X. Chen, L. Jiao, W. Li, X. Fu, Efficient multi-user computation offloading for mobile-edge cloud computing. IEEE/ACM Trans. Netw. **24**, 2795–2808 (2016)
16. S. Kosta, A. Andrius, P. Hui, R. Mortier, X. Zhang, ThinkAir: Dynamic resource allocation and parallel execution in the cloud for mobile code offloading, in *Proceedings of IEEE International Conference on Computer Communications (INFOCOM)* (2012), pp. 945–953
17. Y.-H. Kao, B. Krishnamachari, M.-R. Ra, F. Bai, Hermes: Latency optimal task assignment for resource-constrained mobile computing, in *Proceedings of IEEE International Conference on Computer Communications (INFOCOM)* (2015), pp. 1894–1902
18. M.-H. Chen, B. Liang, M. Dong, Joint offloading decision and resource allocation for multi-user multi-task mobile cloud, in *Proceedings of IEEE International Conference on Communications (ICC)* (2016), pp. 1–6
19. E. Meskar, T.D. Todd, D. Zhao, G. Karakostas, Energy aware offloading for competing users on a shared communication channel. IEEE Trans. Mob. Comput. **16**, 87–96 (2017)
20. S. Jošilo, G. Dán, A game theoretic analysis of selfish mobile computation offloading, in *Proceedings of IEEE International Conference on Computer Communications (INFOCOM)* (2017), pp. 1–9
21. H. Cao, J. Cai, Distributed multiuser computation offloading for cloudlet-based mobile cloud computing: a game-theoretic machine learning approach. IEEE Trans. Vehic. Technol. **68**, 752–764 (2018)
22. H. Lagar-Cavilla, N. Tolia, E.D. Lara, M. Satyanarayanan, D. O'Hallaron, Interactive resource-intensive applications made easy, in *ACM/IFIP/USENIX International Conference on Distributed Systems Platforms and Open Distributed Processing* (2007), pp. 143–163
23. B.-G. Chun, S. Ihm, P. Maniatis, M. Naik, A. Patti, CloneCloud: Elastic execution between mobile device and cloud, in *Proceedings of the Sixth Conference on Computer Systems* (2011), pp. 301–314
24. E. Cuervo, A. Balasubramanian, D.-K. Cho, A. Wolman, S. Saroiu, R. Chandra, P. Bahl, MAUI: Making smartphones last longer with code offload, in *Proceedings of the 8th International Conference on Mobile Systems, Applications, and Services* (2010), pp. 49–62
25. P. Sadeghi, R.A. Kennedy, P.B. Rapajic, R. Shams, Finite-state Markov modeling of fading channels - a survey of principles and applications. IEEE Signal Process. Mag. **25**(5), 57–80 (2008)
26. E.O. Elliott, Estimates of error rates for codes on burst-noise channels. Bell Syst. Tech. J. **42**, 1977–1997 (1963)
27. W. Zhang, Y. Wen, K. Guan, D. Kilper, H. Luo, D.O. Wu, Energy-optimal mobile cloud computing under stochastic wireless channel. IEEE Trans. Wirel. Commun. **12**, 4569–4581 (2013)

28. L.A. Johnston, V. Krishnamurthy, Opportunistic file transfer over a fading channel: A POMDP search theory formulation with optimal threshold policies. IEEE Trans. Wirel. Commun. **5**, 394–405 (2006)
29. W. Zhang, Y. Wen, D.O. Wu, Collaborative task execution in mobile cloud computing under a stochastic wireless channel. IEEE Trans. Wirel. Commun. **14**(1), 81–93 (2014)
30. E.N. Gilbert, Capacity of a burst-noise channel. Bell Syst. Techn. J. **39**(5), 1253–1265 (1960)
31. C.M. Grinstead, J.L. Snell, Markov chains - Chapter 11, in *Grinstead and Snell's Introduction to Probability* (AMS, Providence, 2006), pp. 405–470
32. P. Teymoori, Efficient Mobile Computation Offloading with Hard Task Deadlines and Concurrent Local Execution. PhD Thesis, McMaster University, 2021
33. G. Peskir, A. Shiryaev, *Optimal Stopping and Free-Boundary Problems*. Lectures in Mathematics ETH Zurich (Springer, Dordrecht, 2006)
34. A.P. Miettinen, J.K. Nurminen, Energy efficiency of mobile clients in cloud computing. HotCloud **10**(4), 19 (2010)

Chapter 11
Online Incentive Mechanism Design in Edge Computing

Gang Li and Jun Cai

11.1 Introduction

As a promising technology to further enhance mobile users' Quality of Experience (QoE), Edge Computing (EC) aims to shorten the round-trip latency by moving computational resources closer to mobile users. In reality, from the economical perspective, edge servers may not always be willing to provide computation services without reimbursements since they need to consume their own energy, storage, and computation resources. Thus, incentive mechanisms have to be used to incentivize edge servers to provide such services. Moreover, any practical EC system has to face a dynamic environment. For example, the computation tasks in an EC system are ordinarily generated over time, so that they cannot arrive at the edge server at the same time. Furthermore, in Collaborative Edge Computing (CEC) systems, where vacant resourceful mobile users, called collaborators, are involved in providing computation services, collaborators are not always stationary and can opt-in and opt-out of the network for their own convenience. Therefore, designing online incentive mechanisms becomes pivotal and meaningful for practical edge computing systems. Different from the counterpart of offline incentive mechanisms, in online incentive mechanisms, the allocation of computation resource, task offloading decisions, and reimbursements have to be made right upon the arrival of computation tasks or collaborators. In general, compared with the traditional offline incentive mechanisms, the design of online incentive mechanisms is more challenging due to difficulties in the following facts:

G. Li · J. Cai (✉)
Department of Electrical and Computer Engineering, Concordia University, Montreal, QC, Canada
e-mail: Gang.Li@mail.concordia.ca; Jun.Cai@concordia.ca

L. Cai et al. (eds.), *Broadband Communications, Computing, and Control for Ubiquitous Intelligence*, Wireless Networks,
https://doi.org/10.1007/978-3-030-98064-1_11

- Decisions must be made rationally in the online incentive mechanism to improve performance.
- Since the information arrives on the fly, it is much more difficult to prevent the current arriving mobile users from misreporting their private information without knowing later arrivals' information [1].

The primal–dual theory [2], originally used to solve linear programming problems, where the dual problem is much more tractable than the primal one, provides a promising approach to address these challenges. In this chapter, we will present a comprehensive description of the primal–dual-based online mechanism design method for both linear and nonlinear systems. Then, example applications of the primal–dual-based online mechanism design method in EC systems are discussed.

11.2 Mechanism Design and Auction

This section provides basic concepts and properties in mechanism designs and basic terminologies in auctions. Mechanism design is a subfield of game theory that is used to define game rules so as to achieve a desired outcome. Formally, a mechanism should consist of the following ingredients:

- Players $i \in \mathcal{N}$ with cardinality of $|\mathcal{N}| = N$ and preference types $b_i \in \mathcal{B}_i$
- A strategy space $\mathcal{S} = \mathcal{S}_1 \times \cdots \times \mathcal{S}_N$, where player i chooses strategy $s_i(b_i) \in \mathcal{S}_i$
- Utility $u_i(s_i(b_i), \boldsymbol{s}_{-i}(\boldsymbol{b}_{-i}))$, where $\boldsymbol{s}_{-i}(\boldsymbol{b}_{-i})$ is the set of strategies excluding player i

In the following, for notational convenience, we use $u_i(b_i, \boldsymbol{b}_{-i})$ to denote $u_i(s_i(b_i), \boldsymbol{s}_{-i}(\boldsymbol{b}_{-i}))$. Traditionally, the following two important properties must be satisfied in mechanism design.

Definition 11.1 (Incentive Compatibility (IC)) A mechanism is incentive-compatible if for any type $b_i = v_i$, it is the dominant strategy regardless of other players' type, where v_i is the true type of player i. Then the incentive compatibility requires

$$u_i(v_i, \boldsymbol{b}_{-i}) \geq u_i(b_i, \boldsymbol{b}_{-i}). \tag{11.1}$$

In real-world scenarios, this property is of great importance as it can force each player to play with their true type. Note that incentive compatibility is also known as truthfulness in the literature.

Definition 11.2 (Individual Rationality (IR)) A mechanism is individual-rational if the utility of each player is non-negative, i.e.,

$$u_i(\boldsymbol{b}) \geq 0. \tag{11.2}$$

The property of individual rationality can guarantee that each player can obtain non-negative benefit if they are willing to participate.

Since auctions are known as mechanisms, we refer to the mechanism design as the auction design, and these two concepts are interchangeably used in this chapter. The history of auction theory in wireless networks dates back to the application of spectrum license distribution [3]. Commonly, auctions consist of two major components, which are the allocation rule, i.e., $\mathcal{S}$, and the payment rule (pricing policy), i.e., $\mathcal{P}$. Moreover, in most auctions, players are required to submit their types, and the winners and corresponding rewards will be made based on the designed allocation and payment rules. Some basic terminologies and definitions on auctions are introduced below:

- Auctioneer: An auctioneer is as an intermediate agent that implements allocation and payment rules in the auction. In wireless networks, the auctioneer could often be the Base Station (BS).
- Seller: A seller offers its items for sale. The items could be spectrum, bandwidth, and computation resources in wireless networks.
- Bidder: A bidder is a buyer that wants to buy the items from the seller.
- Bid: The bid typically consists of the items it wants to buy and the maximal price the bidder is willing to pay for these items.

Definition 11.3 (Social Welfare (SW)) The social welfare is defined as the summation of utilities in the auction including all bidders and sellers. After cancelling the payment term, the social welfare in the auction can be expressed as

$$\sum_{i \in \mathcal{N}} v_i x_i, \tag{11.3}$$

where x_i is a binary variable, which means if bidder i wins, $x_i = 1$; otherwise, $x_i = 0$. Note that SW maximization is the most common objective in auctions.

Definition 11.4 (Competitive Ratio (CR)) An online algorithm $\mathcal{A}$ for a minimization problem (or maximization problem) is α-competitive if $\mathcal{A}(I) \leq \alpha OPT_{offline}(I)$ (or $\mathcal{A}(I) \geq \alpha OPT_{offline}(I)$), where $\alpha \geq 1$ is a constant, called CR, I is a sequence of inputs, and $\mathcal{A}(I)$ is the objective value achieved by online algorithm $\mathcal{A}$, while $OPT_{offline}(I)$ is the optimum value achieved by an optimal offline algorithm.

Definition 11.5 (Online Auction (OA)) An online auction, i.e., $M = (\mathcal{S}, \mathcal{P})$, should solve $\mathcal{S}$ and $\mathcal{P}$ along time, and the designed online auction should satisfy the requirements of IC, IR, and have a theoretic CR.

11.3 Primal–Dual-Based Online Incentive Mechanism

In this section, we will introduce the primal–dual-based online incentive mechanism design method, which is one of the most widely used ones in the literature. The wide study of this design method started when an important theorem was proposed in 2000 and 2003 [4]. In this theorem, the basic requirements for designing an online truthful mechanism were specified:

- The necessary payment of any bidder is only related to the allocated items and independent of the bidders' bids.
- The items that the bidder can obtain depend on whether they maximize the utility of this bidder.

Obviously, the first condition implies that the payment rule should be a function of the items only, and the second one defines the allocation rule.

11.3.1 Primal–Dual-Based Method for Linear Systems

In practical systems, such as EC systems, many studied problems with the consideration of economic aspects can usually be formulated as linear optimization problems [5, 6] where the primal–dual method can work effectively. Besides, by using the primal–dual technique, two extra benefits can be attained.

- Some hidden information for the online truthful mechanism in the primal problem can be revealed in the dual problem, which provides a thread for facilitating mechanism design.
- A theoretical CR analysis is necessary for an online truthful mechanism design. By solving the dual problem, the solution gap between the primal problem and its dual problem can be established. Based on the primal–dual theory, the theoretical CR can be derived.

To better elaborate the idea of this primal–dual-based method for linear system (referred to as PDLN), let us start with a simple example where a seller, owing M kinds of items, would like to sell them to N buyers. Assume that each kind of item, say kind j, has a number of C_j duplicates, each buyer can choose at most one kind to purchase one item within this kind. Moreover, let $v_{i,j}$ be the valuation of buyer i to the j-th kind of item, and $x_{i,j}$ be a binary variable, which means buyer i has the j-th kind of item if $x_{i,j} = 1$; otherwise, $x_{i,j} = 0$. By considering the objective to maximize SW, we can formulate this problem as follows:

$$\max_{x_{i,j}} \sum_{i=1}^{N} \sum_{j=1}^{M} v_{i,j} x_{i,j} \qquad \text{Primal}$$

$$
\begin{aligned}
\text{s.t. } C_1 &: \sum_{j=1}^{M} x_{i,j} \le 1, \quad \forall i,\\
C_2 &: \sum_{i=1}^{N} x_{i,j} \le C_j, \quad \forall j,\\
C_3 &: x_{i,j} \in \{0, 1\}.
\end{aligned}
$$

Based on the dual theory, the dual problem is

$$
\max_{u_i,\, p_j} \sum_{i=1}^{N} u_i + \sum_{j=1}^{M} C_j p_j \qquad \text{Dual}
$$

$$
\begin{aligned}
\text{s.t. } C_4 &: u_i + p_j \ge v_{i,j}, \quad \forall (i, j),\\
C_5 &: u_i \ge 0 \; p_j \ge 0,
\end{aligned}
$$

where u_i and p_j are the dual variables of constraints C_1 and C_2, respectively. Note that in linear programming, the complementary slackness condition, which is considered to be equally important as Karush-Kuhn-Tucker (KKT) conditions in nonlinear programming, plays a vital role in not only linear primal–dual theory, but also the online truthful auction design. Based on the problems (Primal) and (Dual), the complementary slackness conditions can be expressed as follows:

$$
\left(\sum_{j=1}^{M} x_{i,j} - 1\right) u_i = 0, \tag{11.4}
$$

$$
\left(\sum_{i=1}^{N} x_{i,j} - C_j\right) p_j = 0. \tag{11.5}
$$

Based on (11.4) and (11.5), we have the following conclusions:

- CS_1: $u_i = v_{i,j} - p_j$, whenever $x_{i,j} = 1$.
- CS_2: $\sum_{i=1}^{N} x_{i,j} = C_j$, whenever $p_j > 0$.
- CS_3: $\sum_{j=1}^{M} x_{i,j} = 1$, whenever $u_i > 0$.

Note that by observing the dual problem, it can be revealed that one of the dual variables, i.e., u_i, can be interpreted as the utility of the requester as long as the physical meaning of another dual variable, i.e., p_j, is defined as the payment. This is reasonable because $v_{i,j}$ is the valuation of buyer i to the j-th kind in C_4, and u_i should be the utility as long as p_j is interpreted to be the payment. Furthermore,

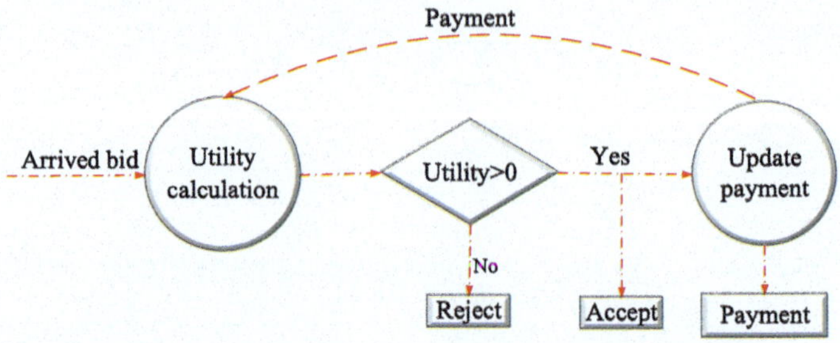

Fig. 11.1 The procedures for online truthful mechanism with linear constraints

based on the conditions from CS_1 to CS_3 and the second basic condition in the aforementioned theorem [4], the allocation rule can be determined by choosing the item whose maximal utility is larger than zero. As shown in Fig. 11.1, at the arrival of the current bid, its utility will be calculated based on a predesigned payment, which will be explained later. This bidder's demand will then be satisfied, and its payment will be determined if its utility is larger than zero; otherwise, its demand will be rejected. This process will continue until a predefined termination rule is satisfied. Note that this method can only obtain one of the feasible solutions to both primal and dual problems, so that there may be a gap between these two problems.

After designing the allocation rule, the payment should also be well designed because it can also directly affect the ultimate optimality. There are three commonly used payment design methods.

The first form of payment increases the payment exponentially after each resource allocation [7], which can be mathematically expressed as follows:

$$p_j = A\left(\frac{B}{A}\right)^{\frac{y}{c_j}}, \tag{11.6}$$

where y represents the accumulated allocated number of j-th kind of items so far. A and B are the minimal and maximal prices per item, respectively. Note that at the very beginning of an auction, the payment equals the minimal price, i.e., $y = 0$. Then, the payment increases exponentially with the accumulated allocated items. When the j-th kind of items is sold out, the payment reaches the maximal value. Under this payment policy, the theoretical CR can be derived as $O(\log(\frac{B}{A}))$. Unfortunately, the actual derivation procedure is quite complex. Moreover, it will be much harder to sell more items as time goes on because the payment may become extremely high. However, one of the advantages of using this payment form is the feasibility, i.e., the designed online truthful mechanism never violates any constraints of the original problem.

In order to sell more items in the later stage, the second form of payment, as indicated in (11.7), is a function of its previous payment and iteratively updates the payment when an item is sold, but with a relatively low increasing rate [8].

$$p_j^{(\ell)} = p_j^{(\ell-1)}\left(1+\frac{x}{\alpha C_j}\right)+\frac{x}{\beta C_j}, \tag{11.7}$$

where the superscript ℓ and the variable x represent the number of price updates and the current allocated number of items in j-th kind, respectively; α and β are the predetermined system parameters. Note that the payment at the beginning of auctions, i.e., $p_j^{(0)}$, is set to be very small, and then it is iteratively updated with each allocated item. Under this payment, the theoretical CR could be a constant. However, this payment form cannot always guarantee feasibility because no considerations for potential violation of constraints are contained in the payment updating. Therefore, a theoretical feasibility analysis has to be provided in the mechanism design, and a preventative step has to be devised in case of infeasibility.

If the central controller or auctioneer knows some a prior information, such as the empiric data about the percentage that the seller would like to sell its items, it is provable that the theoretical performance can be further enhanced [8]. This third form of payment can be expressed as

$$p_j^{(\ell)} = p_j^{(\ell-1)} + \left(\max\{p_j^{(\ell-1)}, p^d\} + \frac{p^d(1-\eta)}{\eta}\right)\frac{x}{C_j}, \tag{11.8}$$

where the parameter η is the prior information, which can be estimated from historic observations; $p^d = f(\eta)$ is a function of η, which can be predetermined. Note that compared to the first and second payment forms, where the payment is always updated exponentially, the third one updates the payment almost linearly till the given percentage of items is sold and then exponentially afterward. The rationality of the third method results from the fact that the first and second payment methods decrease the selling probability of some items in the future auction campaigns too fast to accept more buyers so that the total SW could be degraded. The theoretical CR of the third payment method can reach $O(\frac{1}{\beta\eta})$. In addition, the feasibility analysis should also be provided for the second payment form.

We summarize the core spirit of the PDLN as follows. Initially, all kinds of item have the same price. Then, the prices for different kinds of items increase over time at different "speeds" depending on the "popularity" of the corresponding kinds. Whenever a kind is sold, prices of this kind of item are increased in a multiplicative way. Thus, the algorithm learns the "right" price for the current kind by selling previous ones.

11.3.2 Primal–Dual-Based Method for Nonlinear Systems

The PDLN can only be applied to the problems with both linear objective and linear constraints, while in practical systems, it is common that studied problems include nonlinear constraints. Inspired by the PDLN and the nonlinear primal–

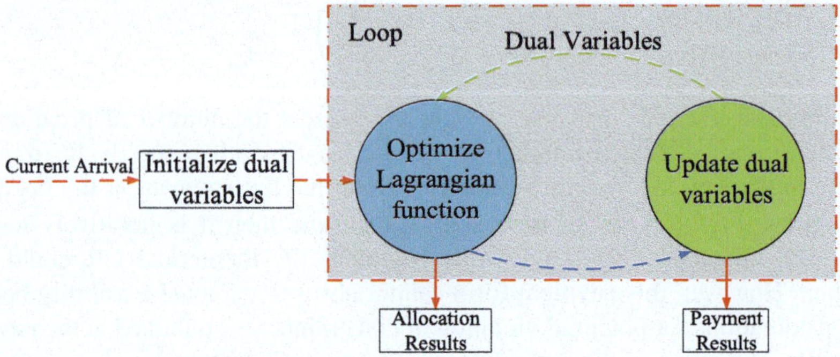

Fig. 11.2 The procedures for online truthful mechanism with nonlinear constraints

dual theory, we briefly introduce another primal–dual-based method for nonlinear systems (PDNON). For explanation purposes, consider a scenario where M sellers are competing to sell their items to one buyer. Each seller has a kind of divisible item and submits a unit price per quantity for kind j, i.e., c_j to this buyer. Furthermore, the total quantities of purchased items satisfy a nonlinear constraint, such as $\sum_{j=1}^{M} g(x_j) \geq \overline{K}$, where x_j denotes the quantity of item purchased in j-th kind. The objective is to minimize the social cost of the buyer, i.e., maximize the social welfare. Note that in this case, the primal problem, which is a Lagrangian function with the nonlinear constraints, defines the allocation rule, while the dual problem is equivalent to maximizing the utility of each seller when the payment rule is defined as the summation of all dual variables. As shown in Fig. 11.2, at the beginning, dual variables corresponding to nonlinear constraints are initialized. After that the constructed Lagrangian function with corresponding dual variables is optimized, and the dual variables in the dual problem will be updated subsequently. Note that a loop is needed to iteratively optimize the Lagrangian function upon the new arrival and learn the right payment to the corresponding seller.

Obviously, compared to the PDLN, PDNON makes it possible to design online truthful mechanisms with nonlinear constraints, especially when allocation variables in the mechanism have real values. It is worth noting that the PDNON can still be applicable for binary allocation variables, but a comprehensive analysis on the feasibility of allocation results should be carried out. Since PDNON consists of a loop process, as shown in Fig. 11.2, to iteratively solve Lagrangian function and update dual variables for each arrival, it has higher complexity and consumes more time in calculation. Moreover, since a loop is needed to derive allocation and payment results for each new arrival, a theoretical analysis on the convergence of this algorithm becomes necessary.

11.4 Application of Primal–Dual Online Incentive Mechanism Design in Edge Computing

In this section, we illustrate how the primal–dual-based mechanism design method can be used to address resource management issue under a dynamic environment in edge computing systems. Specifically, in our system model, upon the arrival of a mobile user who requests task offloading, the BS needs to make a decision right away without knowing any future information. A social welfare maximization problem is formulated, which integrates collaborator selection, communication and computation resource allocation, transmission and computation time scheduling, as well as pricing policy design. Then, an online incentive mechanism integrating computation and communication resource allocation is presented.

11.4.1 System Model Descriptions

In this section, we describe the system under consideration and model the interaction between the BS and arrived requesters as an online auction. After that, the corresponding offline optimization problem is formulated.

11.4.1.1 System Model

We consider a mobile edge computing system as shown in Fig. 11.3. The system consists of a BS integrating edge servers and several mobile users who can also provide computation services, called collaborators. These collaborators are recruited by the BS and are willing to provide computation resources if reimbursements are given. From time to time, there are mobile users, called requesters, who request computing services. The requesters randomly arrive in a sequence, and we denote t_i as the arrival time instant of requester i. Note that the BS does not have any *a priori* information on requesters' arrival times.

Consider a time-slotted structure with a slot length of Δt. For each arrived requester $i \in \mathcal{U}$, where $\mathcal{U}$ denotes the set of all requesters, let $\mathcal{M}_i$ be the set of available collaborators that can provide computation services to it. Note that the set $\mathcal{M}_i$ could be available to requester i by applying the discovery approach [9]. We further define $\mathcal{N}_i = \mathcal{M}_i \cup 0$, where the index 0 represents the BS. Obviously, $\mathcal{N}_i$ consists of all collaborators and the BS that requester i can offload its task to. To simplify notation, we will use the term "task executor" to denote any collaborator or the BS throughout this chapter. We further let $\mathcal{M}$ be the set of all collaborators. Note that since user mobility cannot affect the offloading process in the coverage of one MEC server, it is ignored in our discussion. In fact, if the collaborator moves away after the arrival of the requester's task, the task will fail to be transmitted to the collaborator, and no rewards can be obtained. Thus, there are no

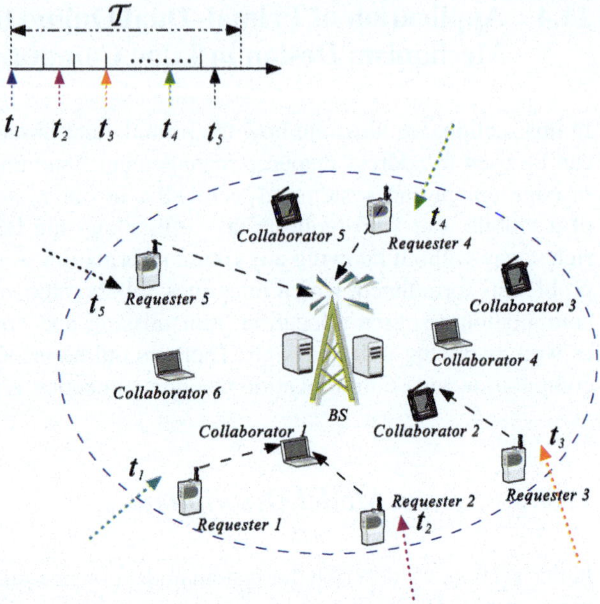

Fig. 11.3 The system model of collaborative task offloading in mobile edge computing where there are 5 requesters arriving at the network in an online fashion and submitting their requests

incentives for collaborators to move. For the movement of requester, if the requester moves around the corresponding collaborator, D2D technologies [10] can be applied to transmit the computational results and the reimbursements between requester and collaborator. Otherwise, results and reimbursements can also be received and transmitted through the cellular link.

The task from requester $i \in \mathcal{U}$ is denoted as $T_i = (s_i, \tau_i)$, where s_i is the size (in bits) of the offloaded task and τ_i is the maximal tolerance delay. Note that the task offloading to the collaborator can be done through a D2D link [10]. Each task i requires Q_i CPU cycles for execution and can be calculated by $Q_i = \kappa_i s_i$ [11], where κ_i is the CPU cycles coefficient. We also define the allocated CPU frequency at task executor j as $f_{i,j}$. Then, the required computation time at task executor j for task i equals

$$I_{i,j}^{C} = \frac{Q_i}{f_{i,j}} = d_{i,j} - a_{i,j}, \tag{11.9}$$

where $a_{i,j}$ and $d_{i,j}$ denote the starting and ending computation time instants, respectively. In addition, given that each requester i is allocated an orthogonal channel with bandwidth $\phi_{i,j}$ for task offloading to task executor j, the transmission rate from requester i to task executor j equals

$$r_{i,j} = \phi_{i,j} \log_2(1 + \gamma_{i,j}), \tag{11.10}$$

where $\gamma_{i,j} = \frac{\Lambda_{i,j}|h_{i,j}|^2}{\sigma^2}$ is the signal-to-noise ratio (SNR), σ^2 is the average power of background noise, and $\Lambda_{i,j}$ and $h_{i,j}$ are the transmission power and the channel gain between requester i and task executor j, respectively. Thus, the transmission time from requester i to task executor j can be calculated as

$$I_{i,j}^T = \frac{s_i}{r_{i,j}} = o_{i,j} - g_{i,j}, \tag{11.11}$$

where $g_{i,j}$ and $o_{i,j}$ denote the starting and ending transmission time instants, respectively. Note that since the available computation and the transmission resources are time-varying, both transmission time and computation time should be optimally determined for each offloading task. Therefore, $g_{i,j}$, $o_{i,j}$, $a_{i,j}$, and $d_{i,j}$ are decision variables. To meet the task delay requirement, we need

$$d_{i,j} - t_i \leq \tau_i. \tag{11.12}$$

In (11.12), similar to other studies in [12, 13], we ignore the time for the task executor to send the computation result back to the requester because the data size of outcomes for many applications is commonly very small. In summary, the whole operation procedure of this system is described as follows:

Step 1. Upon the arrival of requester i, it submits multiple bids to the BS, denoted by $B_{i,j} = (T_i, t_i, v_{i,j})$, $j \in \mathcal{N}_i$, where $v_{i,j}$ is the valuation of requester i to task executor j, which represents its preference to offload the task to task executor j.[1]

Step 2. After collecting the bids from requester i, the BS makes a decision, denoted by a binary variable $x_{i,j}$, whether to accept this requester such that $x_{i,j} = 1$ means requester i is accepted. Otherwise, $x_{i,j} = 0$. The BS further determines what are the optimal transmission and computation time instants for this task.

Step 3. The BS sends the optimal results obtained in Step 2 to requester i and notifies the selected task executor to prepare for the task execution.

Step 4. After the task is completed, requester i will be charged by $p_{i,j}$, which is another decision variable, and the task executor returns computation results to it.

Obviously, the interactions between task executors and requesters can be modelled as an online auction, where the BS is the auctioneer, requesters are buyers, and all the task executors are sellers. Requesters may strategically misreport their private information (i.e., $B_{i,j}$) in order to get more benefits. For example, requester i, who will lose in the auction, may submit false bid $B_{i,j}^{'}$, where $T_i^{'} = T_i, t_i^{'} = t_i$,

[1] Since the collaborators are heterogeneous in terms of available computation resources and geographical locations, which makes the channel conditions between the collaborators and the requesters different, each requestor values the nearby collaborators differently.

and $v'_{i,j} > v_{i,j}$. In this case, this requester has a higher chance to win the auction than if it had reported truthfully. Thus, truthfulness has to be guaranteed through the designed incentive mechanism. Following the previous discussions, the utilities of requester i and the task executor j can be, respectively, expressed as

$$u_i = v_{i,j} - p_{i,j}, \tag{11.13}$$

$$u_j = p_{i,j} - e_{i,j} c_j, \tag{11.14}$$

where $e_{i,j} = Q_i \xi_j f_{i,j}^2$ and c_j are the energy consumption for executing task i and the unit energy cost of task executor j, respectively, and ξ_j is the energy consumption coefficient [14].

11.4.1.2 Problem Formulation

Our target is to design an online auction that satisfies the properties of IR, IC, and has a sound CR. If the information about all tasks is known, we can formulate the corresponding offline optimization problem (MSW) as

$$\max_{\boldsymbol{X},\boldsymbol{G},\boldsymbol{O},\boldsymbol{A},\boldsymbol{D},\boldsymbol{P}} \sum_{i \in \mathcal{U}} \sum_{j \in N_i} w_{i,j} x_{i,j} \qquad \text{MSW}$$

$$\text{s.t. } C_1 : \sum_{j \in N_i} x_{i,j} \leq 1, \quad \forall i \in \mathcal{U};$$

$$C_2 : \sum_{j \in N_i} (o_{i,j} - a_{i,j}) x_{i,j} \leq 0, \quad \forall i \in \mathcal{U};$$

$$C_3 : \text{(11.9), (11.11) and (11.12)};$$

$$C_4 : \sum_{\substack{i \in \mathcal{U}: \\ a_{i,j} \leq t \leq d_{i,j}}} f_{i,j} x_{i,j} \leq F_j, \quad \forall t \in \mathcal{T},\ j \in \mathcal{M} \cup \text{BS};$$

$$C_5 : \sum_{\substack{i \in \mathcal{U}: \\ a_{i,j} \leq t \leq d_{i,j}}} s_i x_{i,j} \leq S_j, \quad \forall t \in \mathcal{T},\ j \in \mathcal{M} \cup \text{BS};$$

$$C_6 : \sum_{\substack{i \in \mathcal{U}: \\ g_{i,j} \leq t \leq o_{i,j}}} \sum_{j \in N_i} \phi_{i,j} x_{i,j} \leq W, \quad \forall t \in \mathcal{T};$$

$$C_7 : \sum_{j \in N_i} (v_{i,j} - p_{i,j}) x_{i,j} \geq 0, \quad \forall i \in \mathcal{U};$$

$$C_8 : \sum_{j \in N_i} (v_{i,j} - p_{i,j}) x_{i,j} \geq \sum_{j \in N_i} (\widetilde{v}_{i,j} - \widetilde{p}_{i,j}) x_{i,j}, \forall i \in \mathcal{U};$$

$$C_9 : x_{i,j} \in \{0, 1\},\ o_{i,j} \in \{t_i, \tau_i\},\ g_{i,j} \in \{t_i, \tau_i\},$$
$$a_{i,j} \in \{t_i, \tau_i\},\ d_{i,j} \in \{t_i, \tau_i\},\ \forall i \in \mathcal{U},\ j \in N_i,$$

where $w_{i,j} = v_{i,j} - e_{i,j}c_j$, F_j, and S_j denote the maximal CPU frequency and storage capacity of the executor j, respectively, W is the whole bandwidth of the system, and $\mathcal{T}$ is the set of all time slots. Decision variables are $\boldsymbol{X} = \{x_{i,j}\}_{i\in\mathcal{U},j\in N_i}$, $\boldsymbol{G} = \{g_{i,j}\}_{i\in\mathcal{U},j\in N_i}$, $\boldsymbol{O} = \{o_{i,j}\}_{i\in\mathcal{U},j\in N_i}$, $\boldsymbol{A} = \{a_{i,j}\}_{i\in\mathcal{U},j\in N_i}$, $\boldsymbol{D} = \{d_{i,j}\}_{i\in\mathcal{U},j\in N_i}$, and $\boldsymbol{P} = \{p_{i,j}\}_{i\in\mathcal{U},j\in N_i}$. Constraint C_1 ensures that each requester can offload its task to at most one task executor. Constraint C_2 means the transmission process occurs before the computation process for any task. Constraint C_3 represents the time rationality and delay requirement. Constraints C_4 and C_5 indicate constrains on the allocated CPU frequencies and storage resources at any task executor, respectively. Constraint C_6 specifies that the allocated bandwidths cannot exceed W, and constraints C_7 and C_8 are the requirements of IR and IC, respectively. Constraint C_9 defines decision variables $\boldsymbol{G}$, $\boldsymbol{O}$, $\boldsymbol{A}$, $\boldsymbol{D}$, and $\boldsymbol{P}$ to be continuous and $\boldsymbol{X}$ to be binary variables.

Obviously, this formulated offline optimization problem is a mixed integer problem and is usually NP-hard [15]. In addition, this formulation requires complete information on system operation in the future. In the following, we will design Online Mechanism Integrating Allocation Rule and Payment (OMAP) to find solutions on the fly.

11.4.2 The Design of OMAP

In this section, we design an online mechanism to find solutions to the problem (MSW). Note that since the payments are not in the objective function in (MSW) but only in the constraints C_7 and C_8, we can decouple (MSW) into two subproblems without losing optimality: an allocation subproblem (including task executor selection, resource allocation, and time scheduling) and a payment rule subproblem. In the following, we first reformulate the offline problem and then solve the allocation problem. After that, a corresponding payment scheme will be designed to not only satisfy IC, but also maintain IR.

11.4.2.1 Problem Reformulation

Since constraints C_4 and C_5 in (MSW) have the same structure, we combine them together as

$$C_{10} : \sum_{\substack{i\in\mathcal{U}: \\ a_{i,j}\le t\le d_{i,j}}} r^k_{i,j}x_{i,j} \le R^k_j, \quad \forall t \in \mathcal{T},\ \forall j \in \mathcal{M},\ \forall k \in \mathcal{K},$$

where

$$r_{i,j}^k = \begin{cases} s_i & \text{if } j \in \mathcal{N}_i \text{ and } k = 1; \\ f_{i,j} & \text{if } j \in \mathcal{N}_i \text{ and } k = 2, \\ 0 & \text{otherwise;} \end{cases} \qquad R_j^k = \begin{cases} S_j & \text{if } j \in \mathcal{M} \text{ and } k = 1; \\ F_j & \text{if } j \in \mathcal{M} \text{ and } k = 2; \\ 0 & \text{otherwise;} \end{cases}$$

$\mathcal{K} = \{1, 2\}$, and $l_{i,j} = l_{i,j}^1 \cup l_{i,j}^2$ denotes all feasible transmission and computation time scheduling pairs from requester i to task executor j with satisfaction of constraints C_2, C_3, and C_9. Let $l_{i,j}^1 = \{l_{i,j}^1(1), l_{i,j}^1(2), \cdots\}$ and $l_{i,j}^2 = \{l_{i,j}^1(1), l_{i,j}^2(2), \cdots\}$ be sets of all the feasible transmission and computation time scheduling, respectively, and each entry $l_{i,j}^1(\ell)$ or $l_{i,j}^2(\ell)$ indicates the ℓ-th feasible scheduling scheme. Let $\mathcal{L}_{i,j}$ be the index set of all feasible solutions from requester i to task executor j. Note that $\mathcal{L}_{i,j}$ has a potentially exponential number of feasible solutions with respect to the decision variables $\boldsymbol{G}$, $\boldsymbol{O}$, $\boldsymbol{A}$, and $\boldsymbol{D}$.

Then, the allocation problem can be rewritten from the original (MSW) as

$$\max_{\hat{X}} \sum_{i \in \mathcal{U}} \sum_{j \in \mathcal{N}_i} \sum_{\ell \in \mathcal{L}_{i,j}} w_{i,j}^\ell x_{i,j}^\ell \qquad \text{EQMSW}$$

$$\text{s.t. } C_{11} \sum_{j \in \mathcal{N}_i} \sum_{\ell \in \mathcal{L}_{i,j}} x_{i,j}^\ell \leq 1, \ \forall i \in \mathcal{U};$$

$$C_{12}: \sum_{i \in \mathcal{U}} \sum_{\ell : t \in l_{i,j}^2(\ell) \in l_{i,j}^2} r_{i,j}^k x_{i,j}^\ell \leq R_j^k, \ \forall t \in \mathcal{T}, \forall j \in \mathcal{M}, k \in \mathcal{K};$$

$$C_{13}: \sum_{i \in \mathcal{U}} \sum_{j \in \mathcal{N}_i} \sum_{\ell : t \in l_{i,j}^1(\ell) \in l_{i,j}^1} \phi_{i,j} x_{i,j}^\ell \leq W, \quad \forall t \in \mathcal{T};$$

$$C_{14}: x_{i,j}^\ell \in \{0, 1\}, \ \forall i \in \mathcal{U}, \ j \in \mathcal{N}_i, \ \ell \in \mathcal{L}_{i,j}.$$

Here, $\hat{\boldsymbol{X}} = \{x_{i,j}^\ell, i \in \mathcal{U}, j \in \mathcal{N}_i, \ell \in \mathcal{L}_{i,j}\}$ are new decision variables; $w_{i,j}^\ell = v_{i,j} - c_j e_{i,j}^\ell$, where $e_{i,j}^\ell$ is the energy consumption at task executor j when the ℓ-th feasible scheduling scheme is selected. In order to devise an online mechanism with sound CR, we resort to its dual problem. The dual problem of (EQMSW) can be formulated as follows by relaxing the constraint C_{14} to allow any values between 0 and 1.

$$\min_{u, \hat{p}} \sum_{i \in \mathcal{U}} u_i + \sum_{t \in \mathcal{T}} W \hat{p}_t + \sum_{t \in \mathcal{T}} \sum_{j \in \mathcal{M}} \sum_{k \in \mathcal{K}} R_j^k \hat{p}_{j,t}^k \qquad \text{EQDP}$$

$$\text{s.t. } C_{15}: u_i + \sum_{t \in l_{i,j}^2(\ell)} \sum_{k \in \mathcal{K}} r_{i,j}^k \hat{p}_{j,t}^k + \sum_{t \in l_{i,j}^1(\ell)} \phi_{i,j} \hat{p}_t \geq w_{i,j}, \ \forall i \in \mathcal{U}, \ j \in \mathcal{N}_i, \ \ell \in \mathcal{L}_{i,j};$$

$$C_{16}: u_i \geq 0, \ \hat{p}_{j,t}^k \geq 0, \ \hat{p}_t \geq 0, \ \forall i \in \mathcal{U}, \ j \in \mathcal{M}_i, \ k \in \mathcal{K},$$

where u_i, $\hat{p}_t$, and $\hat{p}^k_{j,t}$ are the dual variables corresponding to the constraints C_{11}, C_{12}, and C_{13}, respectively. Note that the dual variables $\hat{p}^k_{j,t}$ can be interpreted as the marginal price of task executor j's available computation frequencies and storages resources (i.e., $k = 1$ or 2) at time slot t, while dual variable $\hat{p}_t$ can be regarded as the marginal price of the available bandwidth in the network. Thus, $\sum_{t\in l^2_{i,j}(\ell)} \sum_{k\in\mathcal{K}} r^k_{i,j}\hat{p}^k_{j,t}$ and $\sum_{t\in l^1_{i,j}(\ell)} \phi_{i,j}\hat{p}_t$ represent the total computation cost and the total transmission cost, respectively. Moreover, u_i can be considered as the utility of requester i. In the following sections, we will apply these observations to design an online mechanism to address problem (MSW).

11.4.2.2 OMAP

In our formulated online mechanism, we need to decide whether to accept a new task upon its arrival and which task executor should be assigned as well as how much the requester should be charged. Our basic idea is that if the BS decides to assign the current requester i's task to task executor j, we increase the unit price of task executor j's resource based on the fact that it will have less resources and then apply these updated prices to decide the acceptance of future arrived requesters.

(1) *Allocation Rule* Under the consideration of IC and IR, u_i in constraint C_{13} has to be maximized and greater than zero. In addition, according to the KKT condition [2] in the primal–dual framework, if requester i is accepted (i.e., $x^\ell_{i,j} = 1$), we have

$$u_i = w_{i,j} - \left(\sum_{t\in l^2_{i,j}(\ell)} \sum_{k\in\mathcal{K}} r^k_{i,j}\hat{p}^k_{j,t} + \sum_{t\in l^1_{i,j}(\ell)} \phi_{i,j}\hat{p}_t \right). \tag{11.15}$$

Combining these two requirements together, u_i can be written as

$$u_i = \max\left\{0, \max_{\substack{j\in\mathcal{N}_i,\\ \ell\in\mathcal{L}_{i,j}}} \left\{ w_{i,j} - \left(\sum_{t\in l^2_{i,j}(\ell)} \sum_{k\in\mathcal{K}} r^k_{i,j}\hat{p}^k_{j,t} + \sum_{t\in l^1_{i,j}(\ell)} \phi_{i,j}\hat{p}_t \right)\right\}\right\}. \tag{11.16}$$

From (11.16), we can design the following allocation rule. Upon the arrival of requester i, we choose a task executor in the set $\mathcal{N}_i$ and a scheduling scheme in set $l_{i,j}$ so that u_i is maximized. We denote such best task executor and the scheduling scheme as j^* and $l_{i,j}(\ell^*)$, respectively. Note that the scheduling scheme $l_{i,j}(\ell^*)$, which maximizes the utility of requester i, is referred to as the optimal scheduling scheme at the collaborator j. If at the optimum, u_i in (11.15) is larger than zero, requester i's task is accepted; otherwise, it is rejected. Note that we also refer to the above allocation rule as the acceptance condition in this chapter.

(2) *Payment Design* As indicated before, the marginal prices increase with the acceptance of requesters, and the designed updating rule is vital to the achievable competitive ratio of our online auction that will be discussed later. The designed marginal price updating rule should follow the following three requirements: (1) at the beginning of the auction, the price should be set sufficiently low in order to allow the acceptance of coming requesters; (2) after allocating resources for each accepted requester, prices should be increased rapidly to save resources for the future requesters with high valuations; and (3) if some resources of any task executor are run out at certain time slot, the prices should be set high enough so that no requesters' tasks can be accepted. By considering all these requirements, for any task executor j, we design the marginal prices updating rule as follows:

$$\hat{p}_{j,t}^{k} = \hat{p}_{j,t}^{k}\left(1 + \frac{r_{i,j}^{k}}{R_{j}^{k}}\right) + \frac{r_{i,j}^{k}}{\Gamma_{i,j} R_{j}^{k}},\ \forall t \in [g_{i,j}, o_{i,j}],\ \forall k \in \mathcal{K}, \tag{11.17}$$

$$\hat{p}_{t} = \hat{p}_{t}\left(1 + \frac{\phi_{i,j}}{W}\right) + \frac{\phi_{i,j}}{\Phi_{i,j} W},\ \forall t \in [a_{i,j}, d_{i,j}], \tag{11.18}$$

where $\Gamma_{i,j} = \frac{\sum_{k\in\mathcal{K}} r_{i,j}^{k} I_{i,j}^{C}}{w_{min}}$, $\Phi_{i,j} = \frac{I_{i,j}^{T}\phi_{i,j}}{w_{min}}$, and w_{min} is the minimal valuable of $w_{i,j}$, which can be estimated from the historical data, and both $\Gamma_{i,j}$ and $\Phi_{i,j}$ can be calculated based on the outputs of the allocation rule. Thus, the price for a requester i to pay can be determined by

$$\begin{cases} p_{i,j} = p_{i,j}^{1} + p_{i,j}^{2} = \sum_{t\in l_{i,j}^{2}(\ell)} \sum_{k\in\mathcal{K}} r_{i,j}^{k} \hat{p}_{j,t}^{k} + \sum_{t\in l_{i,j}^{1}(\ell)} \phi_{i,j}\hat{p}_{t} + e_{i,j}c_{j}, & \\ & \text{if } i \text{ is accepted;} \\ p_{i,j} = 0, & \text{if } i \text{ is rejected.} \end{cases}$$

(3) *Scheduling Design* To implement Algorithm 2, the maximization problem in (11.16) needs to be solved. Since we may confront exponential numbers of feasible solutions, it is inefficient to find the best solution through exclusive searching. To address this issue, we propose a new polynomial time method as follows.

From (11.16), the original optimization problem can be equivalently converted into one that minimizes the summation of $p_{i,j}^{1}$, $p_{i,j}^{2}$, and $e_{i,j}c_{j}$. Note that since we try to arrange a certain number of time slots to complete the transmission and computation processes for a task, the newly formulated problem for requester i offloading task to the task executor j becomes

$$\beta_{i,j} = \min_{\substack{y_j(t), z_j(t),\\ N_{\phi_j}, N_{f_{i,j}}}} \sum_{t\in[t_i, t_i+\tau_i]} \left\{\frac{h(t)}{N_{\phi_j}} y_j(t) + \left(\frac{c_1(t)}{N_{f_{i,j}}} + c_2(t)\right) z_j(t)\right\} + \frac{c_3}{N_{f_{i,j}}^{2}}$$

$$\text{s.t. } C_{15}: y_j(t) < z_j(t),\ \forall t \in [t_i, t_i + \tau_i]; \tag{TSP}$$

$$C_{16}: \sum_{t\in[t_i,t_i+\tau_i]} y_j(t) = N_{\phi_j};$$

$$C_{17}: \sum_{t\in[t_i,t_i+\tau_i]} z_j(t) = N_{f_{i,j}};$$

$$C_{18}: y_j(t) \in \{0,1\},\ z_j(t) \in \{0,1\},\ t \in [t_i, t_i + \tau_i],$$

where $h(t) = \frac{s_i \hat{p}_t}{\Delta t \log 2(1+\gamma_{i,j})}$, $c_1(t) = \frac{Q_i \hat{p}^1_{j,t}}{\Delta t}$, $c_2(t) = s_i \hat{p}^2_{j,t}$, and $c_3 = \frac{c_j Q_i^3 \xi_j}{\Delta t^2}$ for any pair i and j; $y_j(t)$ and $z_j(t)$ are two new binary scheduling decision variables. If $y_j(t)$ or $z_j(t)$ equals 1, it means requester i transmits the task to task executor j or task executor j executes the task at time slot t, respectively; N_{ϕ_j} and $N_{f_{i,j}}$ denote the total required transmission and computation time slots at task executor j, respectively. Due to the integral decision variables and the nonlinear objective, it is nontrivial to solve problem (TSP) directly. Instead, we decouple it by letting the optimal dividing time slot between transmission period and computation period be $\bar{t}_{i,j} \in [t_i, t_i + \tau_i]$. Then, the scheduling problem (TSP) can be equivalently transformed into two subproblems as

$$\beta^1_{i,j} = \min_{y_j(t), N_{\phi_j}} \sum_{t\in[t_i,\bar{t}_{i,j}]} \frac{h(t)}{N_{\phi_j}} y_j(t) \tag{SubP1}$$

$$\text{s.t. } C_{16}, \text{ and } y_j(t) \in \{0,1\}$$

$$\beta^2_{i,j} = \min_{z_j(t), N_{f_{i,j}}} \sum_{t\in(\bar{t}_{i,j},t_i+\tau_i]} \left(\frac{c_1(t)}{N_{f_{i,j}}} + c_2(t)\right) z_j(t) + \frac{c_3}{N^2_{f_{i,j}}} \tag{SubP2}$$

$$\text{s.t. } C_{17}, \text{ and } z_j(t) \in \{0,1\}.$$

Lemma 11.1 *The optimal solution of subproblem* (SubP1) *is obtained when* $N_{\phi_j} = 1$, $t^* = \arg\min_{t\in[t_i,\bar{t}_{i,j}]} h(t)$.

Proof The proof by contradiction method is applied to prove our statement. We first sort $h(t)$ during the period of $[\bar{t}_{i,j}, t_i + \tau_i]$ in a non-decreasing order into $h(t^1) \le h(t^2) \le h(t^3) \le \cdots$. According to Lemma 11.1, we choose $h(t^1)$ as the optimal solution of (SubP1). On the other hand, if there exist $N_{\phi_j} = N$ continuous transmission time slots, for example $h(t^{n_1}), h(t^{n_2}), \cdots, h(t^{n_N})$, whose $\beta^1_{i,j}$ is smaller than $h(t^1)$, then, we have

$$\frac{h(t^{n_1}) + h(t^{n_2}) + \cdots h(t^{n_N})}{N} < h(t^1). \tag{11.19}$$

However, this contradicts the fact that $h(t^1) \leq h(t^{n_v})$, $v = 1, 2, \ldots, N$. Thus, our conclusion holds for (SubP1). This completes the proof. □

Lemma 11.2 *Let* $\beta_{i,j}^{2,1}(N) = \sum_{t\in[1,+\infty]} (\frac{c_1(t)}{N} + c_2(t))z_j(t)$ *be the value under the optimal scheduling when* $N_{f_{i,j}} = N$*, and let* $\beta_{i,j}^{2,2}(N) = \frac{c_3}{N^2}$*. Then,* $\beta_{i,j}^{2,1}(N)$ *is an increasing function with respective to N, and there exists at most one intersection point between* $\beta_{i,j}^{2,1}(N)$ *and* $\beta_{i,j}^{2,2}(N)$*.*

Proof This statement is obtained by using an analytical approach. We first compare objective values of $\beta_{i,j}^{2,1}(N)$ and $\beta_{i,j}^{2,1}(N+1)$. Let $t^{n_1}, t^{n_2}, \cdots t^{n_N}$ be the best N numbers of continuous time slots, which means when $N_{f_{i,j}} = N$, the objective of (SubP1) is minimized by selecting these time slots. Likewise, denote $t^{m_1}, t^{m_2}, \cdots t^{m_{(N+1)}}$ as the optimal continuous time slots when $N_{f_{i,j}} = N+1$. Then, we have

$$\beta_{i,j}^{2,1}(N) - \beta_{i,j}^{2,1}(N+1) = \frac{\sum\limits_{v=1}^{N} c_1(t^{n_v})}{N} + \sum_{v=1}^{N} c_2(t^{n_v}) - \left(\frac{\sum\limits_{v=1}^{N+1} c_1(t^{m_v})}{N+1} + \sum_{v=1}^{N+1} c_2(t^{m_v}) \right)$$

$$\Rightarrow (N+1)\left(\beta_{i,j}^{2,1}(N) - \beta_{i,j}^{2,1}(N+1)\right)$$

$$= (N+1)\left(\frac{\sum\limits_{v=1}^{N} c_1(t^{n_v})}{N} + \sum_{v=1}^{N} c_2(t^{n_v}) \right) - \left(\sum_{v=1}^{N+1} c_1(t^{m_v}) + (N+1)\sum_{v=1}^{N+1} c_2(t^{m_v}) \right)$$

$$= \underbrace{(N+1)\left(\frac{\sum\limits_{v=1}^{N} c_1(t^{n_v})}{N} + \sum_{v=1}^{N} c_2(t^{n_v}) \right)}_{①} - \underbrace{N\left(\frac{\sum\limits_{v=1}^{N} c_1(t^{m_v})}{N} + \sum_{v=1}^{N} c_2(t^{m_v}) \right)}_{②} -$$

$$\underbrace{\left(\frac{N c_1(t^{m_{N+1}})}{N} + \sum_{v=1}^{N} c_2(t^{m_v}) \right)}_{③} - (N+1)c_2(t^{m_{N+1}}). \tag{11.20}$$

Since ① is the optimal objective value when $N_{f_{i,j}} = N$, we have $(N+1) \times ① < N \times ② + ③$. Thus, we have $\beta_{i,j}^{2,1}(N) < \beta_{i,j}^{2,1}(N+1)$, which means $\beta_{i,j}^{2,1}(N)$ is an increasing function with respective to N. Moreover, $\beta_{i,j}^{2,2}(N)$ is a decreasing function with respective to N and $\beta_{i,j}^{2,2}(+\infty) = 0 < \beta_{i,j}^{2,1}(+\infty)$. Thus, $\beta_{i,j}^{2,1}(N)$ and $\beta_{i,j}^{2,2}(N)$ have one intersection point only when $\beta_{i,j}^{2,1}(N) = \beta_{i,j}^{2,2}(N)$. This completes the proof. □

Based on Lemma 11.1, the allocated transmission bandwidth for requester i is always $\phi_{i,j} = \frac{s_i}{\Delta t \log_2(1+\gamma_{i,j})}$. According to Lemma 11.2, there must exist a $\overline{N}$, which can minimize the value of $\beta_{i,j}^{2,1}(N) + \beta_{i,j}^{2,2}(N)$. Note that $\beta_{i,j}^{2,1}(N) + \beta_{i,j}^{2,2}(N)$ decreases when $N < \overline{N}$ but increases when $N > \overline{N}$. If there are M_2 available time slots during $(\overline{t}_{i,j}, t_i + \tau_i]$, we apply the following strategies to get the optimal solution of subproblem (SubP2):

- If $\beta_{i,j}^{2}(1) > \beta_{i,j}^{2}(M_2-1) > \beta_{i,j}^{2}(M_2)$, we choose $\beta_{i,j}^{2}(M_2)$ and the corresponding scheduling scheme, denoted as the set π^*, as the optimal solution.
- Otherwise, we apply sequential search to compare the values of $\beta_{i,j}^{2}(N+1)$ and $\beta_{i,j}^{2}(N)$ till $\beta_{i,j}^{2}(N+1) > \beta_{i,j}^{2}(N)$. We then choose $\beta_{i,j}^{2}(N)$ and the corresponding scheduling scheme, denoted as the set π^*, as the optimal solution.

Obviously, for the worst case, we only need $\frac{(\overline{N}+1)(2M_2-\overline{N})}{2}+\overline{N}$ comparisons to reach the optimal solution, which is much more computationally efficient compared to the brute force approach. The detailed procedures for solving the scheduling problem are summarized in Algorithm 1. Obviously, Algorithm 1 can find the globally optimal solution for the scheduling problem (TSP).

Algorithm 1 Online auction for scheduling problem

Require: s_i, Δt, $\hat{p}_t$, $\hat{p}_{j,t}^k$, t_i, τ_i, and $\gamma_{i,j}$
Ensure: Optimal schedule $l_{i,j}(\ell)$ and minimum $\beta_{i,j}$ for requester i offloading task to requester j or BS

```
l¹_{i,j}(ℓ) = ∅, l²_{i,j}(ℓ) = ∅, and β_{i,j} = +∞
while t̄_{i,j} ∈ [t_i, t_i + τ_i] do
    t* = arg min_{t∈[t_i, t̄_{i,j}]} h(t) and get β¹_{i,j}   ▷ Solve (SubP1)
    Apply the above strategies for (SubP2) in [t̄_{i,j}, t_i + τ_i] and get β²_{i,j} as well as π*
    if β_{i,j} > β¹_{i,j} + β²_{i,j} then
        β_{i,j} = β¹_{i,j} + β²_{i,j}
        l¹_{i,j}(ℓ) = l¹_{i,j}(ℓ) ∪ t* and l²_{i,j}(ℓ) = l²_{i,j}(ℓ) ∪ π*
        l_{i,j}(ℓ) = l¹_{i,j}(ℓ) ∪ l²_{i,j}(ℓ)
    end if
    Move t̄_{i,j} to the next time slot in [t_i, t_i + τ_i]
end while
return l_{i,j}(ℓ) and β_{i,j}
```

We summarize the OMAP in Algorithm 2.

Algorithm 2 The OMAP for MEC

Require: $w_{i,j}$, s_i, Δt, $\hat{p}_t$, $\hat{p}^k_{j,t}$, t_i, and τ_i
Ensure: Optimal schedule $l_{i,j}(\ell^*)$, j^* and payment $p_{i,j}$
  $x^{\ell}_{i,j} = 0,\ \forall i \in \mathcal{U},\ \forall j \in \mathcal{N}_i,\ \ell \in \mathcal{L}_{i,j}$
  $u_i = 0$;   ▷the utility of requester i
  $j^* = \emptyset$ and $l_{i,j}(\ell^*) = \emptyset$
  while the arrival of requester i's task **do**
    for $j \in \mathcal{N}_i$ **do**
      Run Algorithm 1 to get the best scheduling scheme $l_{i,j}(\ell)$ and minimum $\beta_{i,j}$
      if $w_{i,j} - \beta_{i,j} > u_i$ **then**
        $u_i = w_{i,j} - \beta_{i,j}$
        $j^* = j$
        $l_{i,j}(\ell^*) = l_{i,j}(\ell)$
      end if
    end for
    if $u_i > 0$ **then**
      Accept requester i and set $x^{\ell^*}_{i,j^*} = 1$
      Allocate the collaborator or BS and implement schedule scheme according to j^* and $l_{i,j}(\ell^*)$
      Charge requester i at price $p_{i,j}$
      Update $\hat{p}^k_{j,t}$ and $\hat{p}_t$ based on (11.17) and (11.18)
    else
      Reject requester i and set $x^{\ell}_{i,j} = 0$ and $p_{i,j} = 0$
    end if
  end while

11.4.3 Performance Analyses

In this section, we will theoretically analyze the OMAP in terms of competitive ratio, feasibility of primal and dual solutions, CE, IC, and IR.

Lemma 11.3 *The competitive ratio of OMAP is 3.*

Proof Assume that the requester i offloads its task to task executor j, and we define $\Delta P^{(i)}$ and $\Delta D^{(i)}$ as the increment of objective values in primal and its dual problems after requester i has been served, respectively. Then, we have

$$\Delta D^{(i)} = u_i + \sum_{t \in \mathcal{T}} W \Delta \hat{p}_t + \sum_{t \in \mathcal{T}} \sum_{k \in \mathcal{K}} R^k_j \Delta \hat{p}^k_{j,t}$$

$$= w_{i,j} - \sum_{a_{i,j} \le t \le d_{i,j}} \phi_{i,j} \hat{p}_t - \sum_{a_{i,j} \le t \le d_{i,j}} \sum_{k \in \mathcal{K}} r^k_{i,j} \hat{p}^k_{j,t} + \sum_{t \in \mathcal{T}} W \Delta \hat{p}_t + \sum_{t \in \mathcal{T}} \sum_{k \in \mathcal{K}} R^k_j \Delta \hat{p}^k_{j,t}$$

$$= w_{i,j} - \sum_{a_{i,j} \le t \le d_{i,j}} \phi_{i,j} \hat{p}_t - \sum_{a_{i,j} \le t \le d_{i,j}} \sum_{k \in \mathcal{K}} r^k_{i,j} \hat{p}^k_{j,t} + \sum_{a_{i,j} \le t \le d_{i,j}} \left(\phi_{i,j} \hat{p}_t + \frac{\phi_{i,j}}{\Phi_{i,j}} \right)$$

$$+ \sum_{a_{i,j} \le t \le d_{i,j}} \sum_{k \in \mathcal{K}} (r^k_{i,j} \hat{p}^k_{j,t} + \frac{r^k_{i,j}}{\Gamma_{i,j}})$$

$$= w_{i,j} + \sum_{a_{i,j} \le t \le d_{i,j}} \frac{\phi_{i,j}}{\Phi_{i,j}} + \sum_{a_{i,j} \le t \le d_{i,j}} \sum_{k \in \mathcal{K}} \frac{r^k_{i,j}}{\Gamma_{i,j}} \le 3w_{i,j} = 3\Delta P^{(i)}.$$

Let $\mathcal{U}^*$ be the set of the offloaded requesters, and $\overline{P}$ and $\overline{D}$ be solutions of primal and its dual problems by OMAP, respectively. Then, we must have

$$\overline{P} = \sum_{i \in \mathcal{U}^*} \Delta P^{(i)} = 3 \sum_{i \in \mathcal{U}^*} \Delta D^{(i)} = 3\overline{D}.$$

From the linear dual theory, we have

$$\frac{P^*}{\overline{P}} \le \frac{\overline{D}}{\overline{P}} = 3,$$

where P^* is the optimal solution to the primal problem (EQMSW). This completes the proof. □

Lemma 11.4 *OMAP produces almost feasible solutions to offline problem* (EQMSW) *if* $W \gg 1$; $R^k_j \gg 1, \forall j$; $r^k_{i,j} \ll R^k_j$, *and* $\phi_{i,j} \ll W, \forall i, j$.

Proof Let Γ_{max}, Φ_{max}, and w_{max} be the maximum values of $\Gamma_{i,j}$, $\Phi_{i,j}$, and $w_{i,j}$, respectively, and r_{min} and ϕ_{min} be the minimum values of $r^k_{i,j}$ and $\phi_{i,j}$, respectively.

We first show that $\hat{p}^k_{j,t}$ can be bounded by the following expression:

$$\hat{p}^k_{j,t} \ge \frac{\left(1 + \frac{1}{R^k_j}\right)^{\sum\limits_{i \in \mathcal{U}'} \sum\limits_{\ell : t \in l^2_{i,j}(\ell) \in l^2_{i,j}} r^k_{i,j} x^\ell_{i,j}} - 1}{\Gamma_{max}}, \tag{11.21}$$

where $\mathcal{U}'$ denotes the set of all accepted requesters before requester i. We prove the above inequality through mathematical deduction.

Define $\hat{p}^k_{j,t}(i)$ as the value of $\hat{p}^k_{j,t}$ before the arrival of requester i. At beginning, we have $x^\ell_{i,j} = 0 \; \forall \, j, \ell$ and $\hat{p}^k_{j,t}(1) = 0$, so that inequality (11.21) holds. We then consider the following two cases:

- Case 1: Requester i is rejected by the BS. In this case, we have $x^\ell_{i,j} = 0$ and $p(i+1) = p(i)$. Obviously, the inequality (11.21) still holds, which does not affect the validation of $p^k_{j,t}(i+1)$.
- Case 2: Requester i is accepted by the BS. In this case, we have

$$\hat{p}_{j,t}^{k}(i+1) = \hat{p}_{j,t}^{k}(i)\left(1+\frac{r_{i,j}^{k}}{R_j^k}\right) + \frac{r_{i,j}^{k}}{\Gamma_{i,j}R_j^k} \geq \hat{p}_{j,t}^{k}(i)\left(1+\frac{r_{i,j}^{k}}{R_j^k}\right) + \frac{r_{i,j}^{k}}{\Gamma_{max}R_j^k}$$

$$\geq \frac{\left(1+\frac{1}{R_j^k}\right)^{\sum\limits_{i\in\mathcal{U}'}\sum\limits_{\ell:t\in l_{i,j}^2(\ell)\in l_{i,j}^2} r_{i,j}^k x_{i,j}^\ell} - 1}{\Gamma_{max}}\left(1+\frac{r_{i,j}^{k}}{R_j^k}\right) + \frac{r_{i,j}^{k}}{\Gamma_{max}R_j^k}$$

$$= \frac{\left(1+\frac{1}{R_j^k}\right)^{\sum\limits_{i\in\mathcal{U}'}\sum\limits_{\ell:t\in l_{i,j}^2(\ell)\in l_{i,j}^2} r_{i,j}^k x_{i,j}^\ell}\left(1+\frac{r_{i,j}^{k}}{R_j^k}\right)}{\Gamma_{max}} - \frac{1}{\Gamma_{max}}$$

$$\approx \frac{\left(1+\frac{1}{R_j^k}\right)^{\sum\limits_{i\in\mathcal{U}'}\sum\limits_{\ell:t\in l_{i,j}^2(\ell)\in l_{i,j}^2} r_{i,j}^k x_{i,j}^\ell}\left(1+\frac{1}{R_j^k}\right)^{r_{i,j}^k}}{\Gamma_{max}}$$

$$= \frac{\left(1+\frac{1}{R_j^k}\right)^{\sum\limits_{i\in\mathcal{U}''}\sum\limits_{\ell:t\in l_{i,j}^2(\ell)\in l_{i,j}^2} r_{i,j}^k x_{i,j}^\ell}}{\Gamma_{max}}, \tag{11.22}$$

where $\mathcal{U}'' = \mathcal{U}' \cup i$ and the approximation holds because $R_j^k \gg 1$ and $r_{i,j}^k \ll R_j^k$, and $(1+a)^x \approx 1 + ax$ when a and x are small enough.

Therefore, inequality (11.21) holds no matter whether requester i is accepted or not. However, $\hat{p}_{j,t}^k(+\infty) < \frac{w_{max}}{r_{min}}(1+1)+1 = 2\frac{w_{max}}{r_{min}}+1$ because of the conditions $w_{i,j} > \beta_{i,j}$ and $r_{i,j}^k \ll R_j^k$. By reconsidering the inequality (11.21), we have

$$\frac{\sum\limits_{i\in\mathcal{U}'}\sum\limits_{\ell:t\in l_{i,j}^2(\ell)\in l_{i,j}^2} r_{i,j}^k x_{i,j}^\ell}{R_j^k} \leq \frac{\log\left(\Gamma_{max}\left(2\frac{w_{max}}{r_{min}}+1\right)+1\right)}{R_j^k\log\left(1+\frac{1}{R_j^k}\right)} \approx \log\left(\Gamma_{max}\left(2\frac{w_{max}}{r_{min}}+1\right)+1\right), \tag{11.23}$$

where the last approximation holds when $R_j^k \gg 1$. Inequality (11.23) indicates that the constraint C_{10} in problem (EQMSW) may be violated by at most $\log(\Gamma_{max}(2\frac{w_{max}}{r_{min}}+1)+1)$.

To verify that the solution meets constraint C_{13} in problem (EQMSW), we can follow the similar procedure to demonstrate that before the arrival of requester i, the value of $\hat{p}_t$ can be bounded as

$$\hat{p}_t(i) \geq \frac{(1+\frac{1}{W})^{\sum\limits_{i\in\mathcal{U}'}\sum\limits_{j\in N_i}\sum\limits_{\ell:t\in l_{i,j}^1(\ell)\in l_{i,j}^1}\phi_{i,j}x_{i,j}^\ell} - 1}{\Phi_{max}}. \tag{11.24}$$

However, we have $\hat{p}_t(+\infty) < \frac{w_{max}}{\phi_{min}}(1+1)+1 = 2\frac{w_{max}}{\phi_{min}}+1$ because of the conditions $w_{i,j} > \beta_{i,j}$ and $\phi_{i,j} \ll W$. Combining similar inequalities as (11.22)–(11.24), we have

$$\frac{\sum_{i\in\mathcal{U}}\sum_{j\in N_i}\sum_{\ell:t\in l^1_{i,j}(\ell)\in l^1_{i,j}}\phi_{i,j}x^{\ell}_{i,j}}{W} \le \frac{\log(\Phi_{max}(2\frac{w_{max}}{\phi_{min}}+1)+1)}{W\log(1+\frac{1}{W})}$$

$$\approx \log\left(\Phi_{max}\left(2\frac{w_{max}}{\phi_{min}}+1\right)+1\right), \tag{11.25}$$

where the last approximation holds when $W \gg 1$. It indicates that the constraint C_6 in problem (EQMSW) may be violated by at most $\log(\Phi_{max}(2\frac{w_{max}}{\phi_{min}}+1)+1)$. This completes the proof. □

Lemma 11.5 *OMAP produces a feasible solution to dual problem* (EQDP).

Proof We consider the following two cases:

- Case 1: Requester i is rejected, which means $w_{i,j^*} - \beta_{i,j^*} \le 0$ for the best selected task executor j^* and $u_i = 0$ according to the acceptance condition (11.16). Thus, constraint C_{15} holds in this case.
- Case 2: Requester i is accepted, which means $u_i = w_{i,j^*} - \beta_{i,j^*} > 0$ for the best selected task executor j^*. Thus, constraint C_{15} still holds in this case.

Therefore, constraint C_{15} in problem (EQDP) always holds no matter whether requester i is accepted or not. This completes the proof. □

Lemma 11.6 *OMAP runs in polynomial time.*

Proof The computational complexity of OMAP is evaluated in terms of computation times with respect to the number of requesters and collaborators. Recall that OMAP consists of Algorithms 2 and 1. For Algorithm 1, given that there are a total of M time slots during the period of $[t_i, t_i + \tau_i]$, the computational complexity of Algorithm 1 can be calculated as $O(M(M - M + 1 + \frac{M(M-1)}{2} + M - 1)) = O(M^2\frac{(M+1)}{2})$. Therefore, the computational complexity of Algorithm 2 is $O(|\mathcal{U}| \times |N_{max}| \times M^2\frac{(M+1)}{2})$. Note that this is the worst case computational complexity. Obviously, Algorithm 2 runs in polynomial time, which completes the proof. □

Lemma 11.7 *OMAP can guarantee truthfulness (IC) and individual rationality (IR).*

Proof We first prove the truthfulness in the requesters' bidding values. Note that the marginal prices $\hat{p}^k_{j,t}$ and $\hat{p}_t$ depend only on the past accepted requesters and are independent of the bidding values of current requester i. Furthermore, OMAP always assigns the requested resource to that requester only when the utility of that requester is maximized among all its bidding values and is greater than zero given the current marginal prices. Therefore, OMAP can be treated as a sequential posted price mechanism [16] or iterative auction [17], where the auctioneer posts the price and the bidders choose the best bidding values to maximize their utilities. In this way, the bidders cannot gain more utilities by misreporting their bidding values.

Next, we demonstrate the truthfulness in arrival time t_i. If a requester reports the arrival time $t_i^{'}$ earlier than the actual value (i.e., $t_i^{'} < t_i$), this requester cannot increase its utility or even suffers from the failure to complete its task when $t_i^{'} < t_{i-1}$ or the transmission time is scheduled within the period of $[t_i^{'}, t_i]$. When the requester declares its arrival time later than t_i (i.e., $t_i^{'} > t_i$), the mechanism will find the optimal transmission and computation times after $t_i^{'}$, while in fact, such optimal times may happen in $[t_i, t_i^{'}]$, which results in an increased payment and a decreased utility. Thus, the requesters will not misreport their arrival time.

Third, it is obvious that the requesters will not misreport their offloaded tasks (i.e., T_i) due to the fact that this can result in the failure in completing their tasks.

Finally, we verify the individual rationality. According to the acceptance condition (11.16), a requester can be accepted only if one of its maximum biddings can lead to a positive utility; otherwise, that requester is rejected and its utility is zero. Hence, OMAP satisfies individual rationality. This completes the whole proof. □

In summary, based on **Lemmas 11.3, 11.6**, and **11.7**, we can make the following conclusion.

Theorem 11.1 *OMAP has a competitive ratio of 3, runs polynomially, and guarantees truthfulness and individual rationality.*

11.4.4 Numerical Simulations

In this section, numerical simulations are conducted to verify the effectiveness of OMAP. Since the total social welfare, revenue, and utility of requester are the most important economical metrics and the competitive ratio is a vital metric to measure an online mechanism, in this section, we will focus on evaluating these two performance metrics with respect to different numbers of requesters and collaborators. In the simulation, the wireless channels between requesters and task executors (i.e., collaborators or the BS) experience Rayleigh fading and all the channel coefficients are zero-mean, circularly symmetric complex Gaussian (CSCG) random variables with variances $d^{-\upsilon}$, where d is the distance between the transmitter and the receiver and $\upsilon = 4$. Table 11.1 lists the main simulation parameters, some of which have also been employed in [11, 18, 19]. For comparison purpose, the following three online strategies are also simulated as comparison benchmarks:

- Random online mechanism: For each requester, the BS randomly selects the task executor and randomly schedules the transmission and computation times.

Table 11.1 Main simulation parameters

Parameter	Value
Cell radius	500 m
Total bandwidth	40 MHz
Transmission power at requesters	1.5 W
Background noise average power	−60 dBm
Total running time	30 minutes
Time slot length	1 second
Task size	Randomly from 10 to 30 MB
CPU cycles coefficient	330 cycles/Byte
Energy consumption coefficient	10^{-26}
Unit energy cost	$0.1
Valuation	Randomly from [$0.1, $10]
The maximum delay	Randomly from [5, 15] seconds
Computation capacity of BS	10 GHz
Storage capacity of BS	10 GB
Computation capacity of collaborators	2 GHz
Storage capacity of collaborators	5 GB

- Greedy online mechanism: Upon the arrival of a requester, the BS chooses the task executor with the maximal valuation as the winner and schedules one time slot for transmission and $\lceil \tau \rceil - 1$ time slots for computation.
- First-in first-out (FIFO) online mechanism [20]: Arriving tasks are always accepted with a fixed transmission and computation time schedule till the resources are run out.

Figure 11.4 shows the total social welfare achieved by different online mechanisms with respect to different numbers of arrived requesters when there are 30 collaborators, i.e., $|\mathcal{M}| = 30$. From this figure, we can see that the achievable total social welfare increases with the number of requesters. This trend is obvious since with the arrival of more requesters, the BS will admit more before resources are exhausted, which results in the increment on the total social welfare. It is worthwhile to note that OMAP outperforms both the random and FIFO online mechanisms but underperforms the greedy one. This is because OMAP tries to minimize the scheduling problem (TSP) so as to maximize the utility of each requester, while the greedy one only attempts to maximize the total social welfare and ignores the maximality of the individual utility. In addition, according to [21], the simple greedy online mechanism cannot guarantee truthfulness and individual rationality.

Figure 11.5 reevaluates the total social welfare under various numbers of collaborators. In the simulation, the total number of arrived requesters is fixed at 75, i.e., $|\mathcal{U}| = 75$. It can be seen from the figure that the total social welfare increases with the number of collaborators till reaching a saturation when the number of collaborators is large enough (e.g., 55 in our simulation). This is because with the excessive amount of collaborators, each requester is always served by

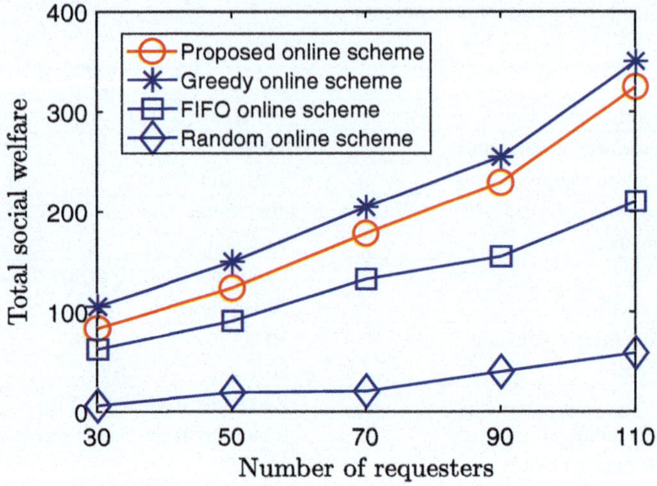

Fig. 11.4 TSW versus the numbers of requesters

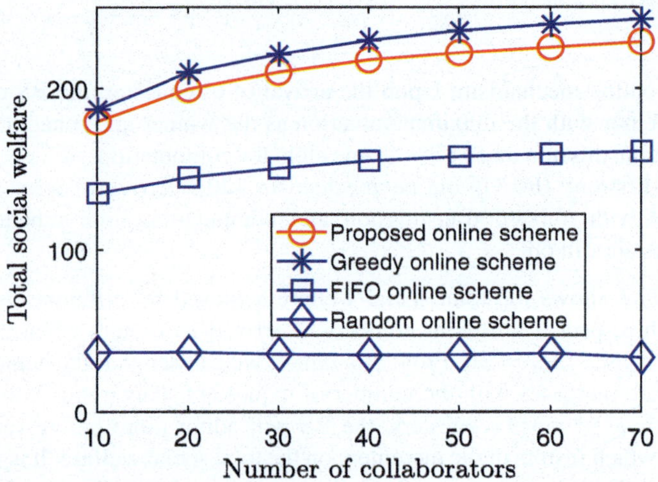

Fig. 11.5 TSW versus the numbers of collaborators

its most effective collaborator, while other collaborators have no effects on the achievable total social welfare. In addition, the total social welfare of random online mechanism is almost a constant. This is because this mechanism selects the collaborator in random so that it treats all collaborators equally regardless of how many collaborators exist. Similar to Fig. 11.4, OMAP outperforms both the random and FIFO online mechanisms but is still inferior to the greedy one. In summary, from both Figs. 11.4 and 11.5, we can conclude that although it is not difficult to design an online algorithm with a sound competitive ratio (or larger social welfare),

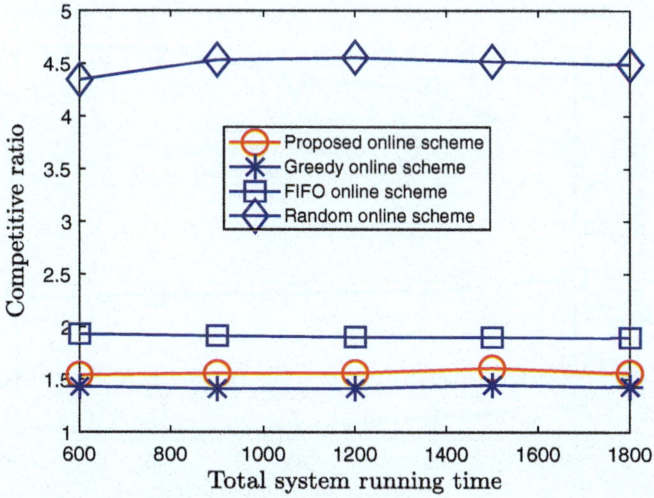

Fig. 11.6 CR versus the total system running time

it is difficult to devise an online mechanism that possesses sound competitive ratio, truthfulness, and individual rationality at the same time. In fact, OMAP sacrifices a little bit of competitive ratio to achieve other economical properties.

Figure 11.6 presents the comparison among different online mechanisms in terms of the competitive ratio by varying the system running time when the number of collaborators is 25. Note that the optimal offline solution is obtained by *Yalmip* optimizer, and the payments are based on the *VCG* mechanism [22]. From Fig. 11.6, we can observe that the CR of OMAP is less than 3, which matches our theoretical analyses. Besides, the competitive ratio almost stays unchanged with different system running times, which demonstrates that OMAP is stable.

Figure 11.7 evaluates the performance of different online mechanisms in terms of competitive ratio with respective to different numbers of collaborators when the number of requesters is 40. For OMAP, FIFO, and greedy online mechanisms, their competitive ratios stay less than 3 and slightly decrease until they stabilize when the number of collaborators becomes large enough. This is because more collaborators can increase the available resources in the system and increase the number of potential alternative collaborators around requesters. In contrast, the competitive ratio of the random online mechanism increases (i.e., the worse performance) with the number of collaborators. According to Fig. 11.5, as the increase of $|\mathcal{M}|$, the social welfare of random online mechanism stays unchanged, while the offline optimal solution increases because more resources are available for allocation. Thus, the competitive ratio of random online mechanism increases.

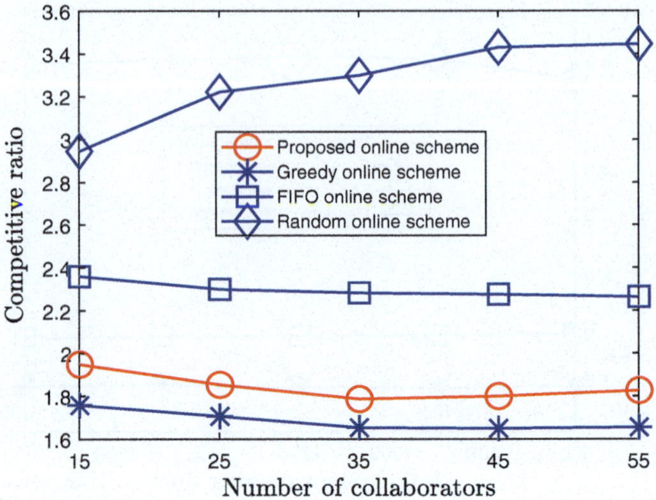

Fig. 11.7 CR versus the numbers of collaborators

11.5 Summary

In this chapter, we introduce the primal–dual-based online mechanism design method and its application in edge computing systems. First, we provide some basic concepts, definitions, and properties that are commonly used in mechanism design. Second, we present the design method for linear online incentive mechanisms based on the primal–dual theory and briefly introduce the design method for nonlinear cases based on the Lagrangian dual method. Finally, we discuss how task offloading in an edge computing system could be modelled as a market and then apply the introduced primal–dual-based online design method to address the formulated mechanism design problem.

References

1. N. Nisan, T. Roughgarden, E. Tardos, V.V. Vazirani, *Algorithmic Game Theory* (Cambridge University Press, Cambridge, 2007)
2. S. Boyd, L. Vandenberghe, *Convex Optimization* (Cambridge University Press, Cambridge, 2004)
3. D. Niyato, N.C. Luong, P. Wang, Z. Han, *Auction Theory for Computer Networks* (Cambridge University Press, Cambridge, 2020)
4. R. Lavi, N. Nisan, Competitive analysis of incentive compatible on-line auctions, in *Proceedings of the 2nd ACM Conference on Electronic Commerce (EC)* (2000), pp. 233–241
5. G. Li, J. Cai, An online incentive mechanism for collaborative task offloading in mobile edge computing. IEEE Trans. Wirl. Commun. **19**(1), 624–636 (2020)

6. H. Li, C. Wu, Z. Li, Socially-optimal online spectrum auctions for secondary wireless communication, in *Proceedings of the 2015 IEEE Conference on Computer Communications (INFOCOM)*, Kowloon (2015), pp. 2047–2055
7. X. Zhang, et al., Online auctions in IaaS Clouds: welfare and profit maximization with server costs. IEEE/ACM Trans. Netw. **25**(2), 1034–1047 (2017)
8. G. Li, J. Cai, An online incentive mechanism for crowdsensing with random task arrivals. IEEE Int. Things J. **7**(4), 1–14 (2020)
9. B. Fan, H. Tian, L. Jiang, A.V. Vasilakos, A social-aware virtual MAC protocol for energy-efficient D2D communications underlying heterogeneous cellular networks. IEEE Trans. Veh. Technol. **67**(9), 8372–8385 (2018)
10. K. Doppler, M. Rinne, C. Wijting, C. Ribeiro, K. Hugl, Device-to device communication as an underlay to LTE-advanced networks. IEEE Commun. Mag. **47**(12), 42–49 (2009)
11. T. Thinh, J. Tang, Q. La, T. Quek, Offloading in mobile edge computing: task allocation and computational frequency scaling. IEEE Trans. Commun. **65**, 3571–3584 (2017)
12. X. Chen, L. Jiao, W. Li, X. Fu, Efficient multi-user computation offloading for mobile-edge cloud computing. IEEE/ACM Trans. Netw. **24**(5), 2795–2808 (2016)
13. D. Huang, P. Wang, D. Niyato, A dynamic offloading algorithm for mobile computing. IEEE Trans. Wirel. Commun. **11**(6), 1991–1995 (2012)
14. Y. Wen, W. Zhang, H. Luo, Energy-optimal mobile application execution: Taming resource-poor mobile devices with cloud clones, in *Proceedings of the IEEE INFOCOM* (2012), pp. 2716–2720
15. J. Borghoff, L.R. Knudsen, M. Stolpe, Bivium as a mixed-integer linear programming problem, in *IMA Cryptography and Coding*. Lecture Notes in Computer Science, vol. 5921 (Springer, Berlin, 2009), pp. 133–152
16. S. Chawla, J. Hartline, D. Malec, B. Sivan, Multi-parameter mechanism design and sequential posted pricing, in *Proceedings of the forty-second ACM symposium on Theory of computing* (2010), pp. 311–320
17. G. Iosifidis, L. Gao, J. Huang, L. Tassiulas, A double-auction mechanism for mobile data-offloading markets. IEEE/ACM Trans. Netw. **23**(5), 1634–1647 (2015)
18. M.H. Chen, B. Liang, M. Dong, Joint offloading and resource allocation for computation and communication in mobile cloud with computing access point, in *IEEE INFOCOM 2017-IEEE Conference on Computer Communications* (2017), pp. 1–9
19. F. Guo, et al., Joint load management and resource allocation in the energy harvesting powered small cell networks with mobile edge computing, in *IEEE INFOCOM 2018-IEEE Conference on Computer Communications Workshops (INFOCOM WKSHPS)* (2018), pp. 299–304
20. D.E. Irwin, L.E. Grit, J.S. Chase, Balancing risk and reward in a market-based task service, in *Proceedings. 13th IEEE International Symposium on High performance Distributed Computing* (2004)
21. R. Lavi, N. Nisan, Competitive analysis of incentive compatible on-line auctions. Theor. Comput. Sci. **310**(1–3), 159–180 (2004)
22. W. Vickrey, Counterspeculation, auctions, and competitive sealed tenders. J. Finance **16**(1), 8–37 (1961)

Chapter 12
Collaborative Deep Neural Network Inference via Mobile Edge Computing

Wen Wu, Yujie Tang, Peng Yang, Weiting Zhang, and Ning Zhang

12.1 Introduction

Advanced neural network techniques and ubiquitous Internet of Things (IoT) devices enable deep neural network (DNN) inference as a key technology in next generation wireless networks. In recent years, DNNs have been applied in many intelligent applications, ranging from facility monitoring, fault diagnosis, to object detection [1, 2]. For example, IoT devices in industrial applications, such as vibration sensors, can sense the industrial operating environment. Then, the sensing data is sent to a pre-trained DNN via wireless communication links, and the DNN processes the sensing data and renders inference results. Such a process

W. Wu (✉)
Pengcheng Laboratory, Shenzhen, P.R. China
e-mail: wuw02@pcl.ac.cn

Y. Tang
School of Computer Science and Technology, Algoma University, Sault Ste. Marie, ON, Canada
e-mail: yujie.tang@algomau.ca

P. Yang
School of Electronic Information and Communications, Huazhong University of Science and Technology, Wuhan, P.R. China
e-mail: yangpeng@hust.edu.cn

W. Zhang
School of Electronic and Information Engineering, Beijing Jiaotong University, Beijing, P.R. China
e-mail: 17111018@bjtu.edu.cn

N. Zhang
Department of Electrical and Computer Engineering, University of Windsor, Windsor, ON, Canada
e-mail: ning.zhang@uwindsor.ca

L. Cai et al. (eds.), *Broadband Communications, Computing, and Control for Ubiquitous Intelligence*, Wireless Networks,
https://doi.org/10.1007/978-3-030-98064-1_12

is referred to as *DNN inference* [3]. A large number of experimental results indicate that DNN inference can achieve high inference accuracy as compared to traditional alternatives, such as decision trees in classification tasks.

Executing DNN inference tasks is computation intensive. Tremendous numbers of multiply-and-accumulation operations are conducted in a DNN inference task [4]. A *device-only* solution that purely executes DNN inference tasks at resource-constrained mobile devices becomes intractable, due to prohibitive energy consumption and a high service delay. For instance, processing an image using AlexNet incurs up to 0.45 W energy consumption even in a tailored energy-efficient chip [5]. An *edge-only* solution that purely offloads large-volume sensing data to resource-rich edge nodes, e.g., access point (AP), suffers from an unpredictable service delay due to time-varying wireless channels [6]. Therefore, neither a device-only nor an edge-only solution can effectively support low-delay DNN inference services.

Collaborative DNN inference, which coordinates resource-constrained mobile devices and the resource-rich AP, is a potential framework to provide low-delay and high-accuracy inference services [7]. Within the collaborative inference framework, sensing data from mobile devices can be either processed locally or offloaded to the AP. At mobile devices, light-weight *compressed* DNNs, i.e., neural networks are compressed without significantly decreasing their performance, are deployed due to constrained on-board computing capability, which saves computing resources at the cost of inference accuracy [8, 9]. At the AP, *uncompressed* DNNs are deployed to provide high-accuracy inference services at the cost of network resources including computing and communication resources. The overall service performance can be enhanced through the task offloading between mobile devices and the AP.

However, the *sampling rate adaption* technique that dynamically configures the sampling rates of mobile devices, is seldom investigated in the collaborative DNN inference framework. The sampling rates of mobile devices can be dynamically adjusted based on mobile devices' real-time channel conditions and the AP's computation workloads. As such, the sensing data from mobile devices can be compressed, thereby reducing not only the offloaded data volume but also the task computation workload. On the one hand, when the mobile device's channel condition is poor or the AP's computation workload is heavy, the sampling rate is decreased to reduce the offloaded data volume and the requested computation workload. As a result, the service delay is reduced at the cost of limited inference accuracy. Our experimental results show that the reduction of inference accuracy is acceptable in harsh network environments. On the other hand, when the mobile device's channel condition is good and the edge computation workload is light, the sampling rate can be increased to help deliver a high-accuracy service with an acceptable service delay. Therefore, sampling rate adaption can effectively reduce the service delay, which should be considered as an important component in the collaborative DNN inference framework.

In this chapter, we present the collaborative DNN inference technology in wireless networks. *Firstly*, we give a comprehensive overview of DNN inference, mobile edge computing (MEC), and machine learning. *Secondly*, we study a detailed case on collaborative DNN inference via device-edge orchestration. The problem is formulated as a constrained Markov decision process (CMDP) taking time-varying channel conditions and random task arrivals into account. Specifically, three decisions, i.e., sampling rates of mobile devices, task offloading, and edge computation resource allocation, are jointly optimized to achieve the minimum average service delay while guaranteeing the long-term accuracy requirements of multiple DNN inference services. *Thirdly*, since traditional RL algorithms target at optimizing a long-term reward without considering policy constraints, it is difficult to directly apply them to solve the formulated CMDP with long-term constraints. To address the issue, we propose a *three-step* solution: (1) the Lyapunov optimization technique is leveraged to transform the CMDP into an MDP; (2) to solve the MDP, a learning-based algorithm is developed based on the deep deterministic policy gradient (DDPG) algorithm; and (3) the edge computing resource allocation can be directly solved via an optimization subroutine, and then the optimization subroutine is incorporated in the learning-based algorithm to reduce the training complexity. Extensive simulations are conducted to validate the effectiveness of the proposed algorithm in reducing the average service delay while preserving the long-term accuracy requirements.

The remainder of this chapter is organized as follows. Section 12.2 presents a comprehensive overview of three key technologies, including DNN inference, MEC, and machine learning. The considered scenario, the system model, the formulated problem, and the proposed learning-based solution are presented in Sect. 12.3. Simulation results are given in Sect. 12.4. Finally, Sect. 12.5 concludes this chapter.

12.2 Background

12.2.1 DNN Inference

Recently, DNN inference for mobile devices has attracted much attention from academia. A device-only solution resorts to on-board computing resources to facilitate DNN inference services. DNN compression techniques are applied to reduce the computational complexity at the mobile devices. Typical techniques include weight pruning [8] and knowledge distillation [10]. The authors in [4] designed a light-weight DNN inference model, which can dynamically compress the model size in order to balance inference accuracy and energy efficiency, taking the widely equipped energy-harvesting functionality in IoT devices into account. In another line of research, by utilizing powerful edge computing servers, edge-assisted DNN inference solutions can provide high-accuracy inference services. The authors in [11] proposed an online video quality and computing resource allocation strategy to maximize video analytic accuracy, thereby facilitating low-delay and accurate DNN-based video analytics. Another important work proposed a novel device-edge collaborative inference scheme [7]. In this work, the DNN model is partitioned and deployed at both the device and the edge, and intermediate results are transferred via wireless links. The above works can offer potential resource allocation solutions to enhance DNN inference performance. In comparison with the existing works, the following case study in this chapter takes the sampling rate adaption of IoT devices into account, aiming at providing accuracy-guaranteed inference services in dynamic network environments.

12.2.2 Mobile Edge Computing

In the current wireless networks, a large volume of computing demands are generated by mobile devices to support emerging applications, such as intelligent path planning, safety applications, and on-board entertainments. Taking the autonomous driving service as an example, when an autonomous vehicle is on the road, a large number of computation-intensive tasks are required to be processed [12]. Processing such computation-intensive tasks by mobile devices requires expensive on-device computing facilities and degrades energy efficiency. As a remedy to these limitations, a potential solution is to explore the MEC paradigm. In the MEC paradigm, mobile devices can offload these computation tasks to nearby radio access networks (RANs) with computation-powerful edge servers for prompt processing. Extensive experiments show that the task processing delay can be significantly reduced by leveraging the MEC paradigm.

Recently, MEC problems have been widely investigated from many perspectives in wireless networks. In high-mobility vehicular networks, the roadside MEC servers judiciously collaborate with each other to provide low-latency services for autonomous vehicles [13]. Also, in the context of vehicular networks, a dynamic RAN slicing framework taking roadside MEC servers into account is proposed to guarantee the quality of service requirements of autonomous driving services [14]. In recent emerging unmanned aerial vehicle (UAV) networks, a UAV endowed with an MEC server is dispatched to collect and then process tasks from a large number of IoT devices in the remote area [15]. In this chapter, the MEC server at the AP is applied to handle the computation-intensive DNN inference task.

12.2.3 Machine Learning

Recently, machine learning (ML) has achieved great success in a number of research fields, ranging from computer vision, gaming, natural language processing, object detection, and traffic prediction [16]. The machine learning methods can be classified into three categories: (1) *supervised learning*, in which the training data structure includes both feature and label. For example, the support vector machine algorithm is supervised learning; (2) *unsupervised learning*, in which the training structure only includes feature without label, e.g., K-means algorithm; and (3) *RL*, in which the data structure is defined by state, action, and reward. As shown in Fig. 12.1, the action can be the control decisions, the state can be the

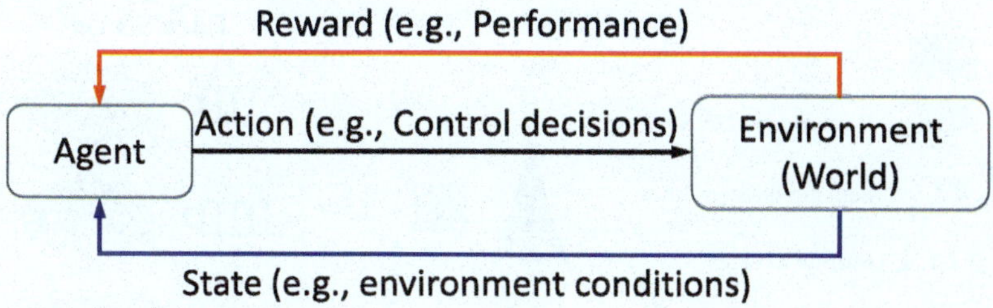

Fig. 12.1 An illustrative example of RL algorithms

environment conditions, and the observed reward from the environment can be system performance. The objective of RL is to learn a good policy in a sequential decision-making problem, such that the learning agent can take appropriate actions based on the current state. A typical example of RL algorithm is the deep Q learning algorithm. Seeing the great benefits of different machine learning methods, it is expected that ML will be widely applied in future wireless networks. The potential ML applications in next generation wireless networks, i.e., 6G networks, are investigated in [17–19], ranging from network slicing, traffic prediction, to digital twin management.

The main benefits of ML methods can be summarized as follows: (1) model-free, which makes ML methods different from traditional model-based approaches. It learns from the data and does not suffer from complicated modelling and strong assumptions; and (2) flexible, which means that ML methods can adaptively adjust the decision based on the current network environment. By training the learning modules properly offline, ML methods can make quick online decisions in highly complex scenarios.

Among different categories of ML algorithms, RL has attracted great attention from both academia and industry in the field of wireless communications. RL algorithms have been widely applied in network resource allocation, such as service migration in vehicular networks [20], network slicing in cellular networks [17], content caching in edge networks [21, 22], and beam alignment in mmWave networks [23, 24]. Hence, RL algorithms can be considered as potential solutions to manage network resources for DNN inference services. In this chapter, we propose a deep RL-based algorithm to deal with resource allocation and sampling rate selection issues in the collaborative DNN inference problem.

12.3 Collaborative DNN Inference via Device-Edge Orchestration

In this section, we introduce a case study on collaborative DNN inference, in which mobile devices and the network edge are orchestrated to provide DNN inference services. The collaborative DNN inference framework is presented in Sect. 12.3.1,

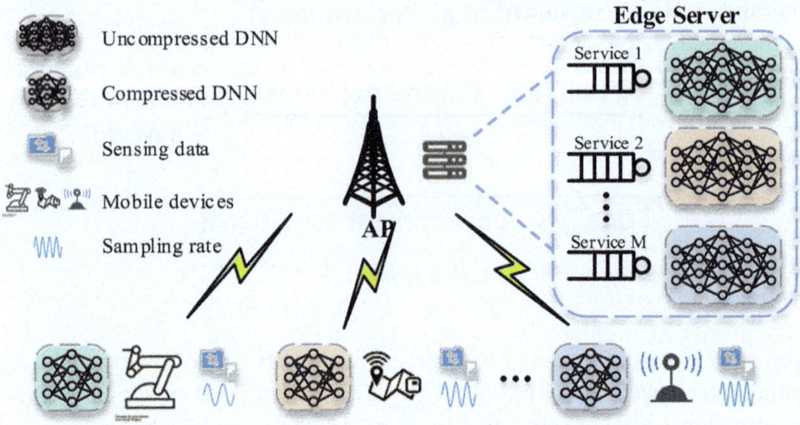

Fig. 12.2 An illustrative example of the collaborative DNN inference framework

and the corresponding detailed performance analysis on service delay and accuracy is provided in Sect. 12.3.2. Based on the system model, the problem is presented in Sect. 12.3.3, which is solved via a learning-based algorithm in Sect. 12.3.4.

12.3.1 Collaborative DNN Inference Framework

We consider a wireless network with one AP to serve multiple types of mobile devices, as illustrated in Fig. 12.2. In the network, the AP collects network information and then conducts resource orchestration decisions. Let $\mathcal{M}$ denote a set of M types of supported inference services, e.g., facility fault diagnosis and facility monitoring services [25]. The set of mobile devices subscribed to service m is denoted by $\mathcal{N}_m$, and the set of all mobile devices is denoted by $\mathcal{N} = \cup_{m \in \mathcal{M}} \mathcal{N}_m$.

Consider industrial facility monitoring services as an example. In a smart factory, wireless sensors are equipped to measure the status of the industrial facility. Vibration sensors can sense the operation condition of a facility with a certain sampling rate, e.g., 24 KHz. Mobile devices send the sensing data to a DNN for a specific inference service, and then DNN processes the sensing data and conducts inference, e.g., fault diagnosis.

In the collaborative inference framework, two kinds of DNNs are deployed:

- **Compressed DNN**, which is deployed on mobile devices. The compressed DNN can be implemented via the weight pruning technique, which prunes less-important weights to reduce computational complexity while maintaining similar inference accuracy [8].

- **Uncompressed DNN**, which is deployed at the AP. As such, M types of uncompressed DNNs share the edge computing resource to serve different kinds of inference requests.

The collaborative DNN inference framework operates in a time-slotted manner. The procedure consists of the following two steps:

- **Step 1: Sampling rate selection**. Mobile devices select their sampling rates based on channel conditions and computation workloads. The set of candidate sampling rates is denoted by $\mathcal{K} = \{\theta_1, \theta_2, ..., \theta_K\}$, where θ_K denotes the raw sampling rate. We assume the sampling rate in $\mathcal{K}$ increases linearly with the index, i.e., $\theta_k = k\theta_K/K$. Let t denote the time index, where $t \in \mathcal{T} = \{1, 2, \ldots, T\}$. Let $\mathbf{X}^t$ denote the sampling rate decision matrix in time slot t, whose element $x_{n,k}^t = 1$ indicates the mobile device $n \in \mathcal{N}$ selects the k-th sampling rate.
- **Step 2: Task processing**. The sensing data from mobile devices within a time slot is deemed as a computation task, which can be either offloaded to the AP or executed locally. Let $\mathbf{o}^t \in \mathbb{R}^{|\mathcal{N}|\times 1}$ denote the offloading decision vector in time slot t, whose element $o_n^t = 0$ indicates offloading the computation task from mobile device n. Otherwise, $o_n^t = 1$ indicates executing the computation task locally.

12.3.2 *Service Delay and Accuracy Analysis of Collaborative DNN Inference*

In this subsection, we analyze the inference delay and accuracy performance in the considered collaborative DNN inference framework.

12.3.2.1 Inference Delay Analysis

In the considered framework, a computation task can be either processed locally or offloaded to the AP. In the following, we analyze the service delay in these two cases, i.e., executing tasks locally or offloading tasks to AP.

Case 1: Executing Tasks Locally The task arrival rate of the n-th mobile device in time slot t is denoted by λ_n^t. We assume that the task arrival follows a general random distribution. Let $\xi_n^t = \lambda_n^t \nu_m, \forall n \in \mathcal{N}_m$ denote the raw data size of the generated tasks at the n-th device. Here, ν_m denotes the raw data size of a task for service m. When the sampling rate is selected, we can represent the data size of the generated task by:

$$\zeta\left(\mathbf{x}_n^t\right) = \sum_{k=1}^{K} \frac{x_{n,k}^t \xi_n^t k}{K}. \tag{12.1}$$

Here, $\mathbf{x}_n^t = \{x_{n,k}^t\}_{k \in \mathcal{K}}$ is the n-th device's sampling rate selection decision vector. If the inference task is processed via a compressed DNN in the local mobile device, the service delay should consist of two parts: the queuing delay in the local computing queue and the task processing delay. The detailed calculation of the two parts is given by:

$$d_{n,l}^t = \frac{o_n^t \eta_{m,c}\left(B_n^t + \zeta\left(\mathbf{x}_n^t\right)\right)}{f_n}, \forall n \in \mathcal{N}_m. \tag{12.2}$$

Here, f_n denotes the n-th mobile device's central processing unit (CPU) frequency, and $\eta_{m,c}$ represents the computation intensity of the compressed DNN for the m-th service. Let B_n^t denote the backlogged computation tasks (in bits) in the local computing queue, which is updated via

$$B_n^{t+1} = \min\left\{\left[B_n^t + o_n^t \zeta\left(\mathbf{x}_n^t\right) - \frac{f_n \tau}{\eta_{m,c}}\right]^+, B_n^{max}\right\}, \tag{12.3}$$

where $[x]^+ = \max\{x, 0\}$. Here, B_n^{max} represents the local computing queue capacity, and τ denotes a time slot duration. It is worth noting that tasks have to be dropped if the local computing queue is full. The amount of the dropped tasks in the local computing queue of device n can be represented by:

$$\Psi_{b,n}^t = \max\left\{B_n^t + o_n^t \zeta\left(\mathbf{x}_n^t\right) - \frac{f_n \tau}{\eta_{m,c}} - B_n^{max}, 0\right\}. \tag{12.4}$$

Here, $\Psi_{n,b}^t > 0$ indicates that a local computing queue overflow event occurs at the n-th device. Then, a corresponding penalty will be incurred to avoid queue overflow.

Case 2: Offloading Tasks to AP If a task is offloaded to the AP, the task will be processed by an uncompressed DNN. The service delay consists of three components: task offloading delay, queuing delay in the edge computing queue, and task processing delay, which are analyzed respectively as follows.[1]

Task Offloading Delay Component The offloading delay of the n-th mobile device is given by:

$$d_{n,o}^t = \frac{\left(1 - o_n^t\right) \zeta\left(\mathbf{x}_n^t\right)}{R_n^t}. \tag{12.5}$$

[1] Note that we assume free transmission backlog in this chapter.

Here, the transmission rate between the n-th mobile device and the AP, R_n^t is represented by:

$$R_n^t = \frac{W}{N} \log_2 \left(1 + \frac{P_T G(H_n^t)}{N_f \sigma^2}\right). \tag{12.6}$$

In the above equation, W, P_T, $G(H_n^t)$, and N_f represent the system bandwidth, transmit power, channel gain, and noise figure, respectively. Here, the background noise is denoted by $\sigma^2 = N_o W/N$, where N_o is thermal noise spectrum density. In this chapter, we assume that channel gain $G(H_n^t)$ varies in terms of channel state H_n^t. Based on extensive real-time measurements, a finite set of channel states $\mathcal{H}$ can be used to model channel state H_n^t [26]. A discrete-time and ergodic Markov chain model can be used to characterize the evolution of channel states. The evolution is given by a transition matrix $\mathbf{P} \in \mathbb{R}^{|\mathcal{H}|\times|\mathcal{H}|}$.

Task Processing Delay Component The tasks from all mobile devices subscribed to the m-th service are placed in the edge computing queue for the m-th service. Here, $\sum_{n\in\mathcal{N}_m} \left(1 - o_n^t\right) \zeta \left(\mathbf{x}_n^t\right)$ represents the amount of aggregated tasks. The computing resource is dynamically allocated among multiple services at the AP based on service task arrivals. The dynamic resource allocation can be implemented via a number of existing containerization techniques, such as Dockers and Kubernetes [27]. The computing resource allocation decision vector in time slot t is denoted by $\mathbf{c}^t \in \mathbb{R}^{M\times 1}$, whose each element $0 \leq c_m^t \leq 1$ represents the portion of the allocated computing resource to the m-th service. As such, the processing delay can be calculated by:

$$d_{n,p}^t = \frac{\eta_{m,u} \left(1 - o_n^t\right) \zeta \left(\mathbf{x}_n^t\right)}{c_m^t f_b}, \forall n \in \mathcal{N}_m. \tag{12.7}$$

Here, f_b represents the computing server's CPU frequency at the AP. The computation intensity of processing the m-th service task by the uncompressed DNN is represented by $\eta_{m,u}$. It is worth noting that $\eta_{m,u} > \eta_{m,c}$. The underlying reason is that the uncompressed DNN consumes more computing resource.

Queuing Delay Component The queuing delay consists of the following two parts:

- The first part is the time taken to process backlogged tasks in the edge computing queue, which is given as follows:

$$d_{n,q}^t = \frac{Q_m^t \eta_{m,u}}{c_m^t f_b}, \forall n \in \mathcal{N}_m. \tag{12.8}$$

In the above equation, Q_m^t represents the edge computing queue backlog for the m-th service in time slot t. The task arrival can be represented by $a_m^t = \sum_{n\in\mathcal{N}_m} \left(1 - o_n^t\right) \zeta \left(\mathbf{x}_n^t\right)$, and hence the edge computing queue backlog is updated according to

$$Q_m^{t+1} = \min\left\{\left[Q_m^t + a_m^t - \frac{c_m^t f_b \tau}{\eta_{m,u}}\right]^+, Q_m^{max}\right\}. \tag{12.9}$$

Similar to that in local computing queues, tasks have to be dropped once the edge computing queue is full, As such, the amount of dropped tasks for the m-th edge computing queue is given as follows:

$$\Psi_{q,m}^t = \max\left\{Q_m^t + a_m^t - \frac{c_m^t f_b \tau}{\eta_{m,u}} - Q_m^{max}, 0\right\}. \tag{12.10}$$

In the above equation, $\Psi_{q,m}^t > 0$ indicates that an edge computing queue overflow event occurs.

- The second part is the average waiting time among all newly arrived tasks until all the tasks of mobile device n are processed, which is given as follows:

$$d_{n,w}^t = \frac{\eta_{m,u} \sum_{i \neq n, i \in \mathcal{N}_m} \left(1 - o_i^t\right) \varsigma\left(\mathbf{x}_i^t\right)}{2 c_m^t f_b}, \tag{12.11}$$

where $\sum_{i \neq n, i \in \mathcal{N}_m} \left(1 - o_i^t\right) \varsigma\left(\mathbf{x}_n^t\right)$ represents the amount of the aggregated tasks excluding the task of mobile device n.

Overall, taking both local execution and task offloading into consideration, the inference delay of the collaborative DNN in time slot t is calculated as follows:

$$\begin{aligned} D^t = &\sum_{n \in \mathcal{N}} \left(d_{n,l}^t + d_{n,o}^t + d_{n,p}^t + d_{n,q}^t + d_{n,w}^t\right) \\ &+ w_p \left(\sum_{n \in \mathcal{N}} \mathbb{1}_{\{\Psi_{b,n}^t > 0\}} + \sum_{m \in \mathcal{M}} \mathbb{1}_{\{\Psi_{q,m}^t > 0\}}\right). \end{aligned} \tag{12.12}$$

Here, $\mathbb{1}_{\{x\}}$ the indicator function, which takes a value of 1 when the event x is true, and $w_p > 0$ is the positive unit penalty cost for queue overflow. In the above equation, the first term indicates the required delay of completing all the tasks in time slot t, and the second term indicates the penalty for the local and edge computing queues overflow events.

12.3.2.2 Inference Accuracy Analysis

The achieved DNN inference accuracy is determined by two factors: the sampling rate of a task and the type of DNN that executes a task. To obtain the inference accuracy, the following two steps are conducted:

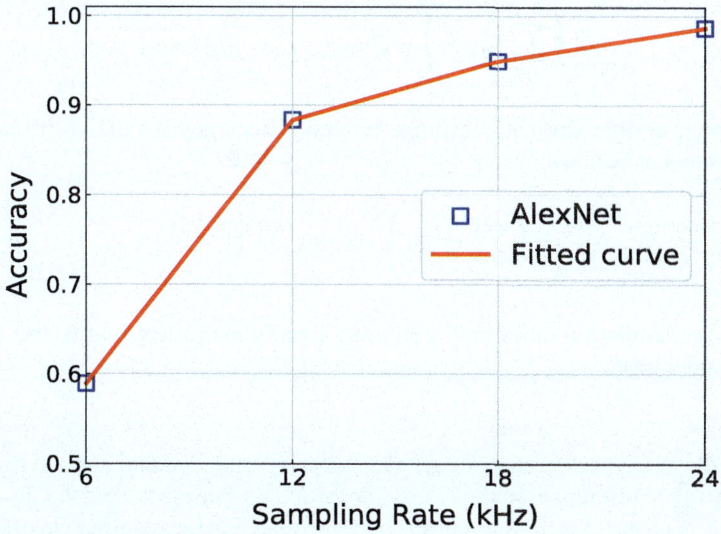

Fig. 12.3 Inference accuracy in terms of different sampling rates on the bearing vibration dataset [29]

- Firstly, we characterize the relationship between the inference accuracy and the sampling rate. The relationship is specified by accuracy function $g(\theta_k)$, $\forall \theta_k \in \mathcal{K}$. To obtain the function, we first implement a DNN inference algorithm, i.e., AlexNet [28]. Then, we use the AlexNet to diagnose facility fault type according to the collected bearing vibration signal [29]. This adopted bearing vibration dataset in the experiment collects the vibration signal of drive end bearings at a sampling rate of 48 KHz, and there are 10 types of possible faults. As shown in Fig. 12.3, inference accuracy grows sub-linearly with the sampling rate. For example, when the sampling rate increases from 18 to 24 KHz, the accuracy increases from 95 to 98.7%. Finally, we measure the accuracy function values in terms of the sampling rates, and the accuracy function is plotted in Fig. 12.3. We can see that the inference accuracy increases with the sampling rate, while the accuracy performance gain decreases at a high sampling rate.
- Secondly, we characterize the relationship between the inference accuracy and the type of DNN via experiments. Here, for the m-th service, the inference accuracy of the compressed DNN is represented by $h_{m,c}$, and that of the uncompressed DNN is represented by $h_{m,u}$. It is worth noting that we have $h_{m,c} < h_{m,u}$. The underlying reason is that an uncompressed DNN achieves higher fault diagnosis accuracy than a compressed DNN.

As the sampling rate selection and the DNN model selection (i.e., task offloading decision) are independent, DNN inference accuracy can be calculated via the product of the accuracy value in terms of the selected sampling rate and the accuracy value in terms of the selected DNN type, i.e.,

$$g\left(\sum_{k\in\mathcal{K}} x_{n,k}^t \theta_k\right)\left(o_n^t h_{m,c} + \left(1-o_n^t\right) h_{m,u}\right).$$

As such, in time slot t, the average inference accuracy for the m-th service can be calculated as follows:

$$A_m^t = \sum_{n\in\mathcal{N}_m} \frac{1}{|\mathcal{N}_m|} g\left(\sum_{k\in\mathcal{K}} x_{n,k}^t \theta_k\right)\cdot\left(o_n^t h_{m,c} + \left(1-o_n^t\right) h_{m,u}\right). \tag{12.13}$$

The above calculation takes both executing locally and offloading to the AP cases into consideration.

The above DNN inference model can be easily extended and applied to cases when other inference methods are adopted. The reason is that the accuracy values in terms of sampling rates and DNN types can be acquired via practical experiments rather than theoretical models.

12.3.3 Joint Sampling Rate Selection and Resource Allocation Problem

12.3.3.1 Constrained Markov Decision Process

In the DNN inference services, not only the service delay is required to be minimized, but also their long-term accuracy requirements should be guaranteed. The CMDP is a class of problems that target at maximizing the long-term reward while satisfying the constraints on the long-term cost [30]. Hence, such problem is suitable to be modeled as a CMDP. We define the action, state, reward, and state transition matrix of the CMDP as follows:

Action The action of the CMDP includes the sampling rate selection, task offloading, and edge computing resource allocation decisions, i.e.,

$$\hat{a}^t = \{\mathbf{X}^t, \mathbf{o}^t, \mathbf{c}^t\}.$$

It is worth noting that the action's components should satisfy following constraints:

- The sampling rate selection decision is constrained by $x_{n,k}^t \in \{0, 1\}$.
- The binary task offloading decision is required, i.e., $o_n^t \in \{0, 1\}$.

- The continuous computing resource allocation decision is constrained by $\sum_{m\in\mathcal{M}} c_m^t \le 1$ and $0 \le c_m^t \le 1$.

The constraint of each action component is satisfied via projecting it into a feasible action set.

State The state of the CMDP includes four components: local computing queues backlog of mobile devices B_n^t, edge computing queues backlog Q_m^t, channel conditions of mobile devices H_n^t, and the raw data size of the generated tasks at mobile devices ξ_n^t. Hence, we have

$$\hat{s}^t = \{\{B_n^t\}_{n\in\mathcal{N}}, \{Q_m^t\}_{m\in\mathcal{M}}, \{H_n^t\}_{n\in\mathcal{N}}, \{\xi_n^t\}_{n\in\mathcal{N}}\}. \quad (12.14)$$

In the above state, both queue backlogs, including $\{B_n^t\}_{n\in\mathcal{N}}$ and $\{Q_m^t\}_{m\in\mathcal{M}}$, adopt a unit in bits. As such, it results in a large state space, especially when the number of mobile devices is large.

Reward The reward of the CMDP is designed to achieve the service delay minimization, as shown in (12.12) in time slot t. In this way, the reward is defined as

$$\hat{r}^t\left(\hat{s}^t, \hat{a}^t\right) = -D^t.$$

State Transition Probability State transition probability of the CMDP is given as follows:

$$\begin{aligned} \Pr\left(\hat{s}^{t+1}|\hat{s}^t, \hat{a}^t\right) = & \prod_{n\in\mathcal{N}} \Pr\left(B_n^{t+1}|B_n^t, x_{n,k}^t, o_n^t\right) \cdot \\ & \prod_{m\in\mathcal{M}} \Pr\left(Q_m^{t+1}|Q_m^t, \mathbf{X}^t, \mathbf{o}^t\right) \cdot \prod_{n\in\mathcal{N}} \Pr\left(H_n^{t+1}|H_n^t\right) \cdot \\ & \prod_{n\in\mathcal{N}} \Pr\left(\xi_n^{t+1}|\xi_n^t\right). \end{aligned} \quad (12.15)$$

The above equality holds since different state terms are independent. Specifically, the first two terms are controlled by the evolution of both local computing queues and edge computing queues, as detailed in (12.3) and (12.9), respectively. The third term is evolved based on the discrete-time Markov chain of channel conditions as mentioned above. The last term is determined by the memoryless task arrival pattern. It is worth noting that each of those state terms only depends on its previous state terms. Such behavior indicates the state transition is Markovian.

In our case, we aim to find a stationary policy $\pi \in \Pi$ that can dynamically configure sampling rates selection $\mathbf{X}^t$, task offloading $\mathbf{o}^t$, and edge computing resource allocation $\mathbf{c}^t$ based on state $\hat{s}^t$. The policy can minimize the service delay and guarantee long-term inference accuracy requirements $\{A_m^{th}\}_{m\in\mathcal{M}}$ simultaneously. To acquire the policy, the optimization problem is formulated as follows:

$$\mathbf{P}_0 : \min_{\pi \in \Pi} \lim_{T \to \infty} \frac{1}{T} \sum_{t=1}^{T} \mathbb{E}_{\pi} \left[D^t \right] \tag{12.16a}$$

$$\text{s.t.} \lim_{T \to \infty} \frac{1}{T} \sum_{t=1}^{T} A_m^t \geq A_m^{th}, \forall m \in \mathcal{M}. \tag{12.16b}$$

The above problem can be deemed as a CMDP.

It is challenging to directly solve the above CMDP via dynamic programming solutions [30]. The reasons are two-fold:

- Firstly, the state transition probability is unknown due to the lack of statistical information on the channel condition variation and task arrival patterns of all mobile devices.
- Secondly, even if the state transition probabilities are known, large action space and state space that grow with respect to the number of mobile devices incur an extremely high computational complexity, which makes dynamic programming solutions intractable.

To solve the CMDP in dynamic environments, we aim to adopt a deep RL-based algorithm. The benefit is that RL-based algorithm can be applied in large-scale networks without requiring statistical information of network dynamics.

However, the existing RL algorithms, such as DDPG, are designed to solve MDP problems without considering policy constraints. Due to the underlying differences between CMDP and MDP, CDMP cannot be solved via traditional RL algorithms.

To solve the problem, we propose a novel learning-based solution for CMDP in the following.

12.3.4 Deep RL-Based Solution

The proposed deep RL-based solution consists of the following three steps:

- **Step 1**: We leverage the Lyapunov optimization technique to deal with the long-term constraints and transform the problem into an MDP, which is suitable to be solved by RL algorithms.

- **Step 2**: We develop a deep RL-based algorithm to solve the MDP.
- **Step 3**: We embed an optimization subroutine in the proposed RL algorithm to directly obtain the optimal edge computation resource allocation.

These three steps are detailed in the following.

12.3.4.1 Markov Decision Process Transformation (Step 1)

To solve problem $\mathbf{P}_0$, the major challenge is to handle the long-term constraints. To address this challenge, we leverage the Lyapunov technique [31, 32].

The basic idea of the step is to construct accuracy deficit queues to characterize the satisfaction status of the long-term accuracy constraints, thereby guiding the learning agent to meet the long-term accuracy constraints.

As such, the problem is transformed in the following way:

- Firstly, inference accuracy *deficit queues* are constructed for all services. The dynamics of the queue evolve as follows:

$$Z_m^{t+1} = \left[A_m^{th} - A_m^t + Z_m^t\right]^+, \forall m \in \mathcal{M}. \tag{12.17}$$

Here, the deviation of the achieved instantaneous accuracy from the long-term accuracy requirement is represented by Z_m^t. Its initial state is set to $Z_m^0 = 0$. Next, we introduce a Lyapunov function to characterize the satisfaction status of the long-term accuracy constraint. The Lyapunov function is defined as [31–33]

$$L\left(Z_m^t\right) = \frac{\left(Z_m^t\right)^2}{2}.$$

In the above equation, a smaller value of $L\left(Z_m^t\right)$ means better long-term accuracy constraint satisfaction.

- Secondly, to guarantee the long-term accuracy constraints, the Lyapunov function should be consistently pushed to a relatively low value. Therefore, a *one-shot Lyapunov drift* is introduced to capture the Lyapunov function's variation across two subsequent time slots [31]. When Z_m^t is given, we define the one-shot Lyapunov drift as follows: $\Delta\left(Z_m^t\right) = L\left(Z_m^{t+1}\right) - L\left(Z_m^t\right)$. We can obtain an upper bound as follows:

$$\begin{aligned}\Delta\left(Z_m^t\right) &= \frac{1}{2}\left(\left(Z_m^{t+1}\right)^2 - \left(Z_m^t\right)^2\right)\\ &\le \frac{1}{2}\left(\left(Z_m^t + A_m^{th} - A_m^t\right)^2 - \left(Z_m^t\right)^2\right)\\ &= \frac{1}{2}\left(A_m^{th} - A_m^t\right)^2 + Z_m^t\left(A_m^{th} - A_m^t\right)\\ &\le C_m + Z_m^t\left(A_m^{th} - A_m^t\right).\end{aligned} \tag{12.18}$$

In the above equation, $C_m = \left(A_m^{th} - A_m^{min}\right)^2/2$ is a constant. Here, A_m^{min} is the lowest inference accuracy, which can be required for service m. Due to the substitution of (12.17), the first inequality holds. The second inequality can be derived due to $A_m\left(\mathbf{X}^t, \mathbf{o}^t\right) \ge A_m^{min}$.

- Thirdly, leveraging the Lyapunov optimization theory, the original CMDP to minimize the service delay and guarantee the long-term accuracy requirements can be transformed to a problem of minimizing a *drift-plus-cost*. The transformed problem is given as follows:

$$\sum_{m\in\mathcal{M}} \Delta\left(Z_m^t\right) + V\cdot D^t \le \sum_{m\in\mathcal{M}} C_m + \sum_{m\in\mathcal{M}} Z_m^t\left(A_m^{th} - A_m^t\right) + V\cdot D^t. \tag{12.19}$$

In the above equation, the inequality holds due to the upper bound in (12.18). Here V represents a positive parameter, which can adjust the tradeoff between the satisfaction status of the long-term accuracy constraints and the service delay minimization. The rationale behind this is that when the long-term accuracy constraint is violated, i.e., $Z_m^t > 0$, it is more urgent to stratify the long-term constraints via improving the instantaneous inference accuracy than to reduce the service delay.

Through this transformation, we reformulate the CMDP problem as a regular MDP problem. The objective of the MDP is to minimize the upper bound of drift-plus-cost as shown in (12.19). In such a reformulated MDP, we should modify the action, state, reward, and state transition matrix since the accuracy deficit queues are incorporated. The modified elements of the MDP are given as follows:

Modified Action The action is the same as that in the CMDP, i.e.,

$$a^t = \hat{a}^t = \{\mathbf{X}^t, \mathbf{o}^t, \mathbf{c}^t\}.$$

Modified State The accuracy deficit queue backlog of services $\{Z_m^t\}_{m\in\mathcal{M}}$ should be incorporated in the state space, as compared to the state of the CMDP. The modified state is given by:

$$s^t = \{\hat{s}^t, \{Z_m^t\}_{m\in\mathcal{M}}\}. \tag{12.20}$$

Modified Reward To minimize the drift-plus-cost in (12.19), the reward is modified as follows:

$$r^t = -V \cdot D^t - \sum_{m \in \mathcal{M}} Z_m^t \left(A_m^{th} - A_m^t \right). \tag{12.21}$$

It is worth noting that we ignore the constant term $\sum_{m \in \mathcal{M}} C_m$ in (12.19) in the reward for simplicity.

Modified State Transition Probability The evolution of state transition probability changes due to the incorporation of accuracy deficit queue backlogs in the state, which is detailed as follows:

$$\Pr\left(s^{t+1}|s^t, a^t\right) = \Pr\left(\hat{s}^{t+1}|\hat{s}^t, \hat{a}^t\right) \cdot \prod_{m \in \mathcal{M}} \Pr\left(Z_m^{t+1}|Z_m^t, \mathbf{X}^t, \mathbf{o}^t\right). \tag{12.22}$$

In the above equation, the second term represents the evolution of the accuracy deficit queue backlog based on (12.17). It is clear that the Markovian property holds for the overall state transition.

Based on the above reformulation and modification, we transform problem $\mathbf{P}_0$ into an MDP problem as follows:

$$\mathbf{P}_1: \min_{\pi \in \Pi} \lim_{T \to \infty} \frac{1}{T} \sum_{t=1}^{T} \mathbb{E}_\pi \left[\sum_{m \in \mathcal{M}} Z_m^t \left(A_m^{th} - A_m^t \right) + V \cdot D^t \right]. \tag{12.23}$$

Similar to solving CMDP as mentioned above, using dynamic programming solutions to solve an MDP faces the curse of dimensionality issue since the state space is large. Therefore, we propose a deep RL-based algorithm to deal with the MDP in the following.

12.3.4.2 Optimization Subroutine for Resource Allocation (Step 3)

For better understanding, we first introduce the third step in the optimization subroutine and then introduce the second step in the RL algorithm design.

As mentioned above, problem $\mathbf{P}_1$ can be solved by RL algorithms. However, we can leverage an inherent property of edge computing resource allocation to reduce the training complexity of RL algorithms. Based on theoretical analysis on (12.23), we find that the edge computing resource allocation and the inference accuracy performance are independent. Specifically, the edge computing resource allocation only impacts the one-shot service delay performance. Therefore, in time slot t, when task offloading and sampling rate selection decisions are given, we can obtain the optimal computing resource allocation decision via solving the following optimization problem:

$$\mathbf{P}_2 : \min_{\mathbf{c}^t} \quad D_t$$

$$\text{s.t.} \quad \sum_{m \in \mathcal{M}} c_m^t \leq 1 \tag{12.24a}$$

$$0 \leq c_m^t \leq 1. \tag{12.24b}$$

Furthermore, an analysis of (12.12) demonstrates that the edge computing resource allocation only impacts the task processing delay and queuing delay at the AP, i.e., $\sum_{n \in \mathcal{N}} \left(d_{n,p}^t + d_{n,q}^t + d_{n,w}^t \right)$. In addition, we find that the aggregated delay from the perspective of all devices is equivalent to the aggregated delay from the perspective of all services. As such, we can rewrite the objective function in $\mathbf{P}_2$ as $\sum_{m \in \mathcal{M}} d_m^t$. As such, we have

$$d_m^t = \sum_{n \in \mathcal{N}_m} \left(\frac{\eta_{m,u} \left(1 - o_n^t\right) \zeta \left(\mathbf{x}_n^t\right)}{c_m^t f_b} + \frac{Q_m^t \eta_{m,u}}{c_m^t f_b} + \frac{\eta_{m,u} \sum_{i \neq n, i \in \mathcal{N}_m} \left(1 - o_i^t\right) \zeta \left(\mathbf{x}_i^t\right)}{2 c_m^t f_b} \right). \tag{12.25}$$

The above equation represents the experienced delay of the m-th service. Through analysis, we show the convexity property of the problem. Then, the following theorem can be used to obtain the optimal edge computation resource allocation in each time slot.

Theorem 12.1 *The optimal edge computing resource allocation for problem* $\mathbf{P}_2$ *is given by:*

$$c_m^{t,\star} = \frac{\sqrt{\Lambda_m^t}}{\sum_{m \in \mathcal{M}} \sqrt{\Lambda_m^t}}, \forall m \in \mathcal{M}, \tag{12.26}$$

where

$$\Lambda_m^t = \sum_{n \in \mathcal{N}_m} \left(\eta_{m,u} \left(1 - o_n^t\right) \zeta \left(\mathbf{x}_n^t\right) + Q_m^t \eta_{m,u} + \frac{\eta_{m,u}}{2} \sum_{i \neq n, i \in \mathcal{N}_m} \left(1 - o_i^t\right) \zeta \left(\mathbf{x}_i^t\right) \right). \tag{12.27}$$

Proof The theorem is proved via the following two steps:

- Firstly, we prove the problem to be a convex optimization problem. For simplicity, t is omitted in the proof. By the definition of Λ_m in (12.27), we can rewrite the objective function as $\sum_{m \in \mathcal{M}} \Lambda_m / (c_m f_b)$. The second-order derivative of the objective function can be derived as $2\Lambda_m / \left(f_b c_m^3 \right) > 0$. In addition, we know that the inequality constraint is linear. Hence, the problem is a convex optimization problem.
- Secondly, we construct a Lagrange function for the problem by ignoring the inequality constraints, which is given as follows:

$$\mathcal{L}(\mathbf{c}, a) = \sum_{m \in \mathcal{M}} \frac{\Lambda_m}{c_m f_b} + a \left(\sum_{m \in \mathcal{M}} c_m - 1 \right). \tag{12.28}$$

Here, a represents the Lagrange multiplier. According to Karush–Kuhn–Tucker conditions for convex optimization [34], the following equation is obtained:

$$\frac{\partial L(\mathbf{c}, a)}{\partial c_m} = -\frac{\Lambda_m}{f_b c_m^2} + a = 0, \forall m \in \mathcal{M}. \tag{12.29}$$

Here, $c_m^\star = \sqrt{\Lambda_m / a f_b}, \forall m \in \mathcal{M}$, can be obtained by solving the above equation. Then, we substitute the above result into the complementary slackness condition $\sum_{m \in \mathcal{M}} c_m^\star - 1 = 0$. Then, the optimal value of a can be given by $a^\star = \left(\sum_{m \in \mathcal{M}} \sqrt{\Lambda_m}\right)^2 / f_b$. Based on the above equation, $a^\star$ takes a positive value, and hence $\{c_m^\star\}_{m \in \mathcal{M}}$ are positive values, which shows that constraint (12.24b), i.e., $c_m^t \geq 0, \forall m \in \mathcal{M}$, is automatically satisfied. We can then prove Theorem 12.1 by substituting $a^\star$ into the complementary slackness condition.

□

This optimization subroutine for the edge computing resource allocation is embedded in the following proposed deep RL-based algorithm. As such, we can reduce the training complexity of the proposed RL algorithm. The reason is that it is no longer necessary to train the neural networks to obtain an optimal edge computing resource allocation policy.

12.3.4.3 Deep RL-Based Algorithm (Step 2)

In the following, we propose a deep RL-based algorithm to solve problem $\mathbf{P}_1$. The proposed algorithm is extended from the well-known DDPG algorithm [35]. However, the DDPG algorithm and the proposed algorithm are different. The main difference is that we embed the above optimization subroutine for computing resource allocation into the RL algorithm to reduce the training complexity. The proposed algorithm can be deployed at the AP that is in charge of collecting the network state information and enforcing the policy to all connected mobile devices.

In the proposed algorithm, the learning agent consists of an actor network that determines the action based on the current state and a critic network that evaluates the determined action based on the reward feedback from the environment. The actor network and the critic network are denoted by $\mu(s|\phi^\mu)$ and $Q(s, a|\phi^Q)$, respectively. The corresponding neural network weights are represented by ϕ^μ and ϕ^Q, respectively. The details of the deep RL-based algorithm are shown in Algorithm 1.

Algorithm 1 Deep RL-based algorithm for sampling rate adaption and resource allocation

Initialize all neural networks and the experience replay memory;
for each episode **do**
  Reset the environment and obtain initial state s_0;
  for time slot $t \in \mathcal{T}$ **do**
    Determine sampling rate selection and task offloading actions $\{\mathbf{X}^t, \mathbf{o}^t\}$ according to s^t;
    Determine edge computing resource allocation action $\mathbf{c}^t$ by (12.26);
    Send joint action $a^t = \{\mathbf{X}^t, \mathbf{o}^t, \mathbf{c}^t\}$ to all mobile devices by the AP;
    Execute the joint action at mobile devices;
    Observe reward r^t and new state s^{t+1};
    Store transition $\{s^t, a^t, r^t, s^{t+1}\}$ in the experience replay memory;
    Sample a random minibatch transitions from the experience replay memory;
    Train the critic and actor network by (12.30) and (12.31), respectively;
    Update target networks by (12.32).
  end for
end for

The proposed algorithm operates in a time-slotted manner, which consists of the following three stages:

- **Stage 1: Obtain experience by interacting with the environment.** The actor network generates the task offloading and sampling rate selection decisions based on the current network state s^t. The decisions are generated with an additive policy exploration noise that follows a Gaussian distribution $\mathcal{N}(0, \sigma^2)$. Additionally, the edge computation resource allocation action is generated by the optimization subroutine. Next, the joint action is executed at all mobile devices, and the corresponding reward r^t is obtained. In addition, we can observe the next state s^{t+1} from the environment. The state transition tuple $\{s^t, a^t, r^t, s^{t+1}\}$ is stored in the experience replay memory for actor and critic network training.
- **Stage 2: Train the actor and critic network based on the stored experience.** A minibatch of transitions are randomly sampled from the experience replay memory to break experience correlation, thereby avoiding the divergence issue caused by DNN. By minimizing the following loss function, the critic network is trained:

$$Loss\left(\phi^Q\right) = \frac{1}{N_b} \sum_{i=1}^{N_b} \left(y_i - Q(s_i, a_i|\phi^Q)\right)^2, \tag{12.30}$$

where

$$y_i = r_i + \gamma Q'(s_{i+1}, \mu'(s_{i+1}|\phi^{\mu'})|\phi^{Q'}).$$

Here, N_b represents the minibatch size $\mu'(s|\phi^{\mu'})$ and $Q'(s, a|\phi^{Q'})$ indicate actor and critic target networks with weights $\phi^{\mu'}$ and $\phi^{Q'}$, respectively. The actor network is trained via the following policy gradient:

$$\nabla_{\phi^{\mu}} \approx \frac{1}{N_b} \sum_{i=1}^{N_b} \nabla_a Q(s_i, a|\phi^Q)|_{s=s_i, a=\mu(s_i)} \nabla_{\theta^{\mu}} \mu(s_i|\phi^{\mu})|_{s_i}. \tag{12.31}$$

- **Stage 3: Update target networks.** The actor and critic target networks are softly updated according to the following equations to ensure network training stability, i.e.,

$$\begin{aligned} \phi^{Q'} &= \delta\phi^{Q} + (1-\delta)\phi^{Q'} \\ \phi^{\mu'} &= \delta\phi^{\mu} + (1-\delta)\phi^{\mu'}. \end{aligned} \tag{12.32}$$

In the above equations, $0 < \delta \ll 1$ is the target network update ratio.

Remark Traditional RL algorithms, e.g., DDPG, can be applied to solve MDP problems, in which learning agents seek to optimize a long-term reward without policy constraints, while they cannot deal with constrained long-term optimization problems [30, 36]. Our proposed deep RL-based algorithm can address long-term constraints within the RL framework by the modification of reward based on the Lyapunov optimization technique. In addition, an optimization subroutine is embedded in our algorithm to further reduce the training complexity.

12.4 Performance Evaluation

12.4.1 Experiment Setup

We consider a smart factory in which mobile devices such as vibration sensors are randomly scattered. Those devices mounted on industrial facilities (e.g., robot arms) capture the operating information. Those sensing data are then either locally processed or offloaded to an AP in the factory.

DNN Inference Services We consider two kinds of DNN inference services:

- *Type I Service*: A facility fault diagnosis service that identifies the type of fault according to the collected bearing vibration signal dataset [29]. Because the period of a time slot is one second, we configure the task data size to be the data volume of a one-second signal, given by the multiplication of the raw sampling rate and the signal quantization parameter. The bearing vibration signal is captured at 48 KHz sampling rate and 16 bit quantization. The resulting task data size is 768 Kb. For this type of service, we set the long-term accuracy threshold to be 0.8.

- *Type II Service*: An extended service from the Type I that diagnoses facility fault based on a low-grade bearing vibration dataset at higher inference accuracy requirement, 0.9. The low-grade dataset senses the vibration at a lower sampling rate of 32 KHz, and the resulting task data size is 512 Kb.

We assume the task arrival rates of both services at each device in each time slot form a uniform distribution. Four potential sampling rates for each device are considered in the simulation, which are 25%, 50%, 75%, and 100% of the raw sampling rate. Accordingly, based on extensive experiments on the dataset [29], the required accuracy to those sampling rates are 0.59, 0.884, 0.950, and 0.987, and the balance parameter, V, is set to be 0.05.

Neural Network Structure To train the proposed deep RL-based algorithm, we set the learning rate of the actor and the critic to be 10^{-4} and 10^{-3}, respectively. The hidden units of both the actor and the critic are set to be (64, 32), while the ReLU function is employed for hidden activation. Note that the Tanh function is used for actor output activation. The training process lasts for 1000 episodes, each of which consists of 200 time slots.

Benchmark We consider the following two benchmark algorithms for performance comparison:

- Delay myopic: Each device dynamically determines the sampling rate and task offloading decisions, to maximize the one-step reward in (12.21) based on the network state.
- Static configuration: Each device follows a fixed configuration on the sampling rate and the task offloading, which satisfy the services' accuracy requirements.

12.4.2 Convergence Performance

Figure 12.4 shows the performance comparison of service delay in the training stage. The average service delay drops as the training continues, which suggests the convergence of the proposed RL-based algorithm. In addition, Fig. 12.5 illustrates the accuracy performance for both services with training episodes. The accuracy performance fluctuates at the beginning of the training. But after around 1000 episodes of training, the average accuracy converges to the required level.

12.4.3 Impact of Task Arrival Rate

After the algorithm is well-trained offline, the performance of the proposed inference algorithm is evaluated in an online scenario. Figure 12.6 gives the comparison on the average service delay with respect to different task arrival rates for $W =$

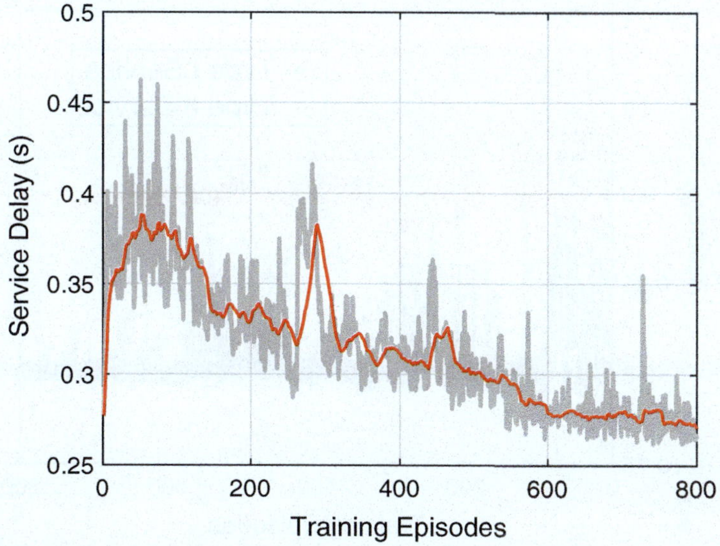

Fig. 12.4 Average delay performance of the proposed algorithm with respect to training episodes in the training stage

20 MHz. Each arrival rate is added up with a 95% confidence interval. It is shown that the service delay grows with the task arrival rate of the constrained communication and computing resources. Meanwhile, the proposed RL-based algorithm gives significantly lower service delay than the benchmark schemes. This is because the proposed algorithm can capture network dynamics, including the pattern of task arrival and channel condition variation, by continuously interacting with the environment. Such knowledge is learnt and utilized by the algorithm to make online decisions that improves long-term performance. In contrast, benchmark schemes only focus on performance in the short term, and they cannot adapt to network dynamics either. In particular, the proposed algorithm reduces the average service delay by 19% and 25%, respectively, as compared with delay myopic and static configuration schemes.

We also give the boxplot accuracy distribution of both services with respect to different task arrival rates in Fig. 12.7. In this figure, the long-term accuracy requirements for both services are 0.8 and 0.9, respectively. The proposed algorithm is able to guarantee the long-term accuracy requirements of both services, with the maximum error probability less than 0.5%.

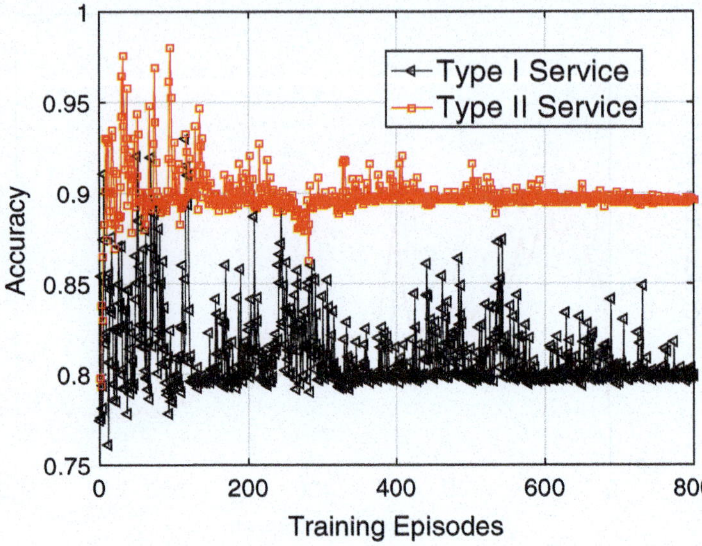

Fig. 12.5 Inference accuracy performance of the proposed learning algorithm with respect to training episodes in the training stage

12.4.4 Impact of Optimization Subroutine

We further evaluate the performance of the proposed algorithm with a fixed computing resource allocation strategy (referred to as proposed-fixed). This strategy allocates edge computing resource based on the average computing demand of two services. As shown in Fig. 12.8, the proposed algorithm provides significant performance gain in case of limited edge computing resource. Specifically, the performance gain in reducing the service delay decreases from 1.98× at 1 GHz CPU frequency to only 1.02× at 1.2 GHz CPU frequency. The underlying reason is that efficient resource allocation weighs higher in resource-constrained scenarios. The simulation curves confirm the effectiveness of the optimization subroutine of computing resource allocation. In light of this optimization subroutine, reducing the training complexity of the proposed RL algorithms can also be considered.

12.5 Conclusion

In this chapter, we have jointly investigated the collaborative DNN inference with sampling rate adaption and resource allocation problem in wireless networks. A deep RL-based algorithm has been devised to capture the pattern of the channel variation and the task arrival, which is then employed to deliver accuracy-guaranteed

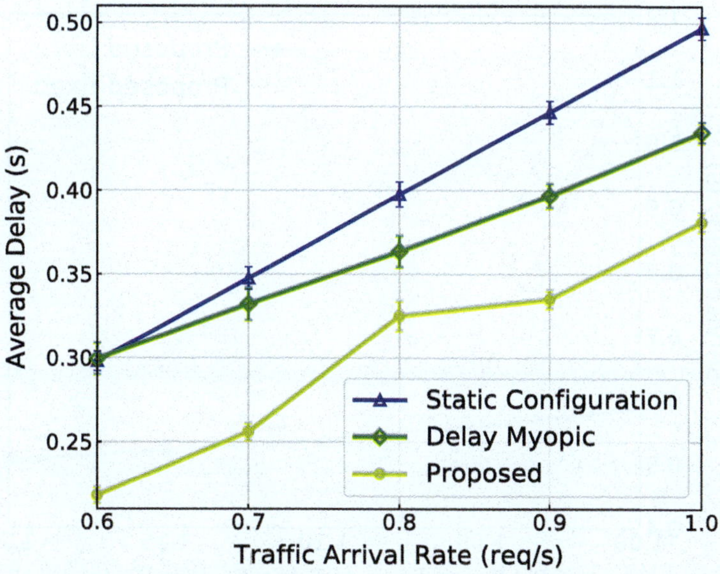

Fig. 12.6 Performance comparison of the service delay in terms of different task arrival rates

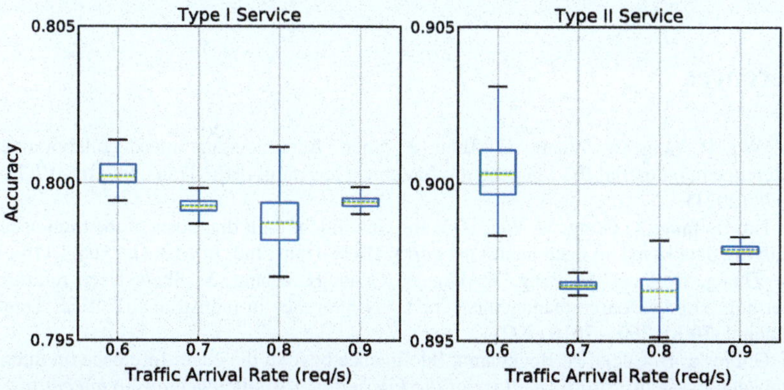

Fig. 12.7 Inference accuracy performance comparison in terms of different task arrival rates

DNN inference services. The proposed algorithm can dynamically reduce the service delay, without requiring prior information of network dynamics.

For DNN inference, further research is required in the following aspects: (1) DNN inference performance should be investigated in the mobile scenarios considering device mobility; and (2) instead of task offloading, the DNN model can be partitioned into a device-side model and a server-side model for collaborative inference. As such, the partition point and resource (computing and spectrum) allocation should be judiciously considered, especially in dynamic network environments.

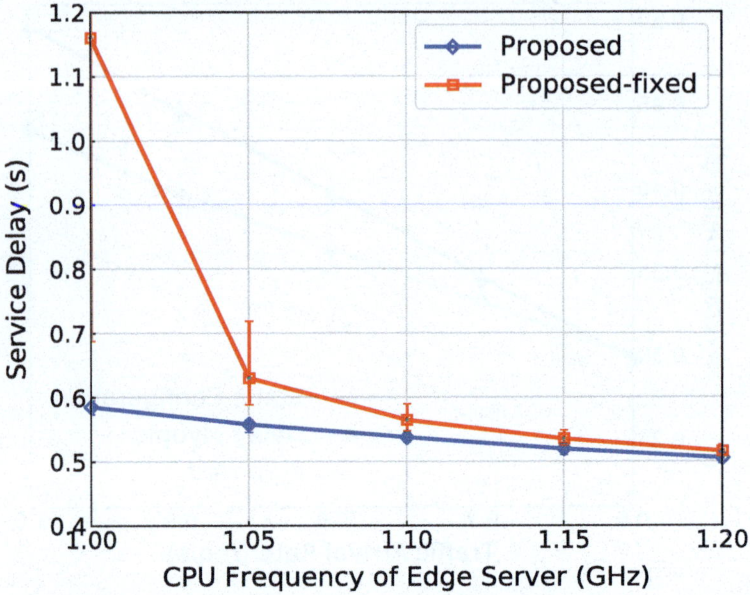

Fig. 12.8 Service delay in terms of CPU frequency of the edge server

References

1. W. Wu, P. Yang, W. Zhang, C. Zhou, X. Shen, Accuracy-guaranteed collaborative DNN inference in industrial IoT via deep reinforcement learning. IEEE Trans. Ind. Inf. **17**(7), 4988–4998 (2021)
2. H. Hu, B. Tang, X. Gong, W. Wei, H. Wang, Intelligent fault diagnosis of the high-speed train with big data based on deep neural networks. IEEE Trans. Ind. Inf. **13**(4), 2106–2116 (2017)
3. W. Zhang, D. Yang, H. Peng, W. Wu, W. Quan, H. Zhang, X. Shen, Deep reinforcement learning based resource management for DNN inference in industrial IoT. IEEE Trans. Veh. Technol. **70**(8), 7605–7618 (2021)
4. G. Gobieski, B. Lucia, N. Beckmann, Intelligence beyond the edge: Inference on intermittent embedded systems, in *Proceedings of the International Conference on Architectural Support for Programming Languages and Operating Systems* (2019), pp. 199–213
5. Y. Chen, T. Krishna, J. S. Emer, V. Sze, Eyeriss: An energy-efficient reconfigurable accelerator for deep convolutional neural networks. IEEE J. Solid-State Circuits **52**(1), 127–138 (2017)
6. D.A. Chekired, L. Khoukhi, H.T. Mouftah, Industrial IoT data scheduling based on hierarchical fog computing: a key for enabling smart factory. IEEE Trans. Ind. Inf. **14**(10), 4590–4602 (2018)
7. E. Li, L. Zeng, Z. Zhou, X. Chen, EdgeAI: On-demand accelerating deep neural network inference via edge computing. IEEE Trans. Wirel. Commun. **19**(1), 447–457 (2020)
8. S. Han, H. Mao, W.J. Dally, Deep compression: Compressing deep neural networks with pruning, trained quantization and Huffman coding (2015). Preprint arXiv:1510.00149
9. S. Teerapittayanon, B. McDanel, H. Kung, Branchynet: Fast inference via early exiting from deep neural networks, in *Proceedings of the IEEE International Conference on Pattern Recognition* (2016), pp. 2464–2469

10. G. Chen, W. Choi, X. Yu, T. Han, M. Chandraker, Learning efficient object detection models with knowledge distillation, in *Proceedings of the Conference on Neural Information Processing Systems* (2017), pp. 742–751
11. P. Yang, F. Lyu, W. Wu, N. Zhang, L. Yu, X. Shen, Edge coordinated query configuration for low-latency and accurate video analytics. IEEE Trans. Ind. Inf. **16**(7), 4855–4864 (2020).
12. W. Zhuang, Q. Ye, F. Lyu, N. Cheng, J. Ren, SDN/NFV-empowered future IoV with enhanced communication, computing, and caching. Proc. IEEE **108**(2), 274–291 (2019)
13. M. Li, J. Gao, L. Zhao, X. Shen, Deep reinforcement learning for collaborative edge computing in vehicular networks. IEEE Trans. Cogn. Commun. Netw. **6**(4), 1122–1135 (2020)
14. W. Wu, N. Chen, C. Zhou, M. Li, X. Shen, W. Zhuang, X. Li, Dynamic RAN slicing for service-oriented vehicular networks via constrained learning. IEEE J. Sel. Areas Commun. **39**(7), 2076–2089 (2021)
15. C. Zhou, W. Wu, H. He, P. Yang, F. Lyu, N. Cheng, X. Shen, Deep reinforcement learning for delay-oriented IoT task scheduling in space-air-ground integrated network. IEEE Trans. Wirel. Commun. **20**(2), 911–925 (2021).
16. Y. Tang, N. Cheng, W. Wu, M. Wang, Y. Dai, X. Shen, Delay-minimization routing for heterogeneous VANETs with machine learning based mobility prediction. IEEE Trans. Veh. Technol. **68**(4), 3967–3979 (2019)
17. X. Shen, J. Gao, W. Wu, K. Lyu, M. Li, W. Zhuang, X. Li, J. Rao, AI-assisted network-slicing based next-generation wireless networks. IEEE Open J. Veh. Technol. **1**, 45–66 (2020)
18. X. Shen, J. Gao, W. Wu, M. Li, C. Zhou, W. Zhuang, Holistic network virtualization and pervasive network intelligence for 6G. IEEE Commun. Surveys Tuts. **24**(1), 1–30 (2022)
19. W. Wu, C. Zhou, M. Li, H. Wu, H. Zhou, N. Zhang, W. Zhuang, X. Shen, AI-native network slicing for 6G networks. IEEE Wirel. Commun. **29**(1), 96–103 (2022)
20. S. Wang, Y. Guo, N. Zhang, P. Yang, A. Zhou, X. Shen, Delay-aware microservice coordination in mobile edge computing: a reinforcement learning approach. IEEE Trans. Mobile Comput. **20**(3), 939–951 (2021)
21. P. Yang, N. Zhang, S. Zhang, L. Yu, J. Zhang, X. Shen, Content popularity prediction towards location-aware mobile edge caching. IEEE Trans. Multimedia **21**(4), 915–929 (2019)
22. W. Wu, N. Zhang, N. Cheng, Y. Tang, Aldubaikhy, K.X. Shen, Beef up mmWave dense cellular networks with D2D-assisted cooperative edge caching. IEEE Trans. Veh. Technol. **68**(4), 3890–3904 (2019)
23. W. Wu, N. Cheng, N. Zhang, P. Yang, W. Zhuang, X. Shen, Fast mmwave beam alignment via correlated bandit learning. IEEE Trans. Wirel. Commun. **18**(12), 5894–5908 (2019)
24. W. Wu, N. Cheng, N. Zhang, P. Yang, K. Aldubaikhy, X. Shen Performance analysis and enhancement of beamforming training in 802.11ad. IEEE Trans. Veh. Technol. **69**(5), 5293–5306 (2020)
25. W. Zhang, D. Yang, Y. Xu, X. Huang, J. Zhang, M. Gidlund, DeepHealth: a self-attention based method for instant intelligent predictive maintenance in industrial Internet of things. IEEE Trans. Ind. Inf. **17**(8), 5461–5473 (2021)
26. L. Lei, Y. Kuang, X. Shen, K. Yang, J. Qiao, Z. Zhong, Optimal reliability in energy harvesting industrial wireless sensor networks. IEEE Trans. Wirel. Commun. **15**(8), 5399–5413 (2016)
27. D. Bernstein, Containers and cloud: From LXC to Docker to Kubernetes. IEEE Cloud Comput. **1**(3), 81–84 (2014)
28. A. Krizhevsky, I. Sutskever, E. Hinton, ImageNet classification with deep convolutional neural networks, in *Proceedings of the Conference on Neural Information Processing Systems* (2012), pp. 1097–1105
29. International Data Corporation. [Online]. Available: https://csegroups.case.edu/bearingdatacenter/pages/download-data-file. Accessed: Jun. 2021
30. E. Altman, *Constrained Markov Decision Processes* (CRC Press, Boca Raton, 1999) Available via DIALOG
31. M.J. Neely, Stochastic network optimization with application to communication and queueing systems. Synth. Lect. Commun. Netw. **3**(1), 1–211 (2010)

32. J. Luo, F. Yu, Q. Chen, L. Tang, Adaptive video streaming with edge caching and video transcoding over software-defined mobile networks: A deep reinforcement learning approach. IEEE Trans. Wirel. Commun. **19**(3), 1577–1592 (2020)
33. J. Xu, L. Chen, P. Zhou, Joint service caching and task offloading for mobile edge computing in dense networks, in *Proceedings of the IEEE International Conference on Computer Communications* (2018), pp. 207–215
34. S. Boyd, S.P. Boyd, L. Vandenberghe, *Convex Optimization* (Cambridge University Press, Cambridge, Vandenberghe). Available via DIALOG
35. T.P. Lillicrap, J.J. Hunt, A. Pritzel, N. Heess, T. Erez, Y. Tassa, D. Silver, D. Wierstra, Continuous control with deep reinforcement learning, in *Proceedings of the International Conference on Learning Representations* (2016)
36. Q. Liang, F. Que, E. Modiano, Accelerated primal-dual policy optimization for safe reinforcement learning (2021). Available via DIALOG, https://arxiv.org/abs/1802.06480

Chapter 13
Automated Data-Driven System for Compliance Monitoring

Humphrey Rutagemwa and François Patenaude

13.1 Introduction

The wireless applications that are becoming an increasing part of our every day lives are based on the electromagnetic spectrum that has been studied since the nineteenth century. While the electromagnetic spectrum spans a very wide range for frequencies, up to the optical domain, only a part has been historically in use for radiocommunication purposes. The traditional radio spectrum band is usually defined between 9 kHz and 30 GHz; however, with new applications emerging, the upper limit could soon be around 300 GHz [1, 2]. As illustrated in Fig. 13.1, the radio spectrum band is itself divided into bands [3] depending on the field of applications. The most common sub-bands are the decade's bands starting with the VLF up to the EHF. Additionally, these sub-bands are divided into sub-sub-bands like in satellite communications (SHF), where they take the letters L, S, C, X, Ku, K, and Ka. The same idea applies to the other sub-bands, where the sub-divisions are often based on the country in which the radio spectrum is used. This makes the spectrum a multi-dimensional natural resource that is finite, however, non-exhaustive.

The electromagnetic spectrum is being used for many communication applications. In general, its use can provide tremendous opportunities for social and economical development. The economic value of commercial spectrum has exploded over the last decades and is now seen as a national resource that requires dedicated management [4].

H. Rutagemwa (✉) · F. Patenaude
Communications Research Centre, Ottawa, ON, Canada
e-mail: humphrey.rutagemwa@ised-isde.gc.ca; francois.patenaude@ised-isde.gc.ca

L. Cai et al. (eds.), *Broadband Communications, Computing, and Control for Ubiquitous Intelligence*, Wireless Networks,
https://doi.org/10.1007/978-3-030-98064-1_13

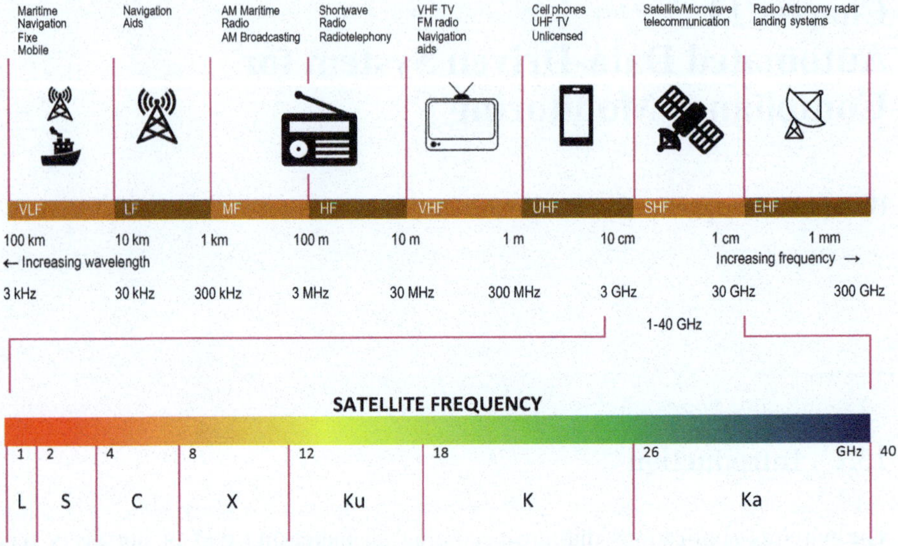

Fig. 13.1 Spectrum overview. Source [3]

13.1.1 Radio Spectrum Management

Radio spectrum management (later referred to as spectrum management) is the overall process of regulating and administering use of the radio spectrum [5]. Spectrum management is predominantly a national responsibility exercised by an organization casually referred to as the regulator. The national spectrum regulators work to enable the conditions for orderly deployment of radiocommunication system within the national territory. The spectrum is made available by considering the stakeholders that include the end-users, the equipment manufacturers, the providers of commercial services, and the providers of public services. The master plans for managing the spectrum are administratively reflected in high-level policies that reflect economic, technical, and social objectives. The economic and social objectives often target the development of a national or continental industry and the promotion of competition to serve various groups in society. The technical objectives are implemented with three key spectrum management functions: spectrum planning, spectrum authorization (assigning, licensing), and spectrum monitoring.

Spectrum planning provides long-term direction and cohesion of policies. The most tangible output of this function is a national radio spectrum allocation chart that defines the different services in the various spectrum sub-bands. Spectrum authorization is the process by which users are granted spectrum resources and permitted to use these resources within specific regulatory conditions, which impose limits on parameters such as geographic coverage, frequency range, and transmission power. Spectrum monitoring serves as the eyes and ears of the spectrum management process [5]. For example, it provides valuable data to facilitate compliance verification and interference investigation.

13.1.2 Spectrum Monitoring for Compliance

Spectrum monitoring to ensure compliance with regulatory requirements is one of the key spectrum management activities that contribute to preventing harmful interference and improving the overall quality of the spectrum. It ensures the integrity of spectrum and radio environments, enabling the orderly operations of spectrum awareness, planning, and licensing activities within the regulatory framework.

The growth of the digital economy, along with the evolution of advanced technology, has led to increasing spectrum demands for more wireless service coverage and dense deployments of intelligent radio devices. As a result, spectrum regulators worldwide are experiencing an increased workload on management activities, including spectrum monitoring activities to verify compliance with license terms and conditions, to detect unlicensed operations and to check unpaid license fees. One of the key challenges faced by the spectrum regulators is how to efficiently and scalably carry compliance activities in a wide range of frequency bands over large geographical areas using limited equipment and human resources.

On the other hand, new advanced technologies for spectrum monitoring, data processing, and data analytics are emerging. A few examples include networked heterogeneous sensors with advanced signal processing, cloud computing and infrastructure, and advanced machine learning algorithms and artificial intelligence. With recent advanced technologies, spectrum regulators have great opportunities to improve their spectrum management approaches and processes.

13.1.3 Chapter Contributions and Organization

This chapter presents an automated data-driven testbed system that leverages advanced technologies to facilitate efficient and scalable spectrum monitoring and compliance activities. It aims to reduce the manual workload on spectrum managers by automatically collecting spectrum data from different sources, identifying compliance issues, and performing analytics to provide actionable insights. The goal of such automated data-driven system is not to replace the judgment and expertise of spectrum managers but to assist in making better evidence-based decisions. It could be beneficial by providing information and insights that will help preventing interference and protecting spectrum integrity, speeding up resolution of the compliance issues, identifying and localizing a wider range of compliance violations, and generating data (evidence) to support potential enforcements.

It should be noted that the objective of this chapter is not to provide a detailed design of operational systems. Rather, it is to present high-level concepts and functions of such systems. Some parts of the research work and live experimental testbed[1] developed at the Communications Research Centre Canada (CRC) are presented through the chapter. The intent is to help explaining and illustrating key concepts of data-driven systems and possible solutions for monitoring spectrum to ensure compliance with regulatory requirements.

The remainder of this chapter is structured as follows: Sect. 13.2 presents an automated data-driven system for supporting compliance monitoring. Section 13.3 describes data sources for spectrum measurements, spectrum management records, and relevant auxiliary data. Section 13.4 presents functions for preprocessing data. Functions for the detection, characterization, and prioritization of violations are discussed in Sect. 13.5. Finally, summary and concluding remarks are provided in Sect. 13.6.

13.2 Automated Data-driven System

This section presents an automated data-driven system for supporting spectrum monitoring and compliance activities in spectrum sharing environments where licensed radio systems may use overlapping frequency channels in the sub-6 GHz bands and operate over large geographical areas. The main functional requirements for the system are to collect spectrum data from various sensors, identify compliance violations, and perform compliance analytics to provide actionable insights.

Figure 13.2 shows a block diagram of the proposed automated data-driven system. It is composed of five functional blocks: data source, signal identification, violation identification, compliance analytics, and user interface. These blocks are described as below.

- **The Data Source Block** provides access to spectrum measurements, spectrum management records, and relevant auxiliary data, which can be obtained from internal or third-party monitoring services. Spectrum measurements include data collected directly from RF sensors or network sniffers. Spectrum management records include data from license/certificate databases and device/equipment databases. Relevant auxiliary data includes historical records of data for weather, events (political, social, economical, environmental), and satellite maps.
- **The Signal Identification Block** preprocesses various data sources to provide more accurate measurements that give a better picture of what is going on. Specifically, it performs the following tasks: decomposing sensed signals into components (modes) from the different emitters, mapping identified signal components to emitters and licenses, and associating measurements and emitters in the databases.

[1] CRC live experimental testbed incorporates spectrum measurements from various sensors located across Canada. However, it is not deployed operationally.

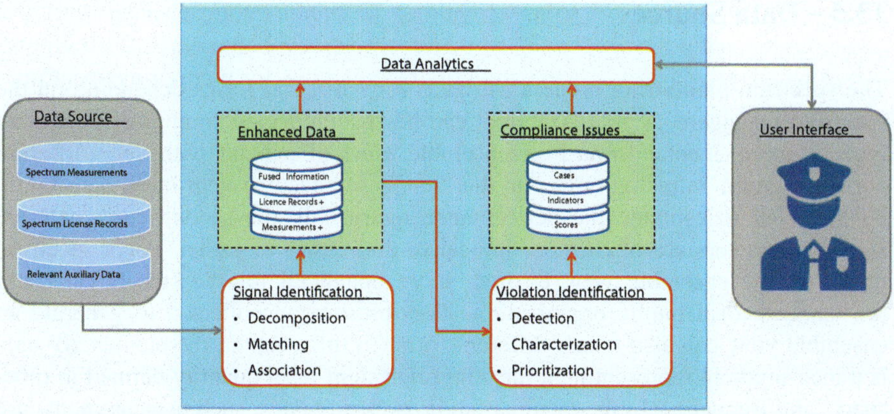

Fig. 13.2 Block diagram of an automated data-driven system for supporting spectrum monitoring and compliance activities

- **The Violation Identification Block** performs the detection, characterization, and prioritization functions. The detection function filters channels with anomalous or unexpected emission characteristics and classifies the type of violations. The characterization function computes confidence, behavior, extent, and impact indicators, which provide a complete and holistic view of compliance violations. The prioritization function considers various factors and computes priority scores that rank the relative importance of compliance violations.
- **The Compliance Analytics Block** analyzes historical data of compliance violations (types of violations, base indicators, and priority scores) and provides actionable insights and foresight to support evidence-based compliance audits and investigations. The functions performed by this block include computing summary statistics, recognizing patterns and trends, predicting the extent of potential and actual impacts, generating prioritized compliance violation alerts, and recommending cause of actions for interference diagnosis and resolutions.
- **The User Interface Block** presents results to the end-users (e.g., spectrum managers) in various ways: graphical user interface (GUI) with dynamic, interactive, and responsive dashboards and drill-down graphs, application programming interface (API) to support remote access and applications on numerous platforms, and data sets and reports in multiple standard file formats.

The rest of this chapter presents detailed descriptions of the data source, signal identification, and violation identification functional blocks.

13.3 Data Sources

The spectrum monitoring process can take very different forms depending on the spectrum management goal. The goal can be specific to a communication system type or more general toward an ensemble view of a band with very different communication infrastructures. In the first case, deep system-level data could be extracted to estimate the performance metrics. In the second case, external signal parameters are often more indicative and easier to gather. For a spectrum regulator, the ensemble view is often more beneficial for the stakeholders and the mandate of a public organization. The information collected to generate an ensemble view can also take a different form. In this chapter, the elementary data is a measurement representing a meaningful system-level quantity defined in time, space, and frequency. In the commercial broadband bands, such measurements are often associated with the key performance indicators (KPIs) used to characterize the system performance. For the land mobile radio (LMR) bands, the traditional measurement is the channels' power.

13.3.1 Spectrum Measurements

For CRC testbed spectrum measurements, signal-to-noise ratio, signal bandwidth, and signal carrier frequencies are also collected using a wideband channel software-defined scanner with probability of false alarm detection [6]. These added dimensions increase the degrees of freedom in several of the analyses presented in the following sections. Note that the instantaneous angle-of-arrival at the sensor could be added if the sensor capability was present. These instantaneous measurements are performed several thousand times per hour per channel, according to a predefined table that contains several thousand channel definitions consisting of channel center frequency, channel bandwidth, and a channel ID. These measurements are done with equipment located at a fixed location for a certain timeframe that can go from a few hours to weeks or months or on the move following roadways. Some examples of spectrum monitoring are [7, 8], where power only is collected to derive radiofrequency occupancy.

The extent of time–frequency–space in the work performed within the CRC testbed is also different than many of these studies. The data collection happened for approximately four years in major cities across Canada in the frequency range from 117 to 960 MHz. The space coverage is defined by the sensor locations and the sensor sensitivity. The sensors' locations are mainly in major Canadian cities, as depicted in Fig. 13.3. The sensors' sensitivity depends primarily on the antenna height and noise figure of the equipment. Because of the larger number of sensors, heterogeneous sensors were used with typical specifications exemplified in Table 13.1.

Such large-scale data collection can only be accomplished with cloud infrastructure to automate the data retrieving, data storage, and data management [9].

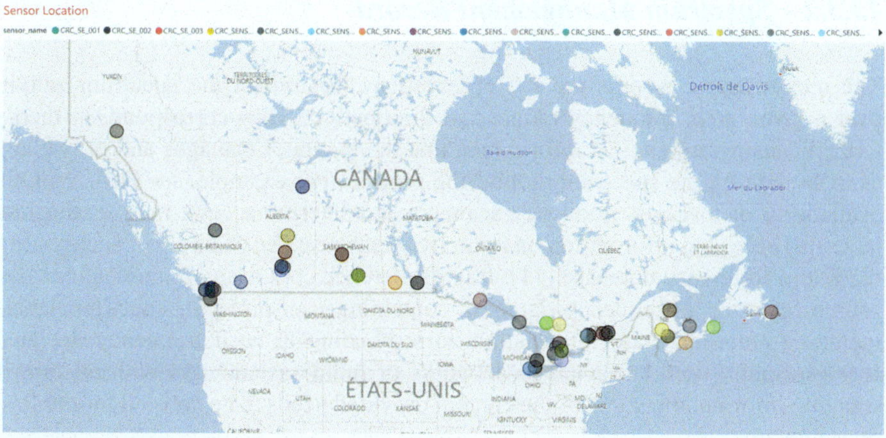

Fig. 13.3 Sensor locations map

Table 13.1 Typical specifications of software-defined heterogeneous wideband scanner

Parameters	Values range
Frequency range	20 MHz to 3 GHz
Tuning resolution	1 Hz
IP3	16–20 dBm
Noise figure	88–20 dB
Instantaneous bandwidth	14–24 MHz
Resolution bandwidth	1–2 kHz
Measurement period	2–4 ms
Instantaneous dynamic range	Greater than 60 dB
Dynamic range	Greater than 100 dB
Revisit period (400 MHz scan)	0.33–4 s
Detection method	Energy detection with constant probability of false alarm or fixed signal-to-noise ratio

The basic data schemas to support the analytics of the use cases are composed of several tables. In that respect, the data is organized in three layers that express the level of aggregation in time. Level 1 (L1) data contains the individual measurements where the time, frequency, and location are attached to the quadruple power, SNR, bandwidth, and carrier estimations measurements. The L1 measurements are then used to build Level 2 (L2) histograms of power, SNR, bandwidth, and carrier estimation measurements over an hour. The L2 histograms also include the burst-on and burst-off distributions in dB-millisecond to characterize the nature of the bursts over the channels. Finally, the Level 3 (L3) statistics parameterized the L2 distributions again over an hour to increase the level of aggregation. Note that measurements are still time–space–frequency organized and not linked to an actual assignee in general.

13.3.2 Spectrum Management Records

The measurement records are the observations that reflect the spectrum activity over a given area, frequency range, and time period. They correspond, in theory, to the licensing/assignment information that the regulator manages and is available as input data for the spectrum monitoring process. For example, see [10]. With the availability of low-cost frequency scanners and Internet access, radio enthusiasts have also collected information on radio frequency use. An example of such a site is the Radio Reference database [11]. RadioReference.com is a communications data source featuring a frequency database containing user-provided spectrum labels, including emission class information corresponding to each frequency detected. It is especially useful to analyze networks of radio systems. As with all crowd-sourcing information, some of the content provided needs to be verified for the level of accuracy.

13.4 Signal Identification

The traditional approach to identify user signals is to analyze the power L2 measurements and identify modes in the histogram structure of fixed sensors of a given channel over a given hour, for example. This approach tends to work well in non-congested areas or when the channel assignment is guided by conservative interference levels with the sensor close to the location of interest. The L2 power histograms approach is suitable for clearly separated power modes. It typically gives a lower bound on the true number of modes. However, it has shortcomings, notably:

- Overlapping powers at the sensors are difficult to distinguish.
- The power samples do not link to the bandwidth and carrier offset information measurements.
- It can be difficult to identify the same signal in space, especially when more than one signal is present.

The overall concept of the proposed signal identification is presented in Fig. 13.4. The components are organized into three categories: the field components representing the sensors (hardware, processing software, and control), the cloud storage to hold all of the spectrum management records and the inferred information, and the cloud analysis where the different use cases get their processing implemented. For the signal identification cloud analysis use case, three main steps are needed. The first step is the mode analysis to decompose the measurements into groups that have similar quantities. This is done on a sensor basis in the cloud but could also happen on the sensor itself. Once each mode is identified on the sensors, the cross-mode, cross-sensor correlation attempts to detect the same signal mode in space (among a group of regional sensors). Signal modes matching in space are concluded to come from the same emitter. Finally, the license-measurement association step is deciding on which signal modes matching in space are associated with which candidate licensee. Note that for each of these steps, information tables are generated to augment the spectrum management information. The final information obtained is the emitter table that holds the actual licensee measurement information aggregated from matching signal modes in space.

13.4.1 Mode Analysis

To better link the measurements to the actual radio environment, the use of L1 data is advantageous. It offers more dimensions to try to separate the spectrum users and ultimately provides the data to build the L2 and L3 data set on a mode basis instead of the channel approach. This section will now look at the classification approach taken to label the L1 samples. The data with a mode label associated with each sample will be denoted L1+ (or augmented data in Fig. 13.4).

In the general field of classification, two main approaches have been used: supervised and unsupervised methods. Supervised classification requires a label for the samples that is not available here. Additionally, supervised classification is useful in the context of classifying new samples in a channel. Since the interest here is more in characterizing the spectrum than predicting it, an unsupervised clustering method is preferred and used to separate the users of a given channel at a given sensor over a defined time period.

Several clustering methods have been developed over the last decades. A good review of these methods can be found in [12]. The constraint in the present context and many other practical situations is that the number of users in a measurement sets

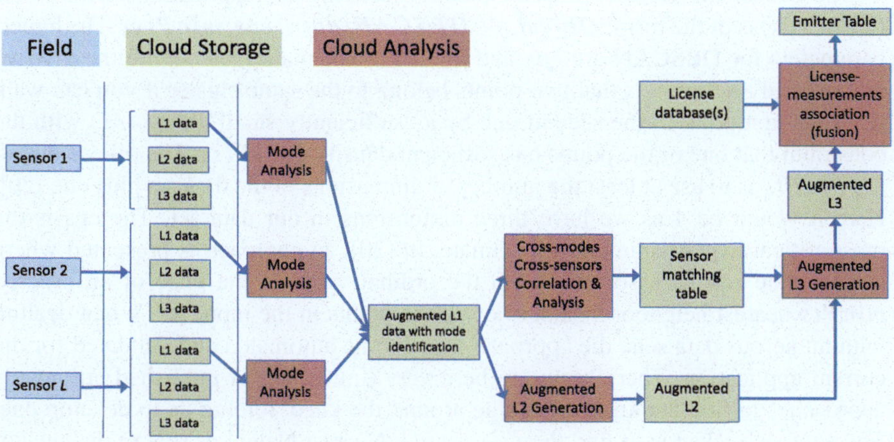

Fig. 13.4 Block diagram of the signal identification steps

is unknown, i.e., the cluster size. Another practical issue here is that the number of channels is large, the number of time periods is large, and the number of sensors is significant. The combination of these makes it impractical to have a manual analysis to set the cluster size. This also implies that the hyper-parameters used at the input of the algorithm need to be fixed or estimated from the measurements itself. With the amount of data to process, it is unlikely that a fixed hyper-parameter value will provide good performance. Practically, some of the measurement points are likely to be outliers or noisy. The algorithm should have a provision to classify samples as outliers so that the statistics of the emitters are not drastically affected. The clusters expected from the measurements are also non-convex as they are generated from a variety of modulated signals and interference patterns. Non-convex clusters are clusters where it is not possible to draw a line between two cluster points without going outside of the cluster. A simplification here is that the number of dimensions is relatively small (on the order of 3–4) and the maximum number of clusters in the measurement set is also small (less than about 5). The time period of interest can be of any value. However, to avoid having to deal with large clustering matrices, the value of one hour is a logical start. The objectives of the clustering on an hourly basis are then to (1) minimize the number of false clusters, (2) minimize the number of missed clusters, and (3) minimize the size of the outlier cluster.

To achieve the above objectives, the selection of the algorithm is based on scalability or suitability for large-scale data sets, the arbitrary shape of data sets, sensitivity to the sequence of input data, and sensitivity to noise/outlier. From [12], the choice reduces to a few algorithms, CURE [13], Moment Method [14], DBCLASD [15], DBSCAN [16], OPTICS [17], STING [18], and GMM [19]. After accounting for the different practical considerations like the number of hyper-parameters, type of hyper-parameters, and availability of code in R [20] and Python [21], it was decided to use DBSCAN. The DBSCAN can deal with border points. Furthermore, only two hyper-parameters need to be provided.

The use of the Density Based Spatial Clustering of Applications with Noise (DBSCAN) is of the form: $Object \leftarrow DBSCAN(data, eps, minPts)$. The hyper-parameters for DBSCAN are eps and $minPts$. The main idea behind the divisive hierarchical algorithm is that two points belong to the same cluster if you can walk from the first point to the second one by a "sufficiently small" step (eps) with the constraint that one of the points has sufficient density ($minPts$). The rule of thumb for $minPts$ is to use at least the number of dimensions of the data set plus one [20]. Here it would be 4 as we have three dimensions in our data set. The eps is the most critical hyper-parameter to estimate. In [20], an approach is presented where the estimate can be associated with the ordinal value of the knee of an ordered $minPts$ nearest neighbor distances of all the points in the input set. When dealing with numerous data sets, the approach needs to be automated and validated for the current application. There seems to be a very limited set of published automated approaches to find the knee or a value around the knee, forcing us to develop one. The approach taken is to find the maximum k Nearest Neighbor (kNN) distance of a line perpendicular to a line going from the first to the last point of the sorted kNN distances.

13.4.2 *Mode-Sensor Matching*

Once the signals at a sensor are identified, it is then necessary to determine if other sensors have also seen the same signal in space. To decide on such cases correlations between all the modes of sensor pairs are performed. The correlation function in a given hour for a given channel between a sensor mode power and another sensor mode power is performed at 1 sample per second, thus with vectors of 3600 samples. The L1+ power needs to be resampled since it is not sample synchronized and not at the same sampling rate. The resampling is done by keeping the max power value for a second or set to a default value if no sample exists within a given second. Once the cross-correlation is performed, the peak value and the lag value of the peak are collected. These values are used to determine which sequences have some commonality. The lag between the sensors should be constant for all signals seen by a sensor pair. Next, the signals are shifted by the lag values to align the power waveform to compute a third parameter called mutual occupancy as a measure of similarity of the waveform. These three parameters (peak level, peak lag, and mutual occupancy) are then analyzed to decide on the commonality. Figures 13.5 and 13.6 show examples of the outputs.

It is to be noted that the lower power modes are incomplete sequences because some of their measurements are hidden by the strongest mode (energy detection). This will have the impact of reducing the correlation peak value. Once the correlation parameters are available, a decision on the commonality or not of a signal mode at two sensors is needed. For an ensemble of channels, the data looks as in Fig. 13.7 for the correlation peak and the mutual occupancy with the blue dots representing the same signals and the green dots different signals at the two sensors. The mutual occupancy M_o is defined as

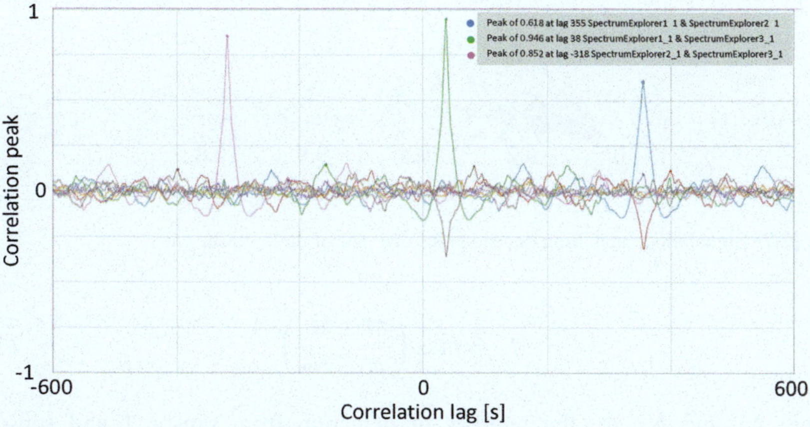

Fig. 13.5 All cross-modes cross-sensors correlation functions

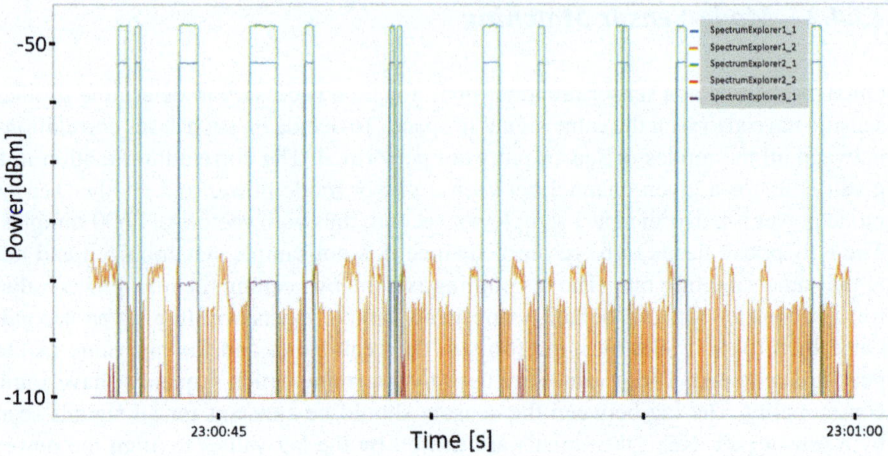

Fig. 13.6 Time-aligned L1 measurements at one sample/sec for three sensors and all modes

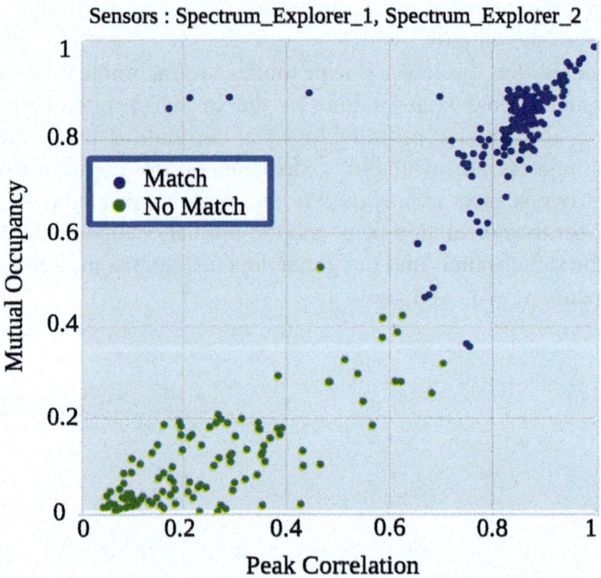

Fig. 13.7 Signal mode matching decisions for one pair of three sensor pairs

$$M_{\mathrm{o}} = \max\left(\frac{M_c}{N_1}, \frac{M_c}{N_2}\right), \tag{13.1}$$

where N_1 and N_2 are the number of detection from sensor 1 and sensor 2, respectively, during a one hour time period. M_{c} is the mutual count computed as

$$M_c = \sum_{i=1}^{3600} \mathbf{1}_I\left[(S_i^1 > T_1)\&(S_i^2 > T_2)\right], \tag{13.2}$$

where S_i^1 and S_i^2 are time-aligned within one second resolution of the sensors measured signal power. T_1 and T_2 are the respective sensor power thresholds computed using [6]. $\mathbf{1}_I[\cdot]$ is the indicator function, that is, $\mathbf{1}_I[\text{True}] = 1$ and $\mathbf{1}_I[\text{False}] = 0$. Figure 13.7 was obtained with a Kmeans classifier with a cluster count of two as an input for the hyper-parameter.

13.4.3 License-Measurements Association

The knowledge of the signals in space provided by the above function enables the association of licenses with the measurements in space. The decision rule adopted here is

$$Associate\ Tx := \underset{Tx\ j}{\arg\min} \left\{ \sum_{i=1}^{L} \left| P_{sensor_i}^{predicted_j} - P_{sensor_i}^{measure} \right| \right\}, \tag{13.3}$$

for $j = 1, 2, ..., C$, where C is the number of candidate transmitters, L is the number of identified signals in space, $P_{sensor_i}^{predicted_j}$ is the predicted power scalar values of transmitter j at sensor I, and $P_{sensor_i}^{measure}$ is the measured power scale value at sensor i. The threshold T ensures that some level of coherency is preserved in terms of power and space.

The value of T depends on factors like sensor calibration and prediction model accuracy. There are many scenarios involved here. In the simplest case, there is one signal mode in space, one licensee, and the summation is less than T. The case where no measurements can be associated with a licensee transmitter results in detecting an illegal transmitter. In that case, a subsequent step is to estimate the location of the illegal transmitter (not represented in Fig. 13.4). The case of no signal mode and licensee may indicate that the location and the sensitivity of the sensors is not adequate to monitor the licensee. This is often the case for low-power emitter licensees. The outcome of the present association is an explicit link between the licenses and the measurements. The resulting table would contain information such as occupancy, estimated bandwidth, carrier offset of the licensees, and other information like potential interference on the channel. In future work, other approaches to consider are using the distribution of power measures (the L2 power measurements) to compare probabilities instead of a sum of absolute power differences.

13.5 Violation Identification

Consider large-scale compliance monitoring operations that require the identification of various types of compliance violations that occur in large geographical areas over long periods of time in many frequency channels. In such situations, the violation identification block shown in Fig. 13.2 can utilize preprocessed data to identify potential violations and generate information to support compliance analytics. The core functions of the violation identification block are detecting, characterizing, and prioritizing compliance violations. The following sections present frameworks and guidelines for developing and implementing these functions.

13.5.1 Detecting Violations

A compliance violation occurs if a spectrum user does not conform to requirements because the transmission is unlicensed or the transmission does not comply with rules and regulations. Typical compliance violations include [5] illegal or unlicensed operation of transmitters or use of frequencies, unauthorized operating periods or locations, illegal emission classes, excessive frequency offset, excessive bandwidth, and excessive power. These violations can occur for various reasons or motives. They can be inadvertent (e.g., equipment failure), reckless (e.g., system misconfiguration), or deliberate (e.g., illegal jamming). In general, compliance violations can be detected or confirmed by comparing parameters derived from spectrum measurements and the corresponding parameters derived from spectrum management databases. In that case, the measurements provide ground-truth data with technical parameters such as center frequency, frequency offset, occupancy, field strength and power-flux density, occupancy bandwidth, direction, polarization, and modulation. The databases provide reference data for licensees, equipment, and stations with technical parameters such as assigned frequency, predicted field strength, emission class, and assigned bandwidth. Detecting compliance violations involves processing spectrum data with various levels, granularities, and abstractions (e.g., measurement traces, statistical distributions, and summary statistics). For large-scale compliance monitoring operations, two key challenges that arise in processing spectrum data are efficiency and scalability problems. A two-stage processing approach is proposed to address these challenges.

- **The first stage:** In this stage, the key function is to detect channels with suspicious compliance issues using methods that are relatively fast or computationally inexpensive. Detecting compliance issues may require analyzing a large number of frequency channels. However, in real-world scenarios, the proportion of channels with actual compliance issues is relatively small. Therefore, the goal in this stage is to perform rapid tests on various technical parameters and reduce the number of frequency channels that will be selected for further analysis. Two general approaches can be applied to achieve this goal. The first approach compares actual measurements themselves and selects frequency channels that show anomalies or unusual patterns. With such an approach, unsupervised machine learning methods (e.g., clustering) or supervised machine learning methods (e.g., binary classification) can be applied to detect anomalous channels. The second approach is to compare parameters derived from measurements with the predicted parameters derived from the spectrum management records. By considering link budget, propagation models or machine learning methods (regression) can be applied to generate predicted parameters. The primary challenge for both approaches is finding fast ways to select suspicious channels while maintaining high precision and accuracy.

- **The second stage:** In this stage, the key function is to confirm and classify types of compliance issues that may exist in the suspicious channels. Note that some suspicious channels may be false positive detections, meaning they do not have compliance issues. This is especially true in congested environments. For example, channels with adjacent channel interference could be falsely detected with excessive bandwidth issues, whereas the ones with co-channel interference could be mistakenly detected with excessive carrier offset issues. Furthermore, even for detections that are true positive, some compliance issues may not be easily classified. For example, excessive field strength derived from measurements may be due to excessive effective radiated power and/or higher antenna height above ground level. Therefore, the goal in this stage is to perform a thorough analysis to determine the most likely cause of observed anomalies or discrepancies and identify types of compliance issues. The most likely causes can be determined by examining several technical parameters across multiple sensors and analyzing the presence of adjacent and co-channel interference. The underlying compliance issues can be identified by applying supervised machine methods such as multi-class or multi-label classifications. In this case, an instance represents suspicious channels. Labels, which may be assigned to instances, represent possible compliance issues. The key challenges here are to model and develop a classifier with a high probability of detection and classification accuracy.

The data processing in the first stage uses a relatively large set of high-level data with coarse granularity. Comparatively, the processing in the second stage uses a relatively small set of low-level data with fine granularity. In general, processing high-level data is less expensive in terms of time and computation compared to processing low-level data. Since the number of channels with actual compliance issues is relatively small, filtering channels in the first stage and then classifying types of compliance issues in the second stage can significantly improve the overall computation efficiency and processing time.

13.5.2 Characterizing Violations

This section identifies and discusses various indicators that show the level of confidence in applied detection and classification methods and quantifies the behavior, extent, and impact of the compliance violations. The indicators can

Table 13.2 List of indicators for characterizing compliance violations

Confidence	Behavioral	Extent	Impact
Reliability	Case rate	Spectral extent	Criticality factor
Sensitivity	Excess margin	Spatial extent	Blocking ratio
Precision	Measurement count	Temporal extent	Capacity loss

provide historical information critical for further analysis and actions. Patterns and trends of indicators can help to diagnose and profile long-term violations. Several indicators can be used to characterize compliance violations. For convenience, these indicators are grouped into four broad categories: confidence indicators, behavioral indicators, extent indicators, and impact indicators. Table 13.2 shows the list of indicators in each category.

13.5.2.1 Confidence Indicators

The process of identifying compliance violations is subjected to random and systematic errors. This is partly due to the randomness of radio environments, errors in sensing spectrum, and uncertainties in detecting compliance violations. Therefore, it is imperative to characterize and evaluate the confidence of the underlying methods or their outputs. Recommended confidence indicators are discussed as follows.

- **Reliability** is the degree to which the results of detected compliance violations can be depended on to be accurate. It indicates the consistency of detection methods.
- **Sensitivity** is the proportion of real compliance violations that are correctly identified. Furthermore, it can be seen as the probability that an emitter is identified as non-compliant, given that the emitter is indeed non-compliant. In general, sensitivity indicates how well an underlying detection or classification method can find all compliance violations. It is computed as (true positive) /(true positive + false negative). In various fields of study, sensitivity is also known as recall, hit rate, true positive rate, or probability of detection.
- **Precision** is the proportion of compliance violation identifications that are correct. In general, precision indicates how accurate the underlying detection or classification methods are in finding non-compliant emitters. It is computed as (true positive)/(true positive + false positive). In various fields of studies, precision is also known as confidence or positive predictive value.

13.5.2.2 Behavioral Indicators

The compliance violations patterns are not always the same in terms of their magnitude and rate. Consequently, different sets of compliance issues may require different levels of attention and urgency. Thus, it is important to characterize and assess the behaviors of compliance violations identified from a set of periodic (e.g., hourly) measurements. Proposed behavioral indicators are described as follows.

- **Case Rate** is the number of periodic measurements with detected compliance issues divided by the number of all periodic measurements.
- **Excess Margin** is the average of the compliance violation margins derived from the periodic measurements. A margin is defined as a degree or amount of which the decision parameter (e.g., bandwidth, power, and carrier offset) exceeds the corresponding decision threshold (e.g., maximum acceptable parameter deviation).
- **Measurement Count** is the number of available periodic measurements observed in a specified time window (e.g., a day). Note that some measurements may not be available for some periods of time due to various reasons, including hardware failures or sensor saturation (i.e., when a sensor measures a value that is larger than its dynamic range). The measurement count provides additional information that shows the validity and practical significance of the overall results.

13.5.2.3 Extent Indicators

Some compliance violations may not immediately interfere with existing spectrum users. Consequently, spectrum managers may not be aware of ongoing violations. However, such unnoticed violations break spectrum integrity and can cause harmful interference to spectrum users assigned in the immediate future. Therefore, it is imperative to characterize and evaluate the extent of the potential impact of violations on spectral, spatial, and temporal dimensions.

- **Spectral Extent** is the average amount of bandwidth that is affected due to compliance violations for a given time period and location area.
- **Spatial Extent** is the average size of the area that is affected due to compliance violations for a given time period and bandwidth.
- **Temporal Extent** is the average of duration of time that is affected due to compliance violations for a given bandwidth and location area.

The three extent indicators could also be combined to produce a spectrum resources indicator, computed as an average of the products of affected bandwidth, affected time, and affected area.

13.5.2.4 Impact Indicators

Compliance violations can cause various levels of interference to radio systems and reduce the radio system capacity in terms of usable bandwidth and coverage. Furthermore, license holders who use radio systems either to support their operations needs or to provide wireless services serve the public interest with various levels of relative importance. Therefore, consequences of failure of their radio systems due to compliance violations can affect the public with various degrees of severity. The proposed impact indicators are described as follows.

- **Criticality Factor** is a relative score that ranks expected consequences of a radio system due to a given type of compliance violation. It is found by multiplying the consequences of failure and the probability of failure of the radio system due to the violation.
- **Blocking Ratio** is the proportion of radio devices affected or interrupted by a compliance violation in a given radio system. It can be found as a ratio of the number of affected receivers and the number of all receivers in the same radio system.
- **Capacity Loss** is the amount of reduced radio capacity due to a given type of compliance violation. It is computed as the product of the amount of reduced bandwidth and the size of overlap coverage (i.e., intersection of service area and affected area).

Criticality factor can be seen as a numeric representation of radio system ranking based on user-defined factors, such as consequences of failure and probability of failure, that identify the relative importance of a particular radio system. Consequences of the failure of a radio system can be established by examining economic, health, safety, security, or social issues arising from the interruption of system services. The probability of failure for a radio system can be estimated by examining the underlying architecture and technologies of the system.

13.5.3 Prioritizing Violations

Numerous non-compliance cases may require directed investigations that involve site inspections or street scans. However, in reality, only a limited number of non-compliance cases can be investigated at one time. This is because of various constraints such as limited budget, monitoring equipment, and human resources. Since not all non-compliance cases are equally urgent or important, prioritizing identified cases becomes a crucial step toward achieving effective compliance monitoring.

Numerous prioritization methods have been proposed and applied in many fields of studies including healthcare (prioritize health problems), software development (prioritize requirements), network security (rank security threats), and project management (prioritize tasks). Across many fields, common methods include [22–24] Sorting, Cumulative Voting, Priority Score, Numerical Assignment (Grouping), and Analytic Hierarchy Process (AHP). These existing methods could potentially be applied to prioritize non-compliance cases; however, produced prioritization results may not be sufficiently robust to support efficient compliance activities. To address the limitations of the existing methods, four aspects that are crucial in prioritizing compliance investigations are identified and incorporated in a proposed framework as follows.

- **V: Value (or Impact).** This aspect estimates the value or impact of resolving a non-compliance case with respect to achieving overall compliance objectives. In general, the value can be estimated by examining the benefit of resolving, and/or penalty of not solving, a compliance issue. For instance, in the spectrum sharing environments, the benefit of resolution could be removing radiofrequency interference and improving spectrum quality, whereas the penalty could be degrading the integrity of the spectrum and depriving the public of potential economic return. The key inputs for quantifying the value aspect are the impact and extent indicators.
- **R: Risk (or Likelihood).** This aspect estimates the probability or likelihood that the identified type of compliance violation is wrong and/or if its associated characteristics are incorrect. It also shows the risk of not being able to resolve identified compliance issues because of erroneous information. The source of errors can be due to various reasons ranging from difficulty in sensing radiofrequency signals to identifying underlying compliance violations. Hence, key indicators that can be used to quantify the risk aspect are confidence indicators and the excess margin indicator.
- **E: Effort (or Cost).** This aspect estimates the effort or cost (in terms of time and resources) that is required to resolve a compliance violation. In general, the amount of time can be estimated by examining the complexity of the issues, whereas the number of resources can be estimated by examining the required human resources, equipment, and finance. The key factors that influence time and resources include type and rate of compliance violation. For instance, compliance violations that occur intermittently tend to take longer to resolve compared to the same type of violation that occurs continuously. Therefore, the key indicators for quantifying effort are type and rate of compliance violations.
- **D: Dependency (or Relationship).** This aspect quantifies the dependency or relationship of the identified compliance violation with the other violations in terms of the actual impact and the extent of the potential impact. The goal is to improve the long-run efficiency in investigations by prioritizing compliance cases that have a higher degree of dependency or relationship. For instance, in spectrum sharing environments, emissions from multiple emitters may impact receivers in different networks. Therefore, prioritizing compliance cases with a higher degree of overlap in impact and extent increases the number of cases that will potentially be examined in a single directed investigation. As a result, the efficiency in resolving compliance issues is increased. The key inputs for quantifying the dependency aspect are the impact and extent overlap degrees.

Figure 13.8 shows the structure of the prioritization framework. For convenience, this framework is referred to as the V-RED framework. The priority score is computed in two steps. In the first step, appropriate values are applied at the input aspect functions, which generate the corresponding values of the aspect score (dotted red box). The goal is to quantify the significance of each aspect to enable

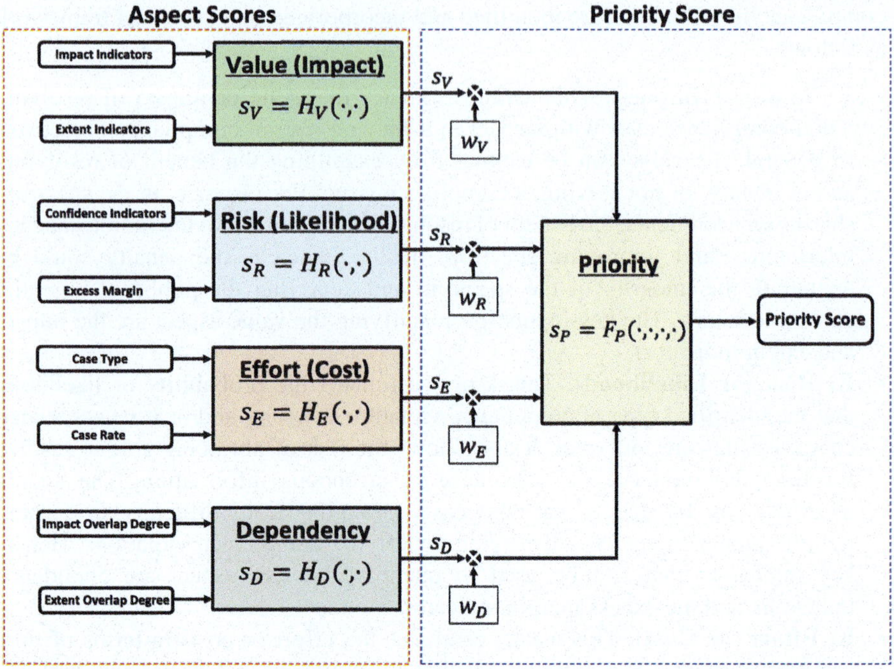

Fig. 13.8 The structure of the prioritization framework

rationale and justification of the prioritization process. In the second step, the aspect scores are multiplied by the aspect weights and applied to a priority function, generating a priority score (dotted blue box). The goal is to generate a priority score that shows the order of importance of the non-compliance cases.

To develop and implement a prioritization solution using the V-RED framework, suitable aspect functions, aspect weights, and a priority function must be defined to reflect a specific use case. In general, aspect weights define the relative importance of each aspect. Therefore, they can be developed by using a pairwise comparison method [25]. Furthermore, the normalized utility functions (output bounded between 0 and 1) can be used to generate aspect scores. Conversely, a linear function can be applied to compute priority scores.

13.6 Summary

Radio spectrum is a finite but non-exhaustive natural resource. It is the foundation for wireless technologies that are increasingly impacting our everyday lives. Spectrum monitoring to ensure compliance with regulatory requirements is a crucial spectrum management function that contributes to the prevention of harmful interference and the protection of the radio environment. The recent advancements of wireless technologies pose a scalability challenge to spectrum regulators. One of the principal issues is how to monitor spectrum compliance in a wide range of frequency bands and over large geographical areas using limited equipment and human resources.

This chapter has presented an automated data-driven system, which leverages newly advanced technologies to facilitate efficient and scalable monitoring of spectrum compliance. The goal is to reduce the manual workload on spectrum managers by automatically collecting spectrum data from different sources, identifying compliance issues, and performing analytics to provide actionable insights. To achieve this, the system performs several functions organized in five functional blocks: data source, signal identification, violation identification, compliance analytics, and user interface.

The data source block provides access to spectrum measurements, spectrum management records, and relevant auxiliary data. The signal identification block decomposes sensed signals into components (modes), maps identified modes to emitters and licenses, and associates measurements and emitters in the databases. The violation identification block detects, characterizes, and prioritizes compliance violations. The compliance analytics block analyzes historical data of compliance violations and provides actionable insights and foresights. The user interface presents results to end-users (e.g., spectrum managers) in various ways, including GUI, APIs, data sets, and reports.

The goal of an automated data-driven system for monitoring spectrum and compliance is not to replace the judgment and expertise of spectrum managers. Rather, it is intended to assist spectrum managers in making better decisions. It could be beneficial as follows:

- **Improve spectrum quality**: Prevent interference and protect spectrum integrity.
- **Increase work throughput:** Allow faster resolution of compliance issues.
- **Increase work outputs:** Identify and localize a wider range of compliance violations.
- **Support enforcement:** Generate data (evidence) to support potential enforcements.

Acknowledgments The authors would like to thank Adrian Florea, Thomas Boyle, David Lu, and Jean-François Roy for their contribution and feedback on the draft version of this chapter.

References

1. Nomenclature of the frequency and wavelengh bands used in telecommunications, Recommendation ITU-R V.431–8 (2015)
2. T.S. Rappaport, Y. Xing, O. Kanhere, S. Ju, A. Madanayake, S. Mandal, A. Alkhateeb, G.C. Trichopoulos, Wireless communications and applications above 100 GHz: opportunities and challenges for 6G and beyond. Special section on millimeter-wave and terahertz propagation, channel modeling and applications. IEEE Access **7**, 78729–78757 (2019)

3. ESA: Satellite frequency bands. https://www.esa.int/applications/telecommunications_integrated_applications/satellite_frequency_bands
4. Economic aspects of spectrum management, Report ITU-R SM.2012-6 (2018)
5. Handbook on Spectrum Monitoring. International Telecommunication Union, Geneva, Switzerland, 2011 (2011) https://www.itu.int/dms_pub/itu-r/opb/hdb/R-HDB-23-2011-PDF-E.pdf
6. S. Wang, R. Inkol, S. Rajan, F. Patenaude, Detection of narrow-band signals through the FFT and polyphase FFT filter banks: noncoherent versus Coherent integration. IEEE Tran. Inst. Meas. **59**(5), 1424–1438 (2010)
7. P. Baltiiski, I. Iliev, B. Kehaiov, V. Poulkov, T. Cooklev, *Long-Term Spectrum Monitoring with Big Data Analysis and Machine Learning for Cloud-Based Radio Access Networks* (Springer Science+Business Media, New York, 2015)
8. E. Wiles, K. Negus, Long-term spectrum monitoring and occupancy from 174 to 1000 MHz in Rural Western Montana, in *12th European Conference on Antennas and Propagation, EuCAP 2018* (2018)
9. S. Campbell, MMCA Sharing Session: Cloud Infrastructure Support. CRC 2020. Available on demand (2020)
10. Spectrum Management System Data, Innovation Science and Economic Development Canada. https://sms-sgs.ic.gc.ca/eic/site/sms-sgs-prod.nsf/eng/h_00010
11. Radio Reference. https://www.radioreference.com
12. D. Xu, Y. Tian, A comprehensive survey of clustering algorithms. Ann. Data Sci. **2**, 165–193 (2015)
13. S. Guha, R. Rastogi, K. Shim, CURE: An Efficient Clustering Algorithm for Large Databases. ACM SIGMOD Rec **27**, 73–84 (1998)
14. R. Yager, D. Filev, Approximate clustering via the mountain method. IEEE Trans. Syst. Man. Cyber. **24**, 1279–1284 (1994)
15. X. Xu, M. Ester, H. Kriegel, J. Sander, A distribution-based clustering algorithm for mining in large spatial databases, in *Proceedings of the Fourteenth International Conference on Data Engineering, ICDE'98* (1998), pp. 324–331
16. M. Ester, H. Kriegel, J. Sander, X. Xu, A density-based algorithm for discovering clusters in large spatial databases with noise, in *Proceedings of 2nd International Conference on Knowledge Discovery and Data Mining, KDD-96* (1996), pp. 226–231
17. M. Ankerst, M. Breunig, H. Kriegel, J. Sander, OPTICS: Ordering points to identify the clustering structure, in *Proceedings on ACM SIGMOD International Conference on Management of Data, SIGMOD'99*, vol. 28 (1999), pp. 49–60
18. W. Wang, J. Yang, R. Muntz, STING: A statistical information grid approach to spatial data mining, in *Proceedings of the 23rd International Conference on Very Large Data Bases, VLDB'97* (1997), pp. 1866–195
19. C. Rasmussen, The infinite Gaussian mixture model, in *Advances in Neural Information Processing Systems*, vol.12 (MIT Press, Cambridge, 1999), pp. 554–560
20. M. Hahsler, M. Piekenbrock, D. Doran, DBSCAN: Fast Density-based Clustering with R. https://rdrr.io/cran/dbscan/f/inst/doc/dbscan.pdf
21. DBSCAN https://pypi.org/project/dbscan
22. M. Yousuf, M.U. Bokhari, M. Zeyauddin, An analysis of software requirements prioritization techniques — A detailed survey, in *3rd International Conference on Computing for Sustainable Global Development (INDIACom)* (2016), pp. 3966–3970
23. P. Berander, P. Andrews, Requirements prioritization, in ed. by A. Aurum, C. Wohlin, *Engineering and Managing Software Requirements* (Springer, Heidelberg, 2005), pp. 69–91
24. J.C.B. Somohano-Murrieta, J.O. Ocharán-Hernández, A.J. Sánchez-García, M. de los Ángeles Arenas-Valdés, Requirements prioritization techniques in the last decade: A systematic literature review, in *2020 8th International Conference in Software Engineering Research and Innovation (CONISOFT)* (2020), pp. 11–20
25. M.A.A. Elsood, H.A. Hefny, E.S. Nasr, A goal-based technique for requirements prioritization, in *International Conference on Informatics and Systems* (2014), pp. 18–24

Chapter 14
AI Driven User Authentication

Hien Nguyen

14.1 Introduction

User authentication is a key aspect of any financial service or security system. The old legacy system of password authentication is deemed insufficient as the user's ID and password can often be compromised by a keylogger virus software or a bad actor's hidden webcam that records all their login information.

AI driven authentication is a framework whereby the system would detect and respond to threats continuously throughout the user's session and not just at login time. If the system detects any abnormal attribute to the user's identity, e.g., changes of mobile device, or location, or changes to biometric data (a person's face, fingerprints, ear features, etc.), then it would prompt the user to verify their identity again.

To promote ease of use for the end users, the new authentication system implements what is called a "frictionless login," wherein password authentication is eliminated. A key component of frictionless authentication is the use of behavioral and physical biometric information for authentication. Behavioral biometric identity is related to the tracking of a user's daily activities as measured by their locality and places they frequent. Physical biometric identity is related to the physical aspects of the human body like facial and fingerprint identity. The combination of behavioral and physical biometric identification is extremely hard to spoof or fake. This has resulted in a highly secure identification system.

It is found that behavioral biometric data provides a more accurate means to determine and even predict identity and identity-linked behaviors such as credit risk

H. Nguyen (✉)
Distilled Identity, Boston, MA, USA

L. Cai et al. (eds.), *Broadband Communications, Computing, and Control for Ubiquitous Intelligence*, Wireless Networks,
https://doi.org/10.1007/978-3-030-98064-1_14

or fraud. This data can be used to obtain a better credit model for financial inclusion and to understand localized economic activity [1].

This chapter describes a general framework for AI driven authentication that has been used in practice. It also focuses on one of the physical biometrics, facial biometric, which is an important part of the authentication system. The remainder of the chapter is organized as follows. Section 14.2 discusses facial recognition techniques, in particular, eigenface detection and a method involving a convolutional neural network. Section 14.3 addresses implementation issues. The chapter is concluded in Sect. 14.4.

14.2 Facial Recognition

Facial recognition is the art of identifying an unknown image to that of a known person. It is achieved by extracting certain facial information and comparing this information to those in the database of facial data in order to determine the identity of a person.

14.2.1 Overview

Facial recognition has the advantage of being highly accurate, with a minimal amount of "friction" to the user. Thus, it is one of the most desirable authentication methods. Before diving deep into facial recognition algorithms, let us look at a high-level view of the overall facial recognition process (see Fig. 14.1).

Facial recognition begins with an input image. The image is processed to detect if there is a face or not. This very first phase of image processing is called "facial detection." There are several facial detection algorithms available, as can be seen in Fig. 14.2. They are different in terms of speed and accuracy. The Haar cascade method was an early method suggested by Voila and Jones [2]. It is still an effective method and in use today. Better face detection algorithms that employ deep learning, which are less affected by changing lighting conditions or partial occlusion of the face, include OpenCV DNN and Multi-Task Cascaded Convolutional Networks (MTCNN) [3]. The choice is really dependent on the application. For example, if the application is to identify an individual at an airport entrance where lighting is good and there is no occlusion of the face, then the Haar cascade will work just fine.

Once a face is detected, it is extracted from the input image. Then the extracted face image goes through a process of facial normalization. The purpose of facial normalization is to minimize the variances in the spatial contrast of the face image. This results in much higher predictive confidence in subsequent recognition processing. Facial normalization often involves face alignment, cropping, contrast normalization, and finally resizing the image to a predefined width and height [4, 5].

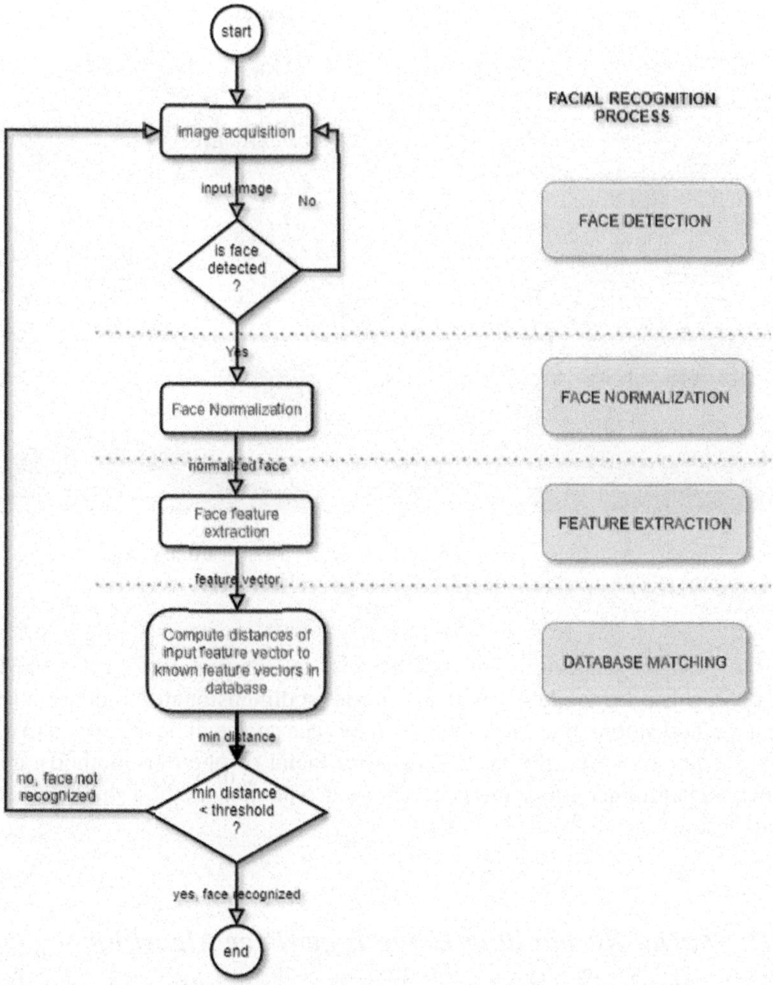

Fig. 14.1 Face recognition flowchart

Facial feature extraction takes a normalized image and produces a feature vector that represents the face. How to construct a feature vector so that all faces of the same person have similar vectors within a small Euclidean distance of each other, and a large distance if the vectors are from two different persons is the key to a successful face recognition algorithm.

Finally, once the facial feature vector is obtained, a database matching process takes place. The facial feature vector is compared to all the known feature vectors in a database of all the known individuals. If the distance between it and another feature vector in the database is less than a threshold, then a match is declared. Otherwise, it is declared a mismatch.

Fig. 14.2 The wonder of image processing. On the left is a very hard to recognize image. On the right is a completely discernible image after histogram equalization

In the past, the majority of classical facial recognition approaches used some methods in subspace modeling with emphasis on dimensionality reduction [6]. The leading method among these is known as eigenface and shall be discussed in the next section. Today, however, the more successful facial recognition methods are those that involve neural networks and deep learning. The new method shall be discussed subsequently.

14.2.2 Facial Recognition Using EigenFace Algorithm

Facial recognition is a computer vision problem. In 1991, Turk and Pentland published a popular paper on face recognition using the eigenface method [7]. The method is actually based on a previous work by Pearson in 1901, which is called Principal Component Analysis (PCA) [8]. PCA is a dimensionality reduction technique that uses eigenvalues and eigenvectors to reduce dimensionality and project a set of training data onto a small feature space.

The eigenface detection method can be separated into two phases: training phase and detection phase.

Training Phase

1. Given a set of M images of a person of dimension $N \times N$, called training images.
2. Convert these images into M vectors of size $N^2 \times 1 : x_1, x_2, x_3, \ldots, x_M$.
3. Normalize these vectors by subtracting the average image Ψ.

$$\Psi = \frac{1}{M}\sum_{i=1}^{M} x_i$$

$$\Phi_i = x_i - \Psi.$$

4. Let A be a matrix $[\Phi_1\Phi_2\Phi_3\ldots\Phi_M]$ of dimension $N^2 \times M$.
5. Let $C = AA^T$ be the covariance matrix for A. Solve for the eigenvalues and eigenvectors of C. However, since C has dimension $N^2 \times N^2$, the computation became expensive.
6. This problem is overcome by taking advantage of the fact that M is much less than N^2. One can compute the eigenvalues and eigenvectors of $C' = A^T A$, whose dimension is $M \times M$. Then, the eigenvectors and eigenvalues of the matrix C can be indirectly computed from that of C'. Let λ_i and $v_i, i = 1, \ldots, M$ denote the eigenvalues and eigenvectors of C'. We then have

$$C'v_i = A^T Av_i = \lambda_i v_i. \tag{14.1}$$

Multiplying (14.1) by A gives

$$AC'v_i = AA^T Av_i = \lambda_i Av_i. \tag{14.2}$$

Replacing $C = AA^T$ gives

$$CAv_i = \lambda_i Av_i. \tag{14.3}$$

From (14.3), it is recognized that the eigenvalue for C and C' are the same, and the eigenvector for C, u_i, is related to that of C' by the relation

$$u_i = Av_i.$$

7. To reduce the computation further, one can choose a set of K eigenvectors, $\{u_i\}$, corresponding to the largest K eigenvalues to form the basis for the linear space spanned by the training images.
8. Compute a set of M vectors, $\{\Omega_i\}$, $i = 1, \ldots, M$, from the training images by projecting Φ_i onto the linear space formed by the K eigenvectors. The components of $\Omega_i = \{\omega_1, \omega_2, \ldots, \omega_K\}$ are the weights of the K eigenvectors in representing each training image.

Note: In order to appreciate how much of a computation reduction this is, let us take a typical example where the image is 512×512 (i.e., $N = 512$) and the number of training images is 10 (i.e., $M = 10$). The dimension of matrix C would be 262144×262144, whereas the dimension of matrix C' is tiny by contrast, i.e., 10×10.

Detection Phase

1. Given an unknown image y, normalize it by subtracting the average image Ψ

$$\Phi = y - \Psi.$$

2. Project the normalized vector Φ onto the K eigenvectors to get the weight vector Ω.
3. Calculate the distance, $\epsilon_k = ||\Omega - \Omega_k||^2$, from Ω to each of the Ω_k vectors in the training set. Then, a face is detected if the minimum distance is smaller than a threshold value Θ_e.

The eigenface algorithm established the first positive step in image recognition, but it is far from ideal. In practice, the images have to be captured in a well-defined area of the face and with proper lighting. Even then the success rate is still far from 100%. In fact, our experiment showed a 60–70% accuracy. This is not good enough, since many authentication applications in banking and high security entry require virtually 100% accuracy.

Then came deep learning face recognition...

Many years ago, as a young undergraduate research assistant, image processing gave me the first shock of seeing how a few math operations can be used to transform an unreadable image into a completely legible image (see Fig. 14.2). Many years later, the second shock that I got was to see how well deep learning image recognition performs in the real world. The success rate is always in the very high 99%, if not 100%. Deep learning for face recognition came about in 2012 with the breakthrough of the AlexNet model [9] by Alex Krizhevsky, Ilya Sutskever, and Hinton. Within a few years, research in deep learning had pushed the success rate from 90 to 99.80% with the FaceNet model.

Figure 14.3 shows the main modules of the face recognition library software that we were using. All of these modules are important in the pipeline of the face recognition process, but we shall focus a bit on the third module "Face feature extraction" using convolutional neural networks (CNN). It is probably not wrong to say that the monumental progress in deep learning face recognition has been largely contributed by one particular algorithm—the Convolutional Neural Network.

14.2.3 Facial Recognition Using CNN

CNN is an architecture that imitates the work of the visual cortex of the human brain. In the 1960s and 1970s, the sensory research work by Dr. Hubel and Dr. Weisel [1] showed that image information is passed through different visual areas of the brain where more primitive information like lines, curves, light, and dark are detected and sent to another layer of the brain that can associate them with a higher-level object like a human face.

CNN network (Fig. 14.4) is generally made up of four layers:

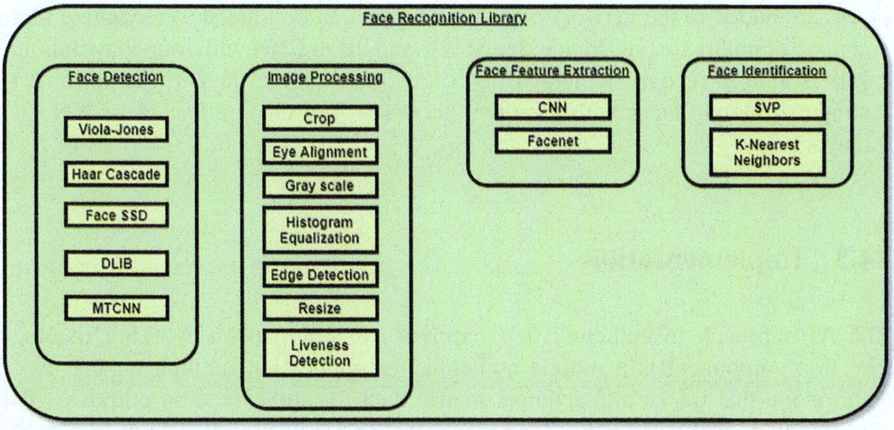

Fig. 14.3 An image recognition software module

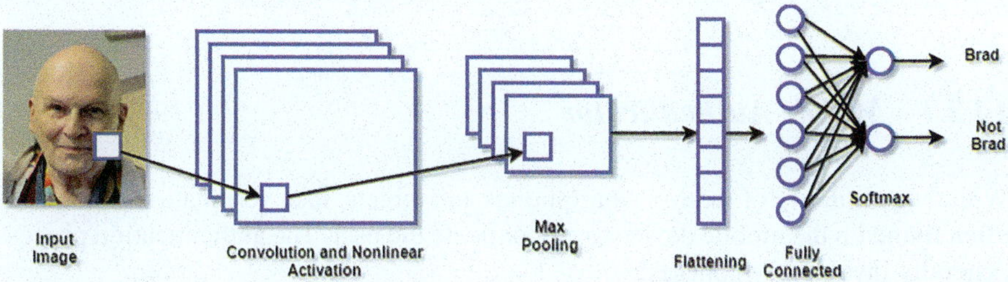

Fig. 14.4 CNN networks

1. Convolutional layer with an activation function: This layer uses filters that perform convolution operations on the input image. A nonlinear activation function is normally applied at the output of the convolution layer to provide an output known as a feature map or activation map.
2. Pooling layer: This layer implements a down sampling operation that is applied after a convolution layer. In particular, max and average pooling are special kinds of pooling operations where the maximum and average value is taken, respectively.
3. Flattening layer: This layer prepares the data for classical neural network operation in the fully connected layer.
4. Fully connected layer: This layer operates on a flattened input where each input is connected to all neurons. Its output is passed through a softmax function to compute the recognition probabilities of all the possible classes.

A CNN network can be viewed as a series of convolutional layers, followed by an activation function, followed by a pooling layer, repeated several times with the flattening and fully connected layer at the end. It was found that the CNN networks are very good at recognizing patterns in an image. As more and more convolutional layers are added to the networks, the later layers can be trained to recognize more and more complex shapes. In practice, it is found that a CNN with four convolutional layers is able to recognize handwritten digits, and with 25 layers it is possible to distinguish human faces with excellent accuracy. The Google FaceNet CNN uses 22 layers.

14.3 Implementation

The AI approach to authentication requires two components working together. The first component is a mobile authenticator application running on the user's smartphone that tracks and acquires metric data for authentication purposes. The second component is the supporting cloud backend that continuously processes the metric data to check for its integrity and alert the mobile authenticator if the user should be re-authenticated.

14.3.1 Mobile Authenticator

With the popularity of today's smartphones and tablets, face verification software is often found on the mobile device to authenticate the user. The authentication process generally involves two phases:

1. Registration phase: In this phase, the application takes several pictures of the user to extract a set of feature vectors.
2. Authentication phase: User authentication is successful when the user is processed to obtain a feature vector that is within an acceptable distance from the registered feature vectors in the database.

In order to make it really hard to break the face recognition system, the phone also collects several metrics from the users that serve as means for cross-checking the user's identity. This data is continuously sent back to the cloud backend for assessing the security of a user. Some of these metrics are:

1. Device profile metric: Each user has a unique mobile device identified by its MAC address, make, and model that can be used to cross check the user's identity.
2. GPS location metric: Each user has a unique daily approximate location between work and home.
3. Cellular location metric: Similar to GPS location, each user has a unique cell tower connection that can be used to authenticate the user.
4. Bluetooth device metric: Each user may have a set of connecting Bluetooth devices. The user may switch phones but the Bluetooth devices may remain the same and can help to establish their identity.

Table 14.1 Device profile metric

Field	Type	Notes
lDeviceID	Long	User device identification
szDeviceModel	Unsigned char[]	Device model name, e.g., Galaxy S7
szBrand	Unsigned char[]	Brand name, e.g., Android, Apple
szOsVersion	Unsigned char[]	Operating system version name, e.g., Android v9
szSerial	Unsigned char[]	Device's serial number, e.g., 604f8c48
szBoard	Unsigned char[]	Device's board name, e.g., msm8996
szManufacturer	Unsigned char[]	Name of manufacturer, e.g., Samsung, LG
iScreenWidth	Integer	Screen width in pixels
iScreenHeight	Integer	Screen height in pixels
fScreenDensityX	Float	Screen x pixel density (dpi)
fScreenDensityY	Float	Screen y pixel density (dpi)
szCollectionTimeUtc	Unsigned char[]	Time at which the metric was collected
fTimezoneUtcOffset	Float	Device's timezone offset in hours from GMU
szTimezoneName	Unsigned char[]	Device's timezone name, e.g., America/New York
szMAC	Unsigned char[]	Device's MAC address, e.g., 00:1B:44:11:3A:B7
szIP	Unsigned char[]	Device's IP address, e.g., 192.168.10.2

5. WiFi network metric: Each user is normally associated with a fixed set of WiFi networks where they live or work. This information is part of his identity and can be used to strengthen the authentication check.
6. Touch motion metric: Each user has a unique finger swiping pattern that can be used to identify the user. The swiping pattern includes swiping speed and touch pressure.

Table 14.1 shows an example of a mobile device metric.

These metrics are collected, with the consent of the users, at regular intervals and help to establish the locality of each user. This is done so that if the authentication software detected that the user was trying to log in to the system from a very foreign place compared to his normal routine, then stricter authentication measures would have to be applied.

Figure 14.5 shows the authentication software application running on a user's phone for tracking and verifying the user's identity.

14.3.2 Supporting Cloud Backend

In order to support all the mobile application data collection and authentication, a substantial backend for data storage and computing backend had to be developed. As an example, if the system is to support a bank with a few million users, then the amount of data gathered could be in terms of terabytes per day and processing this data would take a large computing power. In the last several years, one of the most

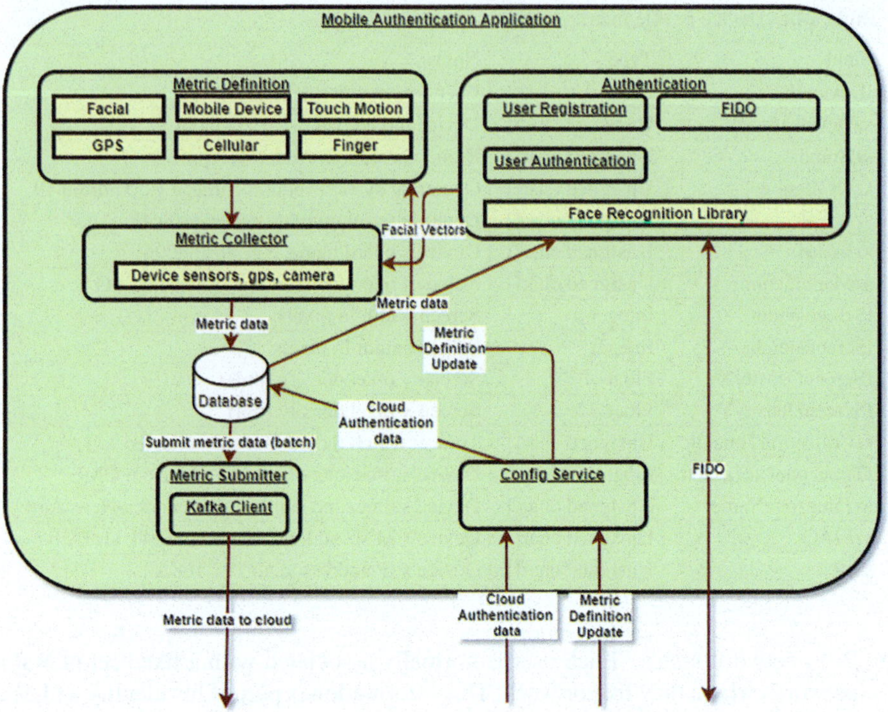

Fig. 14.5 Mobile software application for authenticating the user

impacting developments of large-scale computing is the development of cluster computing. Cluster computing allows computers of various processing power to share work and perform as a single high performing, highly reliable system.

The most well-known of these clusters is the open source project known as Kubernetes, which was started by Google. Kubernetes, also known as k8s, is what Google uses to run billions of the software applications every week. It is the engine, for automating deployment, scaling, and management of software applications.

The development of cluster computing not only allows one to have a highly reliable system, it also allows for the scaling of the computing need according to the ongoing demand.

Figure 14.6 shows an implementation of the authentication backend that works in conjunction with the mobile authenticator to collect data and continuously authenticate the user.

Important software components:

1. Kafka: This is a software framework for storing, reading, and analyzing real-time streaming data. It provides a high-throughput/low-latency messaging framework. Kafka is the real-time streaming service that allows data to be continuously

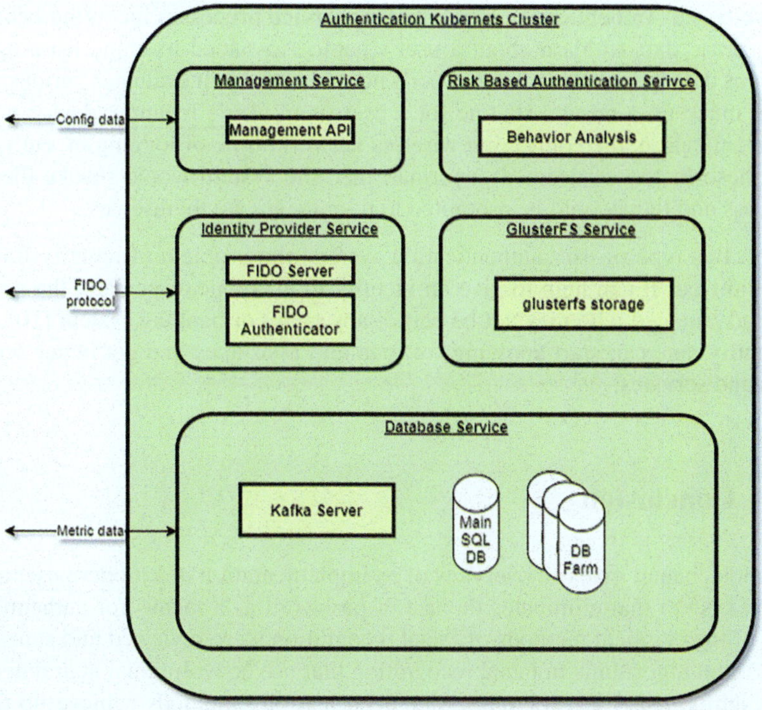

Fig. 14.6 Backend cloud software that stores authenticating data and manages the mobile application

processed and analyzed. This is the subsystem where mobile data is stored for further analysis.

2. DB Farm: Streaming data from Kafka are filtered and moved to SQL databases, where powerful queries can be used to process the data.
3. GlusterFS Service: GlusterFS (GNU Cluster File System) is a distributed scalable file system that aggregates storage components from several servers into one uniform file system. Its special feature is tremendous scalability and the data reliability is guaranteed through redundancy.
4. FIDO Authentication Service: FIDO (Fast ID Online) is a standard, developed by an open industry association—the FIDO Alliance. It is a set of technology-agnostic security specifications for strong authentication using standard public cryptography. It replaces password-only logins with more secure and fast login using facial recognition, fingerprint, and other biometric data. FIDO stores personally identifying information, locally on the user's device to protect it.
5. Management Service: This service allows one to remotely configure the mobile authentication application. It can modify, delete, or add new authentication metrics. If the user is no longer with the organization, then it can disable the authentication service on the mobile device.

6. Risk-based Authentication Service: This service processes incoming behavioral biometric data to learn about a user's habit. Advanced machine learning algorithms detect deviations from a user's normal activity. It evaluates various factors that make up a normal routine for a user (e.g., user's location, device used for authentication, the connecting wireless network, time of logging in, etc.). If any of these factors is deemed abnormal, then the system would revoke the user's access and they would be prompted to re-authenticate themselves.

With this type of user authentication system, the problem of identity fraud can be minimized. It can help to give an identity to everyone, especially the poor and the disadvantaged who may not be part of any credit or banking system [10]. Given an identity, one can start applying for loan and assistance and get better access to goods and services.

14.4 Conclusion

This book chapter gives an overview of an implementation of a modern user authentication system that eliminates the use of password as a means for authentication. Some of the classical methods of facial recognition were reviewed and some of the more recent algorithms in facial recognition that use convolutional neural networks were highlighted. Even though newer deep learning methods achieve up to 99% of accuracy, they are still far from foolproof. For example, a bad actor can still spoof the system with the replay of his photo, video, or using a 3D mask, just to name a few. More sophisticated recognition systems must implement additional countermeasures to combat such spoofing methods [11]. To further enhance the security of the authentication system, several behavioral biometrics are collected to help establish the user's profile and to cross check the user's identity. As humans are creatures of habit, a user's behavioral biometrics data is an effective tool that can be used to minimize the chance that a hacker can break into the system by faking the user's facial identity or fingerprint. The main advantage of an AI based authentication is that it can learn and adapt over time to the user's changing environment and stay ahead of a hacker's attempt at identity theft.

References

1. D.A. Lienhard, David H. Hubel and Torsten N. Wiesel's research on optical development in kittens, in *Embryo Project Encyclopedia*, Oct 2017
2. P. Viola, M.J. Jones, Robust real-time face detection. Int. J. Comput. Vis. **57**, 137–154 (2004)
3. K. Zhang, Z. Zhang, Z. Li, Y. Qiao, Joint face detection and alignment using multitask cascaded convolutional networks. IEEE Signal Process. Lett. **23**(10), 1499–1503 (2016)
4. S.T. Chaudhari, A. Kale, Face normalization: enhancing face recognition, in *3rd International Conference on Emerging Trends in Engineering and Technology, Goa* (2010), pp. 520–525

5. Y. Qian, W. Deng, J. Hu, Unsupervised face normalization with extreme pose and expression in the wild, in *IEEE/CVF Conference on Computer Vision and Pattern Recognition (CVPR), Long Beach, CA* (2019), pp. 9843–9850
6. W. Wójcik, K. Gromaszek, M. Junisbekov, Face recognition: issues, methods and alternative applications, in *Face Recognition - Semi Supervised Classification, Subspace Projection and Evaluation Methods*, Chap. 2, ed. by S. Ramakrishnan (InTech, London, 2016)
7. M. Turk, A. Pentland, Eigenfaces for recognition. J. Cogn. Neurosc. **3**(1), 71–86 (1991)
8. K. Pearson, On lines and planes of closest fit to systems of points in space. Philos. Mag. **2**(11), 559–572 (1901)
9. A. Krizhevsky, I. Sutskever, G.E. Hinton, ImageNet classification with deep convolutional neural networks, in *NIPS'12: Proceedings of the 25th International Conference on Neural Information Processing Systems*, vol. 1, Dec 2012, pp. 1097–1105
10. D. Shrier, http://distilledidentity.com/#video. Last-accessed 10 Nov 2021
11. S. Chakraborty, D. Das, An overview of face liveness detection. Int. J. Inf. Theory (2014). https://doi.org/10.5121/ijit.2014.3202

Chapter 15
Control and Communication Coordination for Industrial Digital Twins of Sintering Process

Cailian Chen, Xiaojing Wen, Xuehan Bai, Lei Xu, Cheng Ren, Jiale Ye, Yehan Ma, and Xinping Guan

15.1 Introduction

The emergence of artificial intelligence and big data technology has greatly accelerated the process of intelligent manufacturing in factories and promoted the industrial production processes toward digitization, networking, and intelligence. However, due to the coupling of each industrial process, big data analysis without the inherent model is not suitable for the complex environment of the factory. Digital twins (DTs) can facilitate the data interaction between physical space and cyberspace by establishing the twin models mapping physical space and adjusting manufacturing parameters timely according to the simulation and prediction of cyberspace, so as to improve production quality [1]. DTs closely integrate model-driven automation with data-driven artificial intelligence technology, providing a new paradigm for the development of intelligent manufacturing.

Establishing digital twin models of industrial processes requires vast amounts of field data, which are generated by various sensors and actuators and need to be transmitted to the edge data center through the field-level industrial network. The coordination of control and communication depends on the timeliness and

C. Chen (✉) · X. Wen · X. Bai · L. Xu · C. Ren · J. Ye · X. Guan
Department of Automation, Shanghai Jiao Tong University, Shanghai, China
e-mail: cailianchen@sjtu.edu.cn; xiaojingwen@sjtu.edu.cn; xuehan_bai@sjtu.edu.cn; xulei1@sjtu.edu.cn; rencheng@sjtu.edu.cn; yejiale@sjtu.edu.cn; xpguan@sjtu.edu.cn

Y. Ma
John Hopcroft Center, Shanghai Jiao Tong University, Shanghai, China
e-mail: yehanma@sjtu.edu.cn

L. Cai et al. (eds.), *Broadband Communications, Computing, and Control for Ubiquitous Intelligence*, Wireless Networks,
https://doi.org/10.1007/978-3-030-98064-1_15

reliability of industrial networks. However, traditional protocols such as Fieldbus and industrial Ethernet cannot guarantee deterministic transmissions for time-sensitive data. Besides, these industrial network protocols are rather sealed and work for specific hardware devices. Therefore, **the first key issue** is how to improve the interconnection among devices and collect industrial field-level data efficiently by exploring deterministic communication technologies.

In order to guarantee the highly real-time transmission for large amounts of time-critical streams between cyber and physical systems, Time-Sensitive Networking (TSN) developed by IEEE 802.1 TSN Task Group[2], extending Ethernet for safety-critical and real-time applications, is the critical technology to guarantee the determinacy. Notably, IEEE 802.1Qbv [3] defines a programmable gating mechanism, i.e., the time-aware shaper (TAS), which employs the time transmission gate and gate control list (GCL) to determine which queues of flows are selected for transmission. Meanwhile, the time of all devices should be synchronized based on IEEE 802.1AS-Rev [4] to guarantee the successful TAS deployment.

However, the characteristics of industrial processes result in new challenges of TSN configurations. Firstly, there are large amounts of tasks in the industrial field that have high requirements of real-time transmissions, such as time-sensitive safety control data flows. Secondly, on-site sensors and actuators are diverse, audio and video data are frequently generated, leading to heterogeneous data streams, which are low time-sensitive (LTS) flows, audio–video bridge (AVB) flows, and so on. Besides, safety data is not time triggered, which should be scheduled with the highest priority. The co-existence of time-triggered (TT) and sporadic flows should be considered for scheduling.

The real-time control and communication coordination depends on scheduling for different flows. The scheduling problem for different priorities is very challenging, even with the specification of the time gating mechanism. For the schedule synthesis problem of TT flows in the TSN network, there are several existing modeling methods, such as integer linear programming (ILP) [5] or mapping the scheduling problem to Satisfiability Modulo Theories (SMT) [6, 7] problem and No-wait Job-shop Scheduling Problem (NW-JSP) [8]. However, the common assumption in these works is the fixed routing paths, which prevents spatial isolations for TT flows. Therefore, it is necessary to improve the schedulability of TT flows through the co-designing of transmission scheduling and route planning. Moreover, in order to improve the scalability (computational efficiency of scheduling method) of the scheduling approaches, the iterated scheduling methods were proposed in [9, 10]. Nonetheless, the stream relevance of these works was based on the intuitive correlations between streams, which lacked data analysis. Therefore, a more objective and precise stream relevance metric needs to be constructed for iterated scheduling in TSN. Except for TT flows, there are other heterogeneous data flows such as AVB flows with bounded end-to-end latency requirements. Gate control list (GCL) schedule synthesis approaches of TT flows taking AVB flows into account were proposed in [11–13].

Because the production process has the characteristics of numerous process parameters, complex mechanisms, significant nonlinearity, and dynamic changes, it is difficult to accurately model the process only by data collection. Therefore, after collecting a large amount of heterogeneous field data, how to build digital twin models in combination with the process mechanism to improve production quality is **the second key issue**. Some successful DT-based applications can predict various physical and chemical components [14–17], as well as predict the performance indicators [18–21], etc. However, existing works ignore the production process and equipment limitations in the actual production process. It is necessary to combine domain knowledge with data-driven intelligent modeling technology to achieve accurate modeling of physical systems and then achieve the prediction of key production indicators.

In addition, there may be thousands of models in the production process. How to achieve on-demand interaction and resource allocation between digital twin models to improve the performance of the product/process in the physical space is **the third key issue**. At present, there have been some results in related research on industrial digital twins. Lu et al. [22] applied the federated learning scheme to build a digital twin edge network. Through the asynchronous model updating scheme, the network not only reduces data transmission overhead but also protects data privacy. In [23], a digital twin model construction method based on federated learning is proposed, and the global model loss function optimization algorithm is given under limited resources. In [24], the construction of the digital twin model is abstracted as a computing task, and a communication and computing resource allocation scheme based on reinforcement learning is proposed to maximize the utilization of network resources. Dong et al. [25] established the digital twin of 5G mobile edge computing network, which applied the twin to train the resource allocation optimization and normalized energy-saving algorithm based on reinforcement learning offline and then updated the scheme to mobile edge computing network. However, the existing research is still at the preliminary stage. How to realize the quality prediction, control decision-making, and network resource optimization of the industrial process based on the constructed digital twin model and then realize the coordination of control and communication still needs in-depth research.

In the rest of this chapter, we propose a scheme of control–communication coordination for industrial digital twins as well as the key technologies of TSN and digital twin modeling. Section 15.2 introduces the network architecture to enhance the real-time control and communication coordination in the environment of the smart factory. For a sintering process described in Sect. 15.3, Sect. 15.4 presents the deterministic and real-time communication protocol based on TSN. Sections 15.5 and 15.6 describe the intelligent modeling of the sintering process and the construction of the corresponding digital twins with the assistance of TSN techniques. Section 15.7 concludes the chapter.

15.2 Control–Communication Coordination Architecture for Industrial Digital Twins

In order to meet stringent production requirements in intelligent manufacturing, various resources in industrial networks need to be efficiently scheduled and managed to achieve production-oriented control and communication coordination. By constructing the digital twin models of process and network, it can provide a new way of resource management for network systems and improve the efficiency of resource utilization. Besides, it can realize the monitoring and management of all factors in the industrial production process and real-time closed-loop control of the whole process and then significantly improve the intelligent level of manufacturing. To this end, we propose a digital twin-assisted control and communication coordination architecture as shown in Fig. 15.1, including Field Industrial Internet of Thing (IIoT), Edge Data Middle Platform, and Industrial Cloud Platform. The whole architecture is optimized in a closed loop through bottom-up information flow and top-down control flow.

In the field IIoT, we deployed a large number of sensing devices (temperature sensors, pressure sensors, infrared thermal imagers, vibration sensors, etc.), transmission nodes (TSN switches), and control devices (PLCs) to monitor the production status of the sintering process. The modeling of the digital twin requires the real-time acquisition of a large number of production data. Therefore, we adopt TSN as the backbone of the architecture to ensure the reliable transmission of field

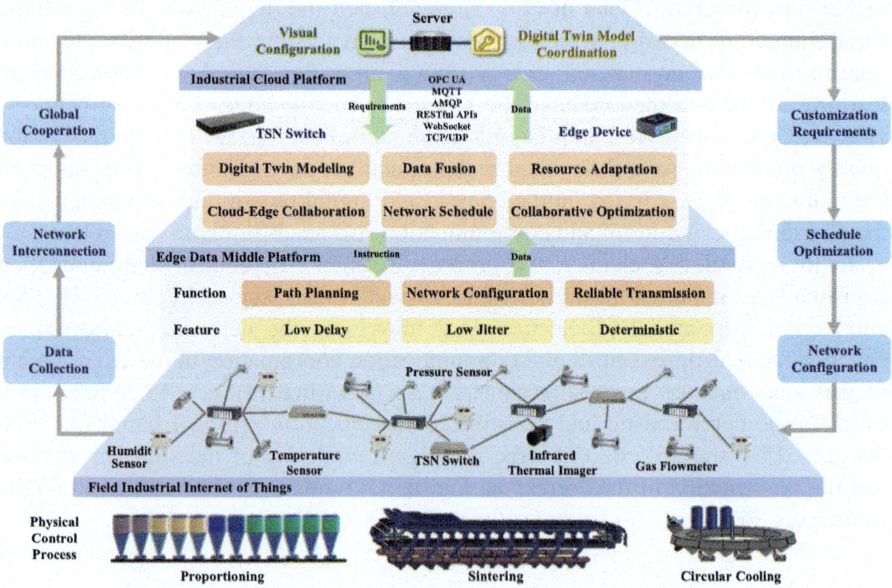

Fig. 15.1 A digital twin-assisted control and communication coordination architecture

perception data, which can meet the network transmission quality requirements of 1-ms delay and 1-μs jitter. At the same time, to ensure the performance of heterogeneous data transmission, some multi-priority data scheduling mechanisms are proposed. After receiving the path planning instruction from the TSN switch at the edge layer, the field layer TSN switch configures the network to achieve reliable transmission on demand. This layer ensures the data penetration of the whole sintering process and the acquisition of high-quality data (low data packet loss rate, accurate data time label, etc.), thus reducing the difficulty of data pretreatment and improving the reliability of algorithms.

The edge data middle platform is deployed with TSN switches for communication and edge devices for control and computing. The management and virtual mapping of production factors such as network, process, and equipment are all completed at this layer. In order to realize the integration and unification of massive scattered and chaotic multi-source data, edge equipment analyzes and processes the data according to the obtained multi-dimensional information and constructs digital twin models of different units based on process mechanism and expert experience to achieve accurate characterization of the production process. At the same time, the quality prediction, quality optimization, and quality traceability models are constructed. The path and network resources are dynamically adjusted according to the requirements of the upper layer. The configuration instructions are issued through the TSN switch to achieve cloud–edge collaboration. Network adaptation reduces the energy consumption of node communication, improves the utilization rate of network resources, and lays a foundation for control performance under dynamic resource conditions. With the support of digital twin technology, the edge data middle platform not only supports more intelligent and flexible system decisions of the top layer but also supports broader and more agile perceptual control of the bottom layer.

In the industrial cloud platform, based on the built digital twin models, a digital twin network for key processes is constructed. Physical entities and various service applications are connected through standardized north–south interfaces. When facing different production needs, the industrial cloud platform can schedule multiple digital twin models on-demand to build a virtual testbed, which can be used to dynamically optimize production resources and adjust and match network optimization schemes. The virtual testbed can efficiently simulate complete process for optimization and then deploy to the industrial site after full verification with the help of optimization algorithms, management methods, and experts' knowledge. Finally, it realizes the re-optimization of process flow, improving production efficiency, and achieves intelligent decision-making and efficient innovation. This layer pays more attention to the comprehensive utilization of field information. With the help of the flexible scheduling of the digital twin model, the resource demand can be predicted accurately, and the communication and control cooperation mechanism between the edge layer and the field layer can be designed. Then, the device states and resource arrangement can be adjusted timely through the feedback of the network operation state to improve and optimize the overall performance. In the bidirectional action, the potential factors of production are connected with the task requirements of the upper layer to meet the demands of intelligent production of the sintering process.

15.3 Sintering Production Line

The process of sintering production is a method of heating powdery materials at a high temperature and sintering them into blocks under incomplete melting conditions. In this process, under the action of a certain high temperature, the surface of iron ore particles softens and melts to produce a certain amount of liquid phase, and they act with other unmelted ore particles. After cooling, the raw materials in the liquid phase solidify and bind into blocks. This process is called sintering.

Obviously, the sintering process is a blocking process in high temperature with complex physical and chemical reactions. The obtained product is called sinter ore and they present irregular pores. The heat energy required for sintering is provided by burning the carbon introduced by material proportioning and the excess air introduced by the exhaust bellows. During this whole process, a series of procedures should be considered such as the materials proportioning, mixing, sintering, crushing, cooling, sifting, and waste gas treatment. Taking the 3 #360 sinter plant of Guangxi Liuzhou Iron & Steel (Group) Company as an example (Fig. 15.2), the sintering production process is briefly introduced.

The first procedure of sintering is mineral proportioning. In order to ensure physical properties, chemical composition stability, and proper permeability of the sintering process, it is necessary to accurately mix multiple iron-containing raw materials and add flux and fuels, according to the quality requirements of sintering product. A well-performed proportioning enables high productivity of blast furnace ironmaking. Generally, the linear programming method can be adopted in sintering proportioning optimization. The optimization output is usually the optimal ratio of chemical composition, such as the TFe, MgO, and R of sinter ore under certain objectives. There are other algorithms for sintering proportioning, such as the Monte Carlo method and genetic algorithm.

The second procedure is the sintering process. Firstly, the sintering feed is carried out, that is, the hearth layer and the sinter mixture are evenly placed on the sintering strand in turn (the hearth layer for sinter is at the bottom of the sinter mixture). Then, ignition and thermal insulation are carried out by coal gas. As the sintering pallet travels toward the end of sinter machine, the sintering is performed from top to bottom under the action of air suction by the sinter bellows. Generally, the materials in the sinter strand can be divided into sinter layer (product), combustion layer, preheating and drying layer, over wet layer, and original sinter ore layer (sinter mixture) from top to bottom. After ignition, five layers appear one after another and move downward. Finally, when the sintering strand moves to the end of the sintering machine, the combustion belt just reaches the bottom of the material layer in the vertical direction of the pallet.

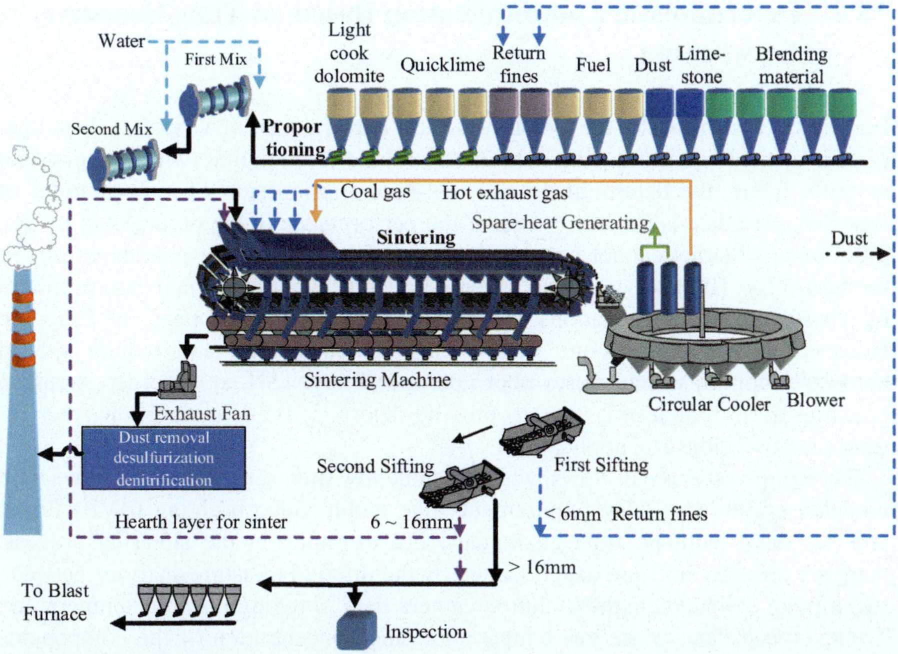

Fig. 15.2 The illustration of the sintering process

At the end of the sintering pallet, the sinter ore falls and is crushed by single roll crusher. After that, considering the high temperature of the sinter ore, it is firstly cooled before being discharged to the belt conveyor. The cooling is undertaken by the circular cooler. After cooling, the sinter ore is transferred to the screening system by belt conveyor. In addition, for the sinter ore as the product of sintering, it is also necessary to collect samples and conduct physical and chemical inspection before they are sent to the blast furnace.

The quality of sinter ore directly affects the quality of iron. Among the evaluation indices of sinter quality, tumble strength (TS), ferrous iron content, total iron, and alkalinity are the most critical. TS reflects the physical strength performance of the sinter, which is used to evaluate the abrasion resistance, collision resistance, and low-temperature reduction pulverization rate of sinter [26]. And the content of FeO of sinter ore reflects the chemical properties, which can be used to evaluate the reducibility of sinter. Higher TS is always accompanied by a high content of FeO, which will reduce the reducibility of the sinter ore and prolong the smelting cycle of sinter ore in the blast furnace. On the contrary, lower TS can increase the powder content of sinter ore and have a negative impact on the blast furnace smelting. Both of them will cause a lot of economic losses.

15.4 Deterministic Communication Based on Time-Sensitive Networking

For digital twin models, the synchronization performance of virtual–real interactions is critical due to frequent data interaction between digital twins and physical systems. Thus, the deterministic and real-time transmission of large amount of time-sensitive data is the key to ensure the performance of the digital twin model. Traditional Ethernet cannot provide deterministic and real-time guarantees due to the best-effort (BE) services, and several real-time Ethernet variants are proposed by corresponding organizations to enhance the real-time property of Ethernet. However, these standards are mostly close to each other and require special hardware support, which causes poor interoperability. TSN, an Ethernet extension breaking these limitations, aims to provide deterministic delay and low jitter for time-sensitive industrial applications.

There are vast types of flows, which include low time-sensitive (LTS) flows with bounded end-to-end delay requirements like audio–video bridging (AVB) flows, and BE flows with no time guarantees. For example, in the sintering process, there are mixture moisture data obtained by the infrared moisture analyzer between two mixing cylinders, high-resolution camera data, sintering machine temperature, Trolley speed data, waste gas temperature, and concentration of the components data, which have different priorities. Although the TSN standard specifies the behavior of the time gating mechanism, the schedule synthesis to achieve the differentiated transmission in the same physical medium under the TSN network for multiple types of flows with different priorities is still challenging.

To improve the schedulability of scheduling instances for TT flows, we propose a co-design approach of transmission scheduling and route planning for TT flows with considering queue assignment [27]. Specifically, the complete co-design constraints of multi-queue scheduling and route planning with deterministic and low delay guarantees are constructed for TT flows, which include queue assignment, conflict avoidance, stream sequence, and real-time and routing constraints. The queue assignment constraint describes the mapping relationships between queues and TT flows on egress port of TSN switches. The conflict avoidance constraint guarantees the deterministic forwarding behavior of TT flows on the egress port, which consists of windows-domain non-overlapping and frame isolation constraints. An illustrative example of conflict avoidance constraint is depicted in Fig. 15.3. The real-time constraint is constructed to ensure that the end-to-end delay of TT flows is lower than the corresponding deadline. The stream sequence constraint describes the transmission sequence of TT flows from talker to listener, i.e., the transmission of TT flows must follow the sequential order on the transmission routing path. The routing no-loop constraint is constructed to avoid the closed loop of the transmission path. Then, the constraint satisfaction problem (CSP) is mapped into SMT problem by an improved mapping method based on Listeners, which solves the indefinite routing problem caused by the introduction of route planning.

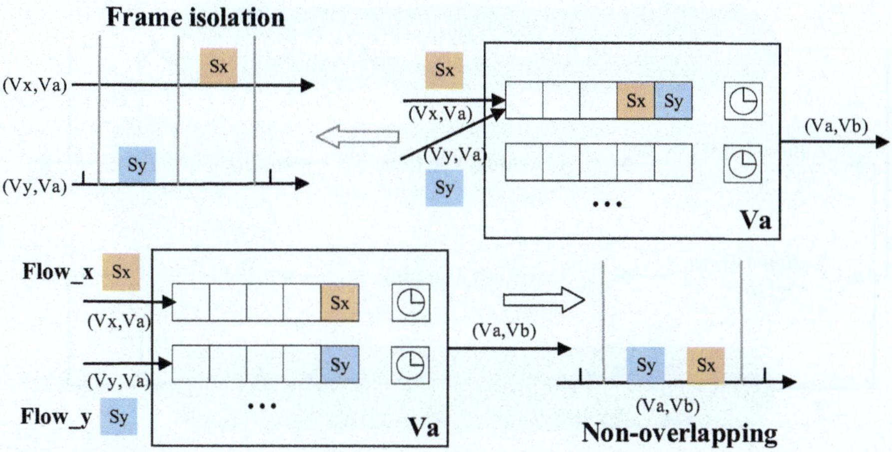

Fig. 15.3 The illustration of conflict avoidance on a shared link

Due to the high computational complexity of TT flows scheduling problems in TSN (i.e., NP-Complete), the above approaches are time consuming and thus are suitable for small and medium scheduling instances. The iterated scheduling methods are proposed for improving scalability (computational efficiency of scheduling method). To construct a more objective and precise stream relevance metric for iterated scheduling in TSN, a semi-supervised learning approach [28] on stream partitioning for iterated scheduling is proposed. Specifically, a large-scale TT stream unlabeled dataset and a small labeled dataset of all possible stream attributes that may influence each flow set's final schedulability are first constructed. The labels are the schedulability under different stream partitioning settings. The stream attributes include a period, a size, a deadline, the number of frames, packets, queues, a sender, a receiver, and a link speed. The scopes of TT stream attributes are selected according to the typical industrial automation applications in the white paper [29] to cover realistic scenarios.

Based on the large-scale TT flow unlabeled dataset, a sparse autoencoder (SAE) is proposed to obtain a sparse and low-rank representation of stream attributes, representing the original input signal to the greatest extent. However, this low-rank representation can only reconstruct the input signal now without any capability of signifying the stream relevance or conflict degrees. Thus, a classifier is added at the SAE top encoding layer and fine-tunes the learned SAE with labeled samples provided in the TT flow labeled dataset. Then, an objective and pervasive evaluation metric on TT flow relevance is constructed based on the low-rank representation of each TT flow without any prior domain knowledge requirements. The evaluation metric on stream relevance provides a data-driven measurement for subsequent stream partitioning. Moreover, the metric can be an additional term on stream relevance or conflicts to guide stream partitioning method such as graph-based approaches and clustering approaches, if accurate domain knowledge is available.

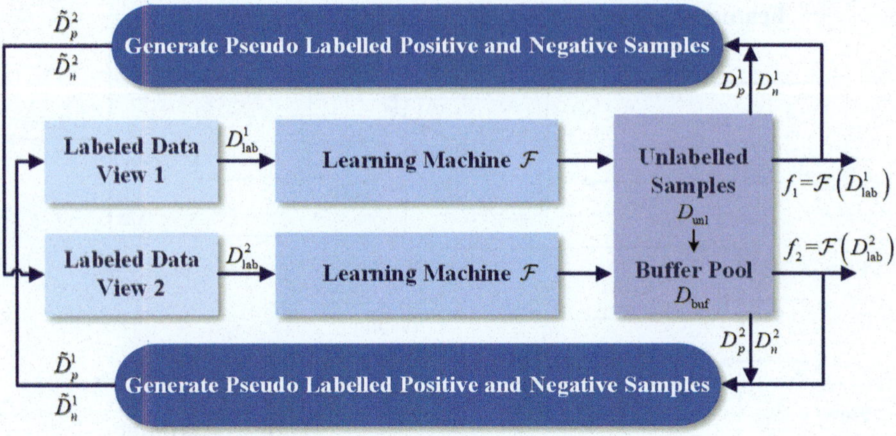

Fig. 15.4 The illustration of the semi-supervised co-training method

To explore a large amount of unlabeled data, a semi-supervised co-training method is proposed to exploit the multi-view relationship of time-triggered stream partitioning, as shown in Fig. 15.4. Each labeled flow set is separated into two views, in particular, divide the streams into two groups on average. To start with, a small portion of labeled training samples is used for learning and stored as the original models. Then the LLR of unlabeled samples is extracted and the pseudo labels are generated by the original models. Next, the original learning models are fine-tuned using an updated training set, which contains labeled training samples and unlabeled training samples with pseudo labels. At last, the second and third steps are iterated until the updated training set becomes stable. The extensive experiments on the TT flow dataset are constructed to demonstrate the effectiveness and performance of the proposed SSL-SP approach, which includes the schedulability comparison, effect of classifiers, dataset size, and parameter setting. The proposed SSL-SP approach obtains the highest schedulability on the TT flow dataset compared to existing iterated scheduling methods.

Nevertheless, the above works focus on the scheduling problem of TT flows from different aspects without considering the impact of TT flow schedule on other flows like AVB flows with bounded end-to-end latency requirements. To achieve the transmission in the same physical medium under TSN network for TT flows, LTS flows, and BE flows, the work [30] proposes a transmission framework by coordinating TAS and cyclic queuing and forwarding (CQF) simultaneously. Under this framework, TT flows are scheduled on the premise of reducing the impact on periodic and aperiodic LTS flows. The flows transmission architecture on the egress port of TSN switch is depicted in Fig. 15.5. The TT flows (i.e., HTS flows), LTS flows, and BE flows are buffered in the egress queue based on the priority mapping table (i.e., assigned internal priority value) and forwarded to the next TSN switch or the nearest edge computing (EC) device based on the predefined GCL schedule.

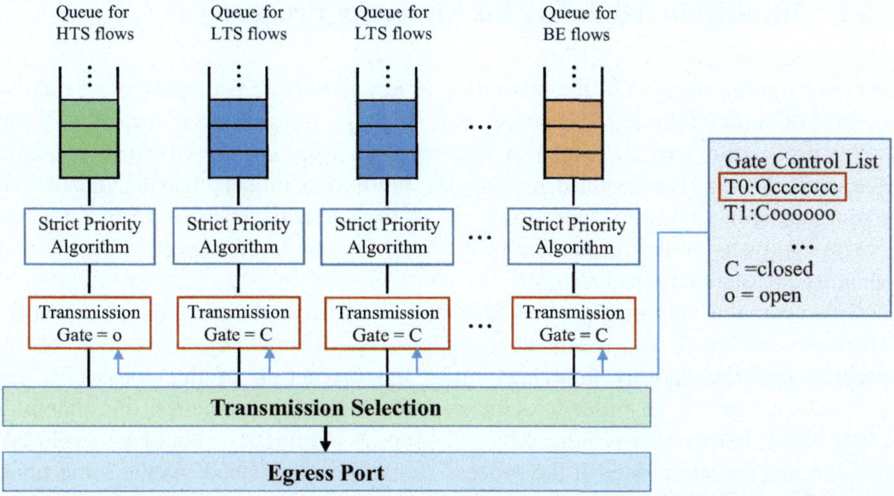

Fig. 15.5 Mixed flows transmission at the egress port of TSN switch

Specifically, the network designer provides the flow sets, including the crucial parameters such as the lower bound and upper bound of the sampling period, the class measurement interval of LTS flows, length of the frames, etc. The network designer then chooses the cycle time and scheduling unit if the flow sets are schedulable. Otherwise, the network designer needs to tune the flows sets by changing the network topography or adding other TSN switches. The network manager then formulates and solves a constrained optimization problem, whereby the objective function is the estimated average delay of LTS flows on the premise of ensuring the constraints of the LTS flows. The decision variable represents the packet injection time of HTS flows. If there exist feasible solutions to the scheduling problem, then the schedules of HTS flows can be generated by the feasible solutions otherwise tune the flows sets. Finally, by merging the schedules of HTS flows and LTS flows, the configurations of the specific egress port on the TSN switch are finished.

To make the coordinated transmission framework practical and achieve fine-grained scheduling of HTS flows, a parameter selection approach is developed by choosing the proper cycle time and the minimum scheduling unit. Moreover, to further reduce the average delay of LTS flows under the coordinated transmission framework, a scheduling problem by planning the packet injection time of each HTS flow is formulated. And then, an injection time grouping algorithm (ITG) is designed by grouping flows of the same period to reduce the computation complexity. Simulation results show the effectiveness of the ITG algorithm.

15.5 Intelligent Modeling for Sintering Process

As described in Sect. 15.3, the prediction of key indicators is crucial to the quality control of sinter. Taking TS as an example, high TS reduces the reducibility of the sinter ore, while low TS results in low strength. Both cases will cause dramatic economic losses. Therefore, designing the prediction model of key indicators to maintain them within an appropriate range in practical production not only has positive impacts on the production of the blast furnace in the aspect of quality and quantity but also helps reduce cost.

However, the sintering process is usually non-linear in view of its industrial chemistry, physical, and mechanical components. Besides, the invisibility of the process and the various time lags make the prediction of the content of TS and FeO nontrivial. In order to achieve accurate quality prediction, the mapping relationship between key indicators and process parameters should be explored, and the mechanism model of the process should be established. At the same time, inspired by machine learning and big data technology, data-driven methods become effective for establishing quality prediction models, which can handle nonlinear approximation problems well. More and more studies begin to focus on data-driven quality prediction in industrial applications.

In terms of TS prediction, some scholars integrated artificial intelligence methods for TS prediction [14–17]. The idea of these hybrid ensemble prediction methods were mainly divided into two types. One is to establish a TS prediction model and then optimize the parameters in the model [14, 15]. The other is to process the data first to weaken some characteristics and then use the processed data to establish prediction models [16, 17]. The data-driven modeling method is not restricted by strict mathematical assumptions and constraints and can reflect the complex relationship between sintering data, which makes the data-driven TS prediction more accurate.

However, the current research work on TS prediction has the following three restrictions. First, the previous works ignore the non-uniform distribution of sintering materials along the width of sintering bed in the practical production process, which makes the low prediction accuracy and cannot satisfy the actual requirements for sub-regional parameters control in sintering production. Second, limited by the capabilities of sensors, the data and features obtained from sintering site are limited. It is difficult for data-driven methods to use data with limited features to establish accurate prediction models. Third, the previous works ignore the inherent time delay in the sintering process and the TS detection process, which means that the data of the same time tag cannot represent the production process of the same batch of sinter materials. Furthermore, the TS prediction model established by the mismatching input and output data will lead to low prediction accuracy. In general, the previous work deviated from the actual production to a certain extent.

To solve the above problems, we proposed a novel TS prediction scheme and a data-driven TS prediction model in [31]. The TS prediction scheme is shown in Fig. 15.6. First, considering the non-uniform distribution of materials in the

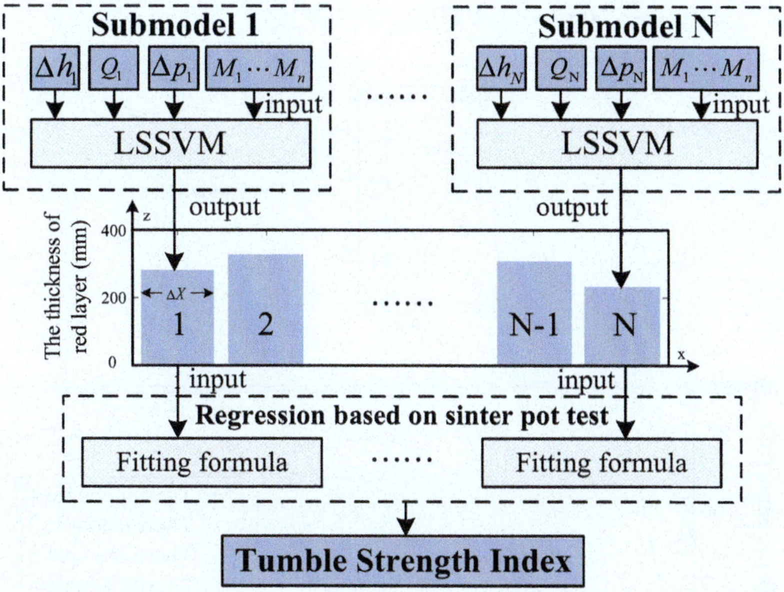

Fig. 15.6 TS prediction scheme

sintering bed and the inherent time delay in the sintering and TS measuring process, we combined local thermal non-equilibrium (LTNE) to establish the TS model (15.1):

$$TS(t + t_f + t_c) = \frac{\int_0^{X_f} \int_{Y_s}^{Y_f} \int_0^{h(t,t_c;x,y)} g(\Delta h, Q, \Delta p, M_1, ..., M_n; x, t)dxdydz}{\int_0^{X_f} \int_{Y_s}^{Y_f} \int_0^{h(t,t_c;x,y)} dxdydz}, \tag{15.1}$$

where t_c is the cooling time, t_f the time gap between the head and the tail, Q the air volume, Δh the height of the mixed materials, Δp the pressure of bellows, and M_1, ..., M_n the composition of sintering mixed materials. The model was more in line with the actual sinter production. Based on the TS model, Q, Δh, Δp, M_1, ..., M_n were selected as the inputs of TS prediction model. Based on the TS model, a more practical data-driven TS prediction scheme for on-site application is proposed. To satisfy the requirements of sub-regional parameters control, the sintering bed is divided into several segments along the width direction, and the TS value in each segment is predicted. This scheme made the TS prediction value worthier for practical sinter production.

To solve the problem of inaccurate modeling caused by limited data features, the thickness of the red layer of the sintering bed tail section was introduced as an intermediate variable, and the sinter pot test data was used to expand the limited feature. By setting up an infrared thermal imaging camera at the tail of the sintering bed, the red layer information was obtained and processed in time, so as to obtain

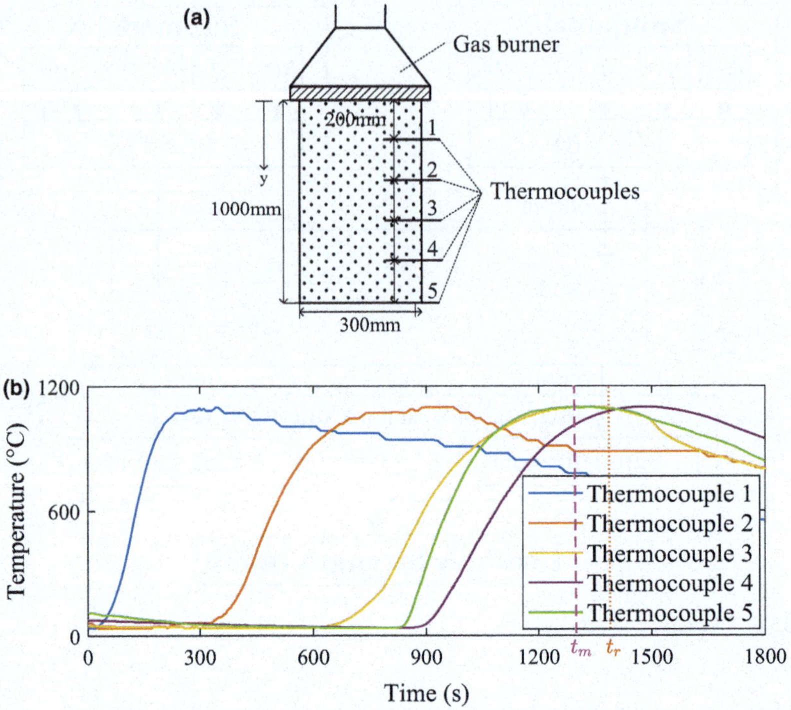

Fig. 15.7 Temperature test in the sinter pot. (**a**) Thermocouple distribution in the sinter pot. (**b**) Temperatures fluctuations in the test

the thickness of the red layer in each segment of the sintering bed tail section. In addition, sinter pot tests were conducted to obtain the fitting formula between the thickness of the red layer and TS. The sinter pot test is designed as a simulation of the industrial sinter process with a standard laboratory process. It overcomes the challenge of exploring the relationship between sintering parameters and the quality of sinter ore under complex and uncertain conditions. The demonstration of the sinter pot is quite simple. The pot is cylindrical with a depth of 1000 mm and a diameter of 300 mm. As shown in Fig. 15.7a, five thermocouples are inserted in the pot at depths of 200 mm, 400 mm, 600 mm, 800 mm, and 1000 mm, respectively. In 8 sinter pot tests, we obtained the TS and temperatures of 5 thermocouples. Figure 15.7b shows the temperature fluctuations in a test.

In order to determine the thickness of the red layer, an interpolation method was adopted to reconstruct the temperature field. Based on the BTP model and practical control strategy in the sintering process, there was 5.9% of the time gap between the beginning and thermocouple 5 appearing maximum temperature for cooling down at the end of the sintering bed. This means that in the sinter pot test, the time to calculate the thickness of the red layer is 5.9% of the time gap after the thermocouple reaches the maximum temperature. Figure 15.8 shows the temperature field changing in the sinter pot and the demonstration of the red layer.

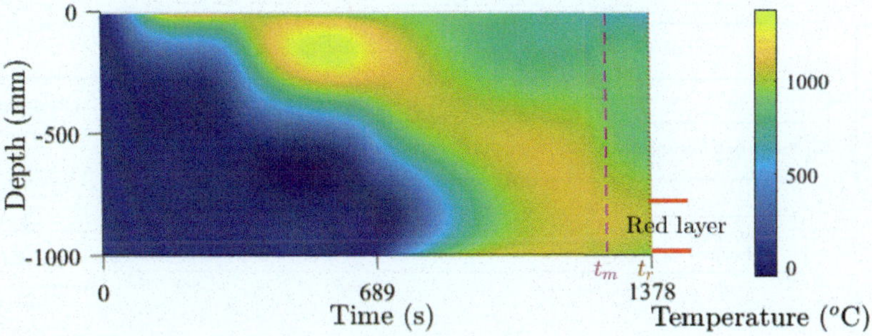

Fig. 15.8 Temperature distribution in the sinter pot

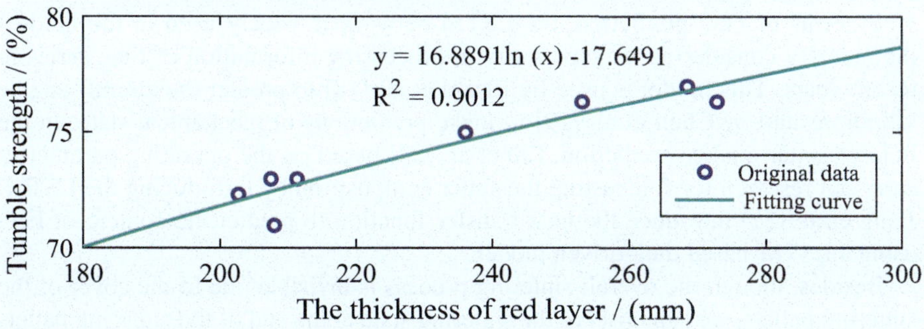

Fig. 15.9 Fitting curve of the thickness of red layer and TS

The fitting curves of the thickness of the red layer and TS are shown in Fig. 15.9. So far, the fitting formula between the thickness of the red layer and TS is expressed as:

$$\theta = 16.8891 \ln(\vartheta) - 17.6491, \tag{15.2}$$

where θ is TS value (%) and ϑ the thickness of red layer (mm).

Further, to solve the problem of time-tag mismatching of input and output data, the LSSVM predictive sub-models were trained by time-matched data of input and red layer thickness. The expanded data features and the time-matched input and output data lead to more accurate predictions. Finally, through the input data and the trained LSSVM prediction sub-models, the prediction value of red layer thickness in each segment was obtained, and then the TS prediction value was calculated by the fitting formula.

The proposed TS prediction scheme was applied to a 3#360 sinter plant of Guangxi Liuzhou Iron & Steel (Group) Company, and the experiments obtained higher prediction accuracy and verified the effectiveness of the proposed method. What's more, it is worth mentioning that the lack of data features and the time-tag mismatching of input and output data are common problems in complex industrial systems. Our proposed scheme provides novel ideas for solving these problems.

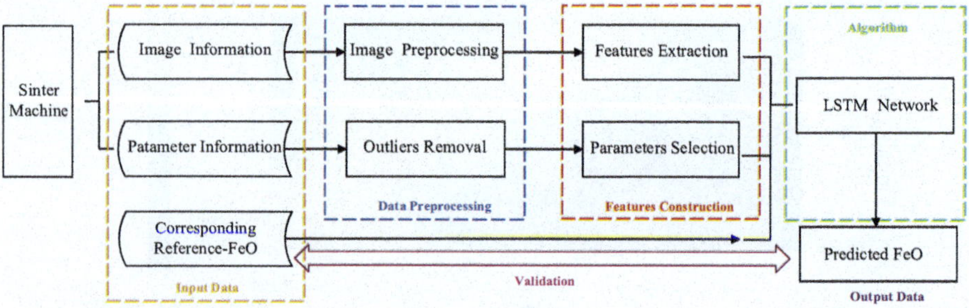

Fig. 15.10 Flow chart of the research scheme routine

In terms of FeO prediction, the LSTM network is widely used in the field of the industry considering its superiority in processing information of time series in recent years. This model is used by Elsaid et al. [18] to predict the aircraft engine vibrations and by Chen et al. [19] to make predictions of mechanical state. In the field of sinter quality prediction, Liu et al. [20] based on the operating parameters carry out research for forecasting the sinter composition by introducing the LSTM. Jiang et al. [21] introduce the heat transfer function to predict the content of FeO using a LSTM-based data-driven model.

Besides, though the overall sintering process is invisible due to the cover of the sintering pallets, we can observe the sintering state at the end of the sintering pallets and capture this information with the camera. These render it accessible to make predictions taking the information from images into consideration. Moreover, the parameters of the sintering process are also used as the features together with the features extracted from images to construct the multi-source feature vector, which can serve to eliminate the errors. They are then as the input of the LSTM network. Furthermore, in order to eliminate the errors caused by the complex large time delay [32], the reference-FeO at the end of the sintering pallets is obtained and it is used as the target output of the LSTM network. Finally, the deep learning model, LSTM network, is utilized to predict the content of FeO with the input of multi-source information and the target output of corresponding reference-FeO.

As shown in Fig. 15.10, we propose a data-driven prediction method for sinter composition FeO based on multi-source information and LSTM network in [33]. The image information, the parameter information of vibration, and temperature are introduced as multi-source features to reflect the content of FeO of sinter ore. Besides, the values of reference-FeO at the end sintering strand are obtained as the target output of the LSTM network. Then, the multi-source information is input to the LSTM network to learn the non-linear relationship between multi-source features and target output. The experimental results show that the data-driven prediction scheme based on multi-source features has better prediction performance, and the absolute error is less than 0.5 compared with reference-FeO, which meets the practical needs of engineering.

Table 15.1 Digital twin models

Category of models		Number of models
Proportioning model	Automatic/manual	24
Mechanism model	Mapping relation	4
Quality prediction	Sintering	9
	Pelletizing	28
	Blast furnace	28
Quality backtracking	Abnormal position	13

Based on the data-driven method, 106 models, such as proportioning optimization, process mechanism, quality prediction, and quality backtracking, were established in Guangxi Liuzhou Iron & Steel (Group) Company, as shown in Table 15.1. The simultaneous evolution of cyberspace and physical space provides decision support for improving quality and efficiency.

15.6 Digital Twins Coordination of the Sintering Process

Based on the coordination architecture in Sect. 15.2, TSN technology, and various data-driven models, control and communication coordination can be realized on the industrial cloud platform. In order to optimize the production process, distributed digital twin models need to be coordinated. In the rest of this chapter, we will introduce control–communication coordination for industrial digital twins of the sintering process, mainly based on TSN and digital twin technology. The development of the digital twin accelerates the intelligent reform of the manufacturing process.

Facing various production demands, a whole process coordination and optimization digital twin network of multiple twins is constructed. Digital twin models are extracted on-demand to simulate the production process in a virtual way, which can gain more efficient network resource allocation. In addition, they predict key process indicators and adjust production process parameters to achieve better production performance.

The network configurations of TSN can be generated by simulating the on-demand interaction of each twin through a digital twin network. If these configurations are directly distributed to the physical network, the normal production process may be affected. In case of production failure, the impact cannot be estimated. Based on the service mapping model of the digital twin network of production collaborative optimization, virtual testing is carried out before parameters arrangement. After network resource allocation and process parameters optimization on the cloud platform, configuration parameters are arranged to the industrial site with less trial and error costs. In this way, the utilization efficiency of production resources is improved.

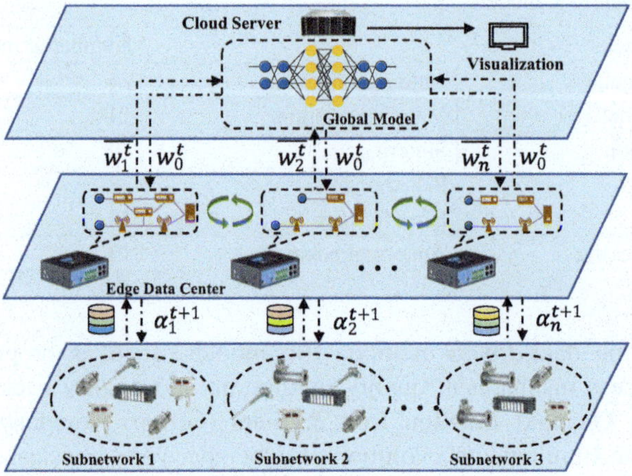

Fig. 15.11 Coordination framework for distributed digital twins

Since the industrial field covers a wide range, different edge devices serve individual regions, and the digital twin models built on each edge device are various. In order to coordinate the local model and the global model, we iteratively update model parameters between each edge device and cloud server through distributed learning technology. Specifically, after each edge device collects massive heterogeneous data, it extracts, converts, loads, cleans, and processes the data to construct digital twin datasets for training various functions in the production process. Then various industrial production functional models, i.e., various forms of deep neural networks, are trained distributed based on these datasets. The digital twins trained on each edge device of the whole production process interact with the industrial cloud platform to serve various production demands as shown in Fig. 15.11. The local model at device i is shown as follows:

$$F(\omega_i^t) = \frac{1}{D_i} \Sigma_{x_i, y_i \in D_i} f(\omega_i^t, x_i, y_i), \tag{15.3}$$

where ω_i^t is the network parameter of device i after t iterations, D_i is the dataset on the edge device i, x_i, y_i is the data sample, and $f(\omega_i^t)$ is the difference between the value output by the network and the true value of the data. $F(\omega_i^t)$ is the loss function obtained by averaging the value for the whole dataset. The network parameter update process is shown in Fig. 15.11 and following formulas:

$$\bar{\omega_i^t} = \omega_i^{t-1} - l_r \nabla_\omega F(\omega_i^{t-1}), \tag{15.4}$$

$$\omega_0^t = \beta_1 \bar{\omega_1^t} + \beta_2 \bar{\omega_2^t} + \cdots + \beta_n \bar{\omega_n^t}, \tag{15.5}$$

where $\bar{\omega_i^t}$ is the local updated network parameters of device i, and l_r denotes the learning rate. In formula (15.5), ω_0^t indicates the global network parameter, and the

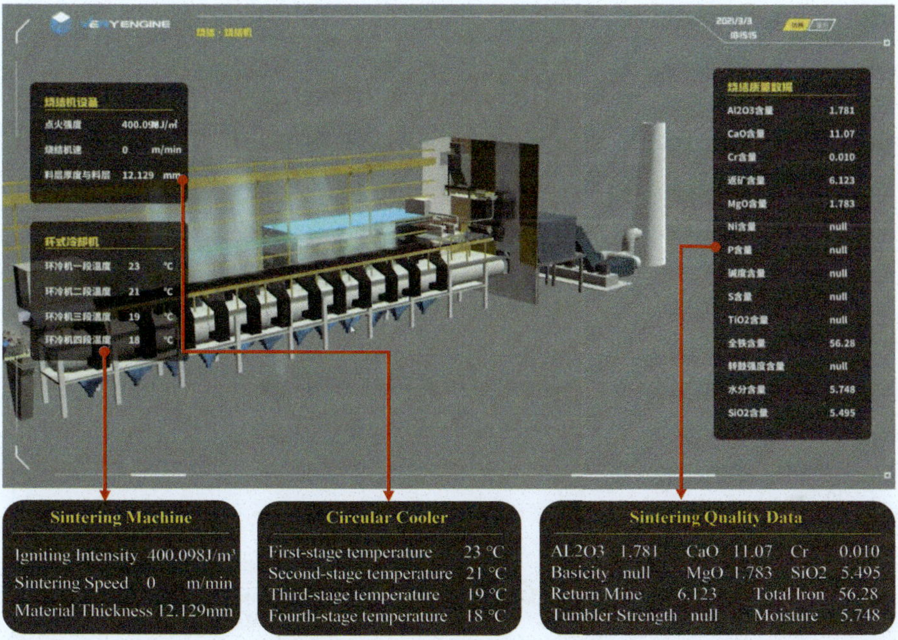

Fig. 15.12 Digital twin platform for sintering quality prediction

update process is a weighted learning process. After T local training iterations, the edge device will upload the model to the cloud server. The industrial cloud platform will integrate all collected models and get a global loss function recorded as $F(\omega_k)$ after k iterations. At the same time, this parameter will be broadcasted to each edge device for update. This process will be repeated until the preset accuracy is met.

Up to now, this technical scheme has been applied in Guangxi Liuzhou Iron & Steel (Group) Company. The quality prediction, quality retrospective, and quality optimization services are provided. In terms of quality prediction service, high-quality and stable production data are obtained through real-time state detection. Then, quality prediction services can be provided based on reliable data. We have built more than 100 prediction models to achieve the prediction of key indicators such as chemical composition and physical properties, as shown in Table 15.1. Based on this, a digital twin platform for sintering quality prediction is formed, as shown in Fig. 15.12, which can provide an important reference for the production site. Figure 15.13 shows the quality prediction curves of TS and FeO drawn based on platform data from November 17 to November 25, 2021. The prediction accuracy is more than 93.75% and over 90.00%, respectively.

At the same time, the forecast service provides early warning. That is, when the forecast value is abnormal, the quality retrospective service is activated to locate the production processes where the abnormality occurs. After that, the occurrence of abnormality is reported to the quality optimization service in time, and the

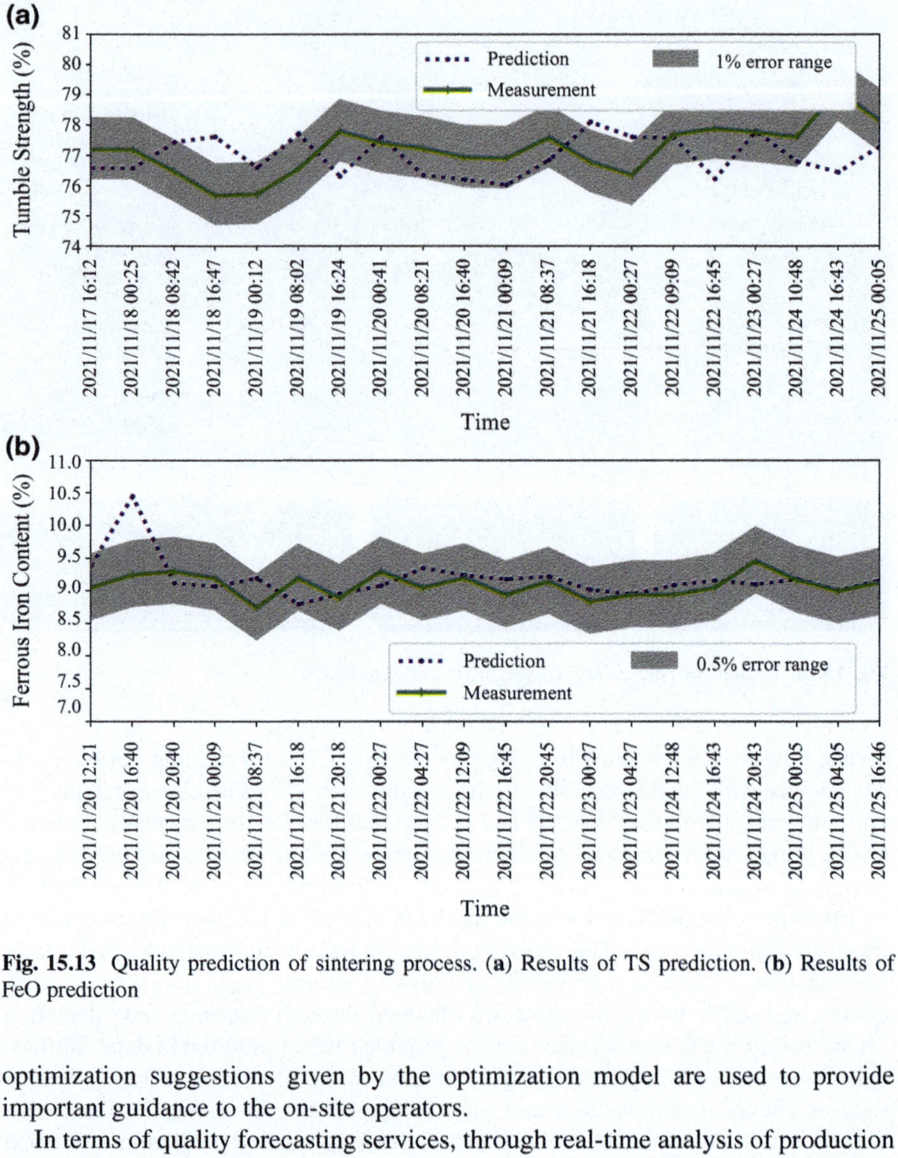

Fig. 15.13 Quality prediction of sintering process. (**a**) Results of TS prediction. (**b**) Results of FeO prediction

optimization suggestions given by the optimization model are used to provide important guidance to the on-site operators.

In terms of quality forecasting services, through real-time analysis of production data such as raw material ratio, operating status (machine speed, valve opening, etc.), observation status (exhaust gas temperature, wind box negative pressure, etc.), the quality prediction of sinter ore is achieved. Accurate prediction of quality indicators provides an important reference for the production site. As shown in Table 15.2, we predict the key indicators 20 minutes in advance, and the prediction accuracy of key indicators is more than 90%.

As for quality optimization, we first analyze the entire pre-iron process and establish a batching optimization model for a single process. Based on this, the

Table 15.2 Quality prediction accuracy and time in advance

Category of models	Number of models	Time scale	Spatial scale	Prediction in advance (hour)	Prediction accuracy
Sintering	9	Minute (forecast period) Hour (forecast advance time)	Equipment/process/ production line	2	>90%
Pelletizing	28	Minute (forecast period) Hour (forecast advance time)	Equipment/process/ production line	2	>90%
Blast furnace	28	Minute scale (changes with the updating frequency of molten iron quality)	Blast furnace/process/ production line	1/3	>92.00%

analysis of the quality correlation between multiple processes is conducted and a proportioning optimization model is established. For the sinter mixture and sinter ore, the corresponding total iron or cost is the optimization target. Constraints are quality conservation constraints, ingredient constraints, inventory constraints, and harmful element constraints of the corresponding process, etc. As for the optimization algorithm, the simple method, gray wolf method, or whale optimization algorithm are utilized to optimize the quality for single process. After that, multi-processes joint batching optimization (mixing material and sinter) is performed. By analyzing the quality relationship between each process and introducing nonlinear convergence factors, Levy flight strategies, etc., joint optimization is achieved under these analyses and constraints. Finally, the proposed proportioning plan satisfies both the requirements of the sinter mixture and the sinter ore.

Furthermore, combined with production cost, process flow, iron grade quality, and other constraints, an economical ore procurement plan was given to substantially save production costs through sintering proportioning optimization. As for quality backtracking, the intelligent fault diagnosis of the production process has been realized by studying the safe intervals and correlation weights of various performance indicators of key processes.

15.7 Summary

To improve the production quality of the sintering process, this chapter firstly introduced a new multi-tier coordination architecture to cooperate control and communication for industrial digital twins, which addressed the field network for smart manufacturing and facilitated the integration of CT, OT, and IT. TSN and its scheduling algorithms were introduced to improve the synchronization and end-to-end performances of communication between cyber and physical systems, such as digital twins and sintering plants. To guarantee the transmission performance of heterogeneous data in the sintering process, some cooperative design methods and mixed transmission framework of route planning considering queue allocation are discussed. This chapter also presented several mechanism and data-driven modeling methods, which laid a solid foundation for more than 100 digital twin functional models. The proposed digital twins' coordination was applied through the on-demand interaction on the industrial cloud platform. The technical scheme is verified over a 3#360 sinter plant of Guangxi Liuzhou Iron & Steel Company and could predict key indicators 20 minutes in advance. The prediction accuracy of TS and FeO is more than 93.75% and 90.00% under normal operating conditions, respectively, which provide important references for improving the quality of the sinter.

Acknowledgments This work was partially supported by the National Key R&D Program of China under the grant 2018YFB1702100, the National Natural Science Foundation of China under the grants 62025305, 61933009, and 62103268, and the Ministry of Industry and Information Technology of China under the grant ZX20200064. Special thanks to Mr. Chugang Shi, the technical director of Sintering Plant, Liuzhou Steel Group, Guangxi, P. R. China, and other technicians for their unreserved supports and constructive comments on the digital modeling for sintering process.

References

1. F. Tao, H. Zhang, A. Liu, A.Y. Nee, Digital twin in industry: state-of-the-art. IEEE Trans. Ind. Inf. **15**(4), 2405–2415 (2018)
2. Time-sensitive networking task group (2017). https://1.ieee802.org/tsn
3. IEEE, 802.1Qbv-2015-IEEE standard for local and metropolitan area networks–bridges and bridged networks-amendment 25: enhancements for scheduled traffic (2015). https://standards.ieee.org/standard/8021Qbv-2015.html
4. 802.1AS-Rev-Timing and synchronization for time-sensitive applications (2017). http://1.ieee802.org/tsn/802.1AS-rev
5. P. Pop, M.L. Raagaard, S.S. Craciunas, W. Steiner, Design optimisation of cyber-physical distributed systems using IEEE time-sensitive networks. IET Cyber-Phys. Syst. Theory Appl. **1**(1), 86–94 (2016)
6. S.S. Craciunas, R.S. Oliver, M. Chmelík, W. Steiner, Scheduling real-time communication in IEEE 802.1Qbv time sensitive networks, in *Proceedings of the 24th International Conference on Real-Time Networks and Systems, ser. RTNS '16* (2016), pp. 183–192
7. R.S. Oliver, S.S. Craciunas, W. Steiner, IEEE 802.1Qbv gate control list synthesis using array theory encoding, in *2018 IEEE Real-Time and Embedded Technology and Applications Symposium (RTAS)* (2018), pp. 13–24
8. F. Dürr, N.G. Nayak, No-wait packet scheduling for IEEE time-sensitive networks (TSN), in *Proceedings of the 24th International Conference on Real-Time Networks and Systems, ser. RTNS '16* (2016), pp. 203–212
9. R. Mahfouzi, A. Aminifar, S. Samii, A. Rezine, P. Eles, Z. Peng, Stability-aware integrated routing and scheduling for control applications in Ethernet networks, in *Proc. Des., Autom. Test Eur. Conf. Exhib.* (2018), pp. 682–687
10. A.A. Atallah, G.B. Hamad, O.A. Mohamed, Routing and scheduling of time-triggered traffic in time-sensitive networks. IEEE Trans. Ind. Inf. **16**(7), 4525–4534 (2020)

11. L. Zhao, P. Pop, Z. Zheng, Q. Li, Timing analysis of AVB traffic in TSN networks using network calculus, in *2018 IEEE Real-Time and Embedded Technology and Applications Symposium (RTAS)* (2018), pp. 25–36
12. V. Gavriluţ, P. Pop, Scheduling in time sensitive networks (TSN) for mixed-criticality industrial applications, in *2018 14th IEEE International Workshop on Factory Communication Systems (WFCS)* (2018), pp. 1–4
13. Y. Huang, S. Wang, B. Wu, T. Huang, Y. Liu, TACQ: enabling zero-jitter for cyclic-queuing and forwarding in time-sensitive networks, in *ICC 2021 - IEEE International Conference on Communications* (2021), pp. 1–6
14. M. Wu, C. Xu, J. She, R. Yokoyama, Intelligent integrated optimization and control system for lead–zinc sintering process. Control Eng. Pract. **17**(2), 280–290 (2009)
15. S. Wang, H. Li, Y. Zhang, Z. Zou, A hybrid ensemble model based on ELM and improved AdaBoost. RT algorithm for predicting the iron ore sintering characters. Comput. Intell. Neurosci. **2019** (2019). https://doi.org/10.1155/2019/4164296
16. Y. Li, C. Yang, Y. Sun, Dynamic time features expanding and extracting method for prediction model of sintering process quality index. IEEE Trans. Ind. Inf. **18**(3), 1737–1745 (2021)
17. X. Fan, J. Feng, X. Chen, Y. Wang, Prediction model and control-guidance expert system of sinter chemical composition. Min. Metall. Eng. **31**(4), 5 (2011)
18. A. ElSaid, B. Wild, J. Higgins, T. Desell, Using LSTM recurrent neural networks to predict excess vibration events in aircraft engines, in *2016 IEEE 12th International Conference on e-Science (e-Science)* (2016), pp. 260–269
19. Z. Chen, Y. Liu, S. Liu, Mechanical state prediction based on LSTM neural network, in *2017 36th Chinese Control Conference (CCC)* (2017), pp. 3876–3881
20. S. Liu, X. Liu, Q. Lyu, F. Li, Comprehensive system based on a DNN and LSTM for predicting sinter composition. Appl. Soft Comput. **95**, 106574 (2020)
21. Z. Jiang, L. Huang, K. Jiang, Y. Xie, Prediction of FeO content in sintering process based on heat transfer mechanism and data-driven model, in *2020 Chinese Automation Congress (CAC)* (2020), pp. 4846–4851
22. Y. Lu, X. Huang, K. Zhang, S. Maharjan, Y. Zhang, Communication-efficient federated learning for digital twin edge networks in industrial IoT. IEEE Trans. Ind. Inf. **17**(8), 5709–5718 (2020)
23. W. Sun, S. Lei, L. Wang, Z. Liu, Y. Zhang, Adaptive federated learning and digital twin for industrial internet of things. IEEE Trans. Ind. Inf. **17**(8), 5605–5614 (2020)
24. Y. Dai, K. Zhang, S. Maharjan, Y. Zhang, Deep reinforcement learning for stochastic computation offloading in digital twin networks. IEEE Trans. Ind. Inf. **17**(7), 4968–4977 (2020)
25. R. Dong, C. She, W. Hardjawana, Y. Li, B. Vucetic, Deep learning for hybrid 5G services in mobile edge computing systems: learn from a digital twin. IEEE Trans. Wirel. Commun. **18**(10), 4692–4707 (2019)
26. J. An, J. Yang, M. Wu, J. She, T. Terano, Decoupling control method with fuzzy theory for top pressure of blast furnace. IEEE Trans. Control Syst. Technol. **27**(6), 2735–2742 (2018)
27. L. Xu, Q. Xu, Y. Zhang, J. Zhang, C. Chen, Co-design approach of scheduling and routing in time sensitive networking, in *2020 IEEE Conference on Industrial Cyberphysical Systems (ICPS)*, vol. 1 (2020), pp. 111–116
28. J. Tu, Q. Xu, L. Xu, C. Chen, SSL-SP: a semi-supervised-learning-based stream partitioning method for iterated scheduling in large-scale time-sensitive networking, in *2021 22nd IEEE International Conference on Industrial Technology (ICIT)* (2021), pp. 1182–1187
29. Integration of 5G with time-sensitive networking for industrial communications, in *5G Alliance for Connected Industries and Automation* (2021), pp. 13–16
30. J. Zhang, Q. Xu, X. Lu, Y. Zhang, C. Chen, Coordinated data transmission in time-sensitive networking for mixed time-sensitive applications, in *IECON 2020 The 46th Annual Conference of the IEEE Industrial Electronics Society* (2020), pp. 3805–3810

31. J. Ye, X. Ding, C. Chen, X. Guan, X. Cao, Tumble strength prediction for sintering: data-driven modeling and scheme design, in *2020 Chinese Automation Congress (CAC)* (IEEE, Piscataway, 2020), pp. 5500–5505
32. S. Zhu, C. Chen, J. Xu, X. Guan, L. Xie, K.H. Johansson, Mitigating quantization effects on distributed sensor fusion: a least squares approach. IEEE Trans. Signal Process. **66**(13), 3459–3474 (2018)
33. X. Bai, C. Chen, W. Liu, H. Zhang, Data-driven prediction of sinter composition based on multi-source information and LSTM network, in *2021 40th Chinese Control Conference (CCC)* (2021), pp. 3311–3316

Index

L. Cai et al. (eds.), *Broadband Communications, Computing, and Control for Ubiquitous Intelligence*, Wireless Networks,
https://doi.org/10.1007/978-3-030-98064-1

The manufacturer's authorised representative in the EU is Springer Nature Customer Service Centre GmbH, Europaplatz 3, 69115 Heidelberg, Germany. If you have any concerns regarding our products, please contact ProductSafety@springernature.com

Printed and bound by CPI Group (UK) Ltd, Croydon, CR0 4YY
28/09/2026
02224077-0009